Bacterial Cell-to-Cell Communication

Many bacterial diseases are caused by organisms growing together as communities or biofilms. These microorganisms have the capacity to coordinately regulate specific sets of genes by sensing and communicating among themselves by means of a variety of signals. This book examines the mechanisms of quorum sensing and cell-to-cell communication in bacteria and the roles that these processes play in regulating virulence, bacterial interactions with host tissues, and microbial development. Recent studies suggest that microbial cell-to-cell communication plays an important role in the pathogenesis of a variety of disease processes. Furthermore, some bacterial signal molecules may possess immunomodulatory activity. Thus, understanding the mechanisms and outcomes of bacterial cell-to-cell communication has important implications for appreciating host–pathogen interactions and ultimately may provide new targets for antimicrobial therapies that block or interfere with these communication networks.

DONALD R. DEMUTH is Professor in the Department of Periodontics, Endodontics and Dental Hygiene at the University of Louisville School of Dentistry.

RICHARD J. LAMONT is Professor of Oral Biology at the University of Florida.

Published titles

Forthcoming titles in the series

CELLULAR MICROBIOLOGY

Over the past decade, the rapid development of an array of techniques in the fields of cellular and molecular biology has transformed whole areas of research across the biological sciences. Microbiology has perhaps been influenced most of all. Our understanding of microbial diversity and evolutionary biology, and of how pathogenic bacteria and viruses interact with their animal and plant hosts at the molecular level, for example, have been revolutionized. Perhaps the most exciting recent advance in microbiology has been the development of the interface discipline of cellular microbiology, a fusion of classical microbiology, microbial molecular biology, and eukaryotic cellular microbiology. Cellular microbiology is revealing how pathogenic bacteria interact with host cells in what is turning out to be a complex evolutionary battle of competing gene products. Molecular and cellular biology are no longer discrete subject areas but vital tools and an integrated part of current microbiological research. As part of this revolution in molecular biology, the genomes of a growing number of pathogenic and model bacteria have been fully sequenced, with immense implications for our future understanding of microorganisms at the molecular level.

Advances in Molecular and Cellular Microbiology is a series edited by researchers active in these exciting and rapidly expanding fields. Each volume will focus on a particular aspect of cellular or molecular microbiology and will provide an overview of the area, as well as examine current research. This series will enable graduate students and researchers to keep up with the rapidly diversifying literature in current microbiological research.

AMCM

ADVANCES IN MOLECULAR AND

Series Editors

Professor Brian Henderson
University College, London

Professor Michael Wilson
University College, London

Professor Sir Anthony Coates
St George's Hospital Medical School, London

Professor Michael Curtis
St Bartholemew's and Royal London Hospital, London

Bacterial Cell-to-Cell Communication

Role in Virulence and Pathogenesis

EDITED BY

Donald R. Demuth

University of Louisville School of Dentistry

and

Richard J. Lamont

University of Florida

CAMBRIDGE
UNIVERSITY PRESS

CAMBRIDGE UNIVERSITY PRESS

Cambridge, New York, Melbourne, Madrid, Cape Town, Singapore, São Paulo

CAMBRIDGE UNIVERSITY PRESS

The Edinburgh Building, Cambridge, CB2 2RU, UK

Published in the United States of America by Cambridge University Press, New York

www.cambridge.org
Information on this title: www.cambridge.org/9780521846387

© Cambridge University Press 2006

First published 2006

Printed in the United Kingdom at the University Press, Cambridge

A catalogue record for this publication is available from the British Library

ISBN-13 978-0-521-84638-7 hardback
ISBN-10 0-521-84638-2 hardback

Contents

viii

CONTENTS

Contributors

Marcie B. Clarke
Department of Microbiology
University of Texas Southwestern Medical Center
5323 Harry Hines Blvd.
Dallas, TX 74390, USA

Dennis G. Cvitkovitch
Oral Microbiology
Faculty of Dentistry
University of Toronto
124 Edward St.
Toronto, ON, Canada M5G 1G6

Donald R. Demuth
Center for Oral Health and Systemic Disease
School of Dentistry
University of Louisville
Louisville, KY 40292, USA

Kangmin Duan
Molecular Microbiology Laboratory
Northwest University
Xián, China

Michael Givskov
Centre for Biomedical Microbiology
BioCentrum Technical University of Denmark
Matematiktorvet, Bldg. 301
DK-2800 Kgs. Lyngby, Denmark

Morten Hentzer
Carlsberg Biosector
Carlsberg A/S
Gl. Carlsberg Vej 10
2500 Valby, Denmark

Barbara H. Iglewski
Department of Microbiology and Immunology
Box 672
University of Rochester School of Medicine and Dentistry
Rochester, NY 14642, USA

Richard J. Lamont
Department of Oral Biology
University of Florida
Gainesville, FL 32610, USA

Celine Levesque
Oral Microbiology
Faculty of Dentistry
University of Toronto
124 Edward St.
Toronto, ON, Canada M5G 1G6

John J. Mekalanos
Department of Microbiology and Molecular Genetics
Harvard Medical School
200 Longwood Ave
Boston, MA 02115, USA

Everett C. Pesci
Department of Microbiology and Immunology
The Brody School of Medicine at East Carolina University
600 Moye Blvd
Greenville, NC 27834, USA

M. Dilani Senadheera
Oral Microbiology
Faculty of Dentistry
University of Toronto
124 Edward St.
Toronto, ON, Canada M5G 1G6

CONTRIBUTORS

Lotte Søgaard-Andersen
Max-Planck Institute for Terrestrial Microbiology
Karl-von-Frisch Str.
35043 Marburg, Germany

Vanessa Sperandio
Department of Microbiology
University of Texas Southwestern Medical Center
5323 Harry Hines Blvd
Dallas, TX 74390, USA

Michael G. Surette
Department of Biochemistry and Molecular Biology
Faculty of Medicine
University of Calgary
Calgary AB, Canada T2N 4N1

Victoria E. Wagner
Department of Microbiology and Immunology
Box 672
University of Rochester School of Medicine and Dentistry
Rochester, NY 14642, USA

Catharine E. White
Department of Microbiology
360A Wing Hall
Cornell University
Ithaca, NY 14850, USA

Stephen C. Winans
Department of Microbiology
360A Wing Hall
Cornell University
Ithaca, NY 14850, USA

Jeremy M. Yarwood
3M Corporate Research Materials Laboratory
3M Center, Bldg. 201-E-01
St. Paul, MN 55144, USA

Jun Zhu
Department of Microbiology
University of Pennsylvania School of Medicine
201B Johnson Pavilion
3610 Hamilton Walk
Philadelphia, PA 19104, USA

CONTRIBUTORS

Preface

It is now well established that a number of bacteria communicate through diffusible signals that may induce and/or regulate a coordinated response by the individual organisms that make up a given population or biofilm. For many of these organisms, it has been suggested that intercellular signaling functions to report population density or to coordinate a response from all cells in a microbial community. Therefore, cell-to-cell communication has been referred to as auto-induction or quorum sensing. The response of bacteria to quorum sensing signals is quite varied and includes, for example, the induction of bioluminescence, the regulation of virulence gene expression, the formation of biofilms, or the induction of horizontal transfer of genetic material. It is also becoming increasingly apparent that some bacteria may communicate via contact-dependent signaling mechanisms, and that the response to direct cell-to-cell contact influences complex behaviors that may contribute to multicellular development or the adaptation to growth in complex biofilms. In the past five to ten years, increased interest and research in the mechanisms of bacterial cell-to-cell communication has revealed surprising complexity both in the signaling processes themselves and in the breadth of the response of recipient cells to the signal molecules. For example, a variety of chemical species, e.g. acyl-homoserine lactones, oligopeptides, furan derivatives (i.e. AI-2), quinolones, butyrolactones, and unsaturated fatty acids are known or have been suggested to function as diffusible signals. Furthermore, some organisms, most notably *Pseudomonas aeruginosa* and species of *Vibrio*, have been shown to produce and respond to multiple diffusible signal molecules. In many cases, the cellular proteins that are required to synthesize and respond to these diffusible signal molecules have been identified; high-resolution crystal structures have been determined for several of these polypeptides. In addition, the response of several organisms to quorum sensing signals has been investigated by using microarray technologies; at least in the case of *P. aeruginosa*, the expression of hundreds of genes is influenced by intercellular signaling. Recent research is also beginning to determine how different signaling pathways are integrated in a given bacterial cell, and how these cell-to-cell signaling mechanisms are linked to the processes that control stationary-phase growth in bacteria.

In contrast to the autoinduction or quorum sensing mechanisms described above, contact-dependent signaling in organisms such as *Myxococcus xanthus* or *Porphyromonas gingivalis* may be initiated by non-diffusible signals such as cell-surface polypeptides. The C-signal of *M. xanthus* is probably the best-characterized cell-surface signal and it plays an essential role in the starvation-induced development of spore-filled fruiting bodies by this organism. Organisms that colonize the human oral cavity may also respond to non-diffusible cell surface signals as well as to diffusible signals. These mechanisms may facilitate the colonization process or adaptation of cell to growth in a multispecies biofilm. Thus, the capacity to respond to non-diffusible cell surface signals may be beneficial for cells that exist in a complex multicellular community, or for cells that must adapt to life in an open-flow environment that might limit the accumulation of a diffusible signal.

In general, the greatly increased interest in bacterial cell-to-cell communication that has occurred in recent years has resulted in a much clearer picture of the signaling mechanisms utilized by bacteria and how organisms integrate various input signals to generate a specific coordinated response as a community. This in turn, has led to efforts to exploit these communication pathways as targets for the development of new antimicrobial therapies. Indeed, chemicals that jam or inhibit bacterial signaling have the potential to be highly effective anti-biofilm agents that may overcome the inherent resistance of biofilm-mediated diseases to treatment with antibiotics.

Our foremost objective in compiling this book is to summarize the rapid advances that have been made, and are continuing to be made, in the field of intercellular communication among bacteria. Owing to the broad scope of new information that is available in this field, we have chosen to target this book on the biomedical community and to focus primarily on the mechanisms of cell-to-cell communication in bacteria of medical importance. A recurring theme throughout many of the chapters is the role that cell-to-cell communication plays in regulating virulence and the development of biofilms by these organisms. The contributors to this volume are leading researchers in the field; each chapter reviews the most recent findings, controversies, and important new information in the author's field of expertise. Our goal is to provide the reader with a comprehensive overview of the various mechanisms and outcomes of bacterial cell-to-cell communication and an appreciation of how these signaling processes may be relevant to the development of microbial communities and to host–pathogen interactions during infection.

We thank the authors for their contributions and for their efforts and cooperation during the assembly of this book. We also thank Katrina Halliday of Cambridge University Press for her assistance in nurturing this project to fruition.

CHAPTER 1

Quorum sensing and regulation of *Pseudomonas aeruginosa* infections

Victoria E. Wagner and Barbara H. Iglewski

University of Rochester School of Medicine and Dentistry, Rochester, NY, USA

①

INTRODUCTION

Pseudomonas aeruginosa is a ubiquitous Gram-negative microorganism that thrives in many environments, from soil and water to animals and people. It is an opportunistic pathogen that can cause respiratory infections, urinary tract infections, gastrointestinal infections, keratitis, otitis media, and bacteremia. *P. aeruginosa* is the fourth most common nosocomial pathogen, accounting for approximately 10% of hospital-acquired infections (www.cdc.gov). Immunocompromised patients, such as those undergoing cancer treatment or those infected with AIDS, burn patients, or cystic fibrosis (CF) patients, are susceptible to *P. aeruginosa* infections. These infections are difficult to treat by using conventional antibiotic therapies, and hence result in significant morbidity and mortality in such patients. The recalcitrant nature of *P. aeruginosa* infections is thought to be due to the organism's intrinsic antibiotic resistance mechanisms and its ability to form communities of bacteria encased in an exopolysaccharide matrix; such communities are known as biofilms.

  P. aeruginosa possesses an impressive arsenal of virulence factors to initiate infection and persist in the host. These include secreted factors, such as elastase, proteases, phospholipase C, hydrogen cyanide, exotoxin A, and exoenzyme S, as well as cell-associated factors, such as lipopolysaccharide (LPS), flagella, and pili. The expression of these factors is tightly regulated. Many factors are expressed in a cell-density-dependent manner known as quorum sensing (QS). Quorum sensing, or cell-to-cell communication, is a means by which bacteria can monitor cell density and coordinate population behavior. The behavior was first identified in *Vibrio fischeri* as a mechanism to induce bioluminescence (20). In *P. aeruginosa*, this

Bacterial Cell-to-Cell Communication: Role in Virulence and Pathogenesis, ed. D. R. Demuth and R. J. Lamont. Published by Cambridge University Press. © Cambridge University Press 2005.

mechanism consists of a regulatory protein and a cognate molecule, termed an autoinducer. At low cell densities, a small amount of autoinducer is present in the extracellular milieu. As cell densities increase, the autoinducer concentration rises in the extracellular environment and a threshold intracellular concentration is exceeded. At this critical concentration, the autoinducer binds to the regulatory protein and this complex acts to induce or repress expression of target genes. The cell-density-dependent regulation of virulence factor production has been suggested as a protective means to prevent host response to invading bacteria before sufficient bacterial numbers have accumulated (15). *P. aeruginosa* possesses two well-studied QS systems, the *las* and *rhl* systems, which have been shown to be important in its pathogenesis. The following sections describe QS in *P. aeruginosa* and its contribution to *P. aeruginosa* virulence, and discuss how QS may represent an attractive target to develop new antimicrobial therapies.

P. AERUGINOSA QUORUM SENSING SYSTEMS

In 1991, Gambello and Iglewski reported that a protein encoded by a gene termed *lasR* could complement elastase production in an elastase-deficient *P. aeruginosa* strain (24). Homology searches showed that LasR belongs to the LuxR family of QS transcriptional regulators (2). Subsequently, two complete QS systems, *las* and *rhl*, have been identified and well studied in *P. aeruginosa*. Each system consists of a transcriptional regulatory protein (LasR in the *las* system and RhlR in the *rhl* system) and cognate autoinducer signal molecules (*N*-(3-oxododecanoyl) homoserine lactone (3-oxo-C_{12}-HSL) in the *las* system, and *N*-butyryl homoserine lactone (C_4-HSL) in the *rhl* system). LasR and RhlR are members of the LuxR family of transcriptional regulators and share 31% and 23% identity, respectively, with the transcriptional activator LuxR in *V. fischeri* (20, 38). Each protein contains an autoinducer binding site and a DNA binding region. Both LasR and RhlR form multimers upon binding to their respective autoinducer (35, 36). This transcriptional regulator protein : autoinducer complex modulates target gene expression, presumably by binding to conserved DNA elements, termed *las* boxes, located upstream of the translational start site of QS-regulated genes (91). The autoinducers are synthesized from *S*-adenolsylmethioinine (SAM), which contributes the homoserine lactone ring and the available cellular acyl–acyl carrier protein (ACP) pool that is used to form the acyl side chain (31). The *lasI* gene encodes the synthase that directs the synthesis of 3-oxo-C_{12}-HSL; *rhlI* encodes the synthase required for C_4-HSL production. The acyl side-chain length has been demonstrated to be important in the specificity of the autoinducer molecule for its cognate transcriptional regulatory protein (49). The length of the acyl side chain is believed to contribute to the differences in transport of the autoinducer molecules from the cell. Diffusion studies have found that, whereas C_4-HSL, with a relatively short acyl side chain, is able to diffuse freely across the cell membrane, 3-oxo-C_{12}-HSL, which has a much longer acyl side chain and is more hydrophobic, is actively pumped from the cell by the multidrug efflux pump MexAB-OprM (51).

The *las* and *rhl* systems are organized in a hierarchical manner in which the *las* system regulates expression of the *rhl* system (Figure 1.1). The LasR-3-oxo-C_{12}-HSL complex was shown to exert both transcriptional control, through activation of *rhlR* transcription, and post-translational control, thought to be mediated by the competitive binding of 3-oxo-C_{12}-HSL to RhlR when the C_4-HSL concentration is low, of the *rhl* system (53). The *las* system also regulates production of elastase (*lasB*), LasA protease (*lasA*), exotoxin A (*toxA*), alkaline proteases (*apr*), and a type two secretion pathway (*xcpR, xcpP*) (3, 23, 24, 48, 50, 68, 86). Expression of *rsaL*, which has been demonstrated to repress *lasI* transcription and alter elastase production when over expressed on a plasmid in *P. aeruginosa*, is also activated by the *las* QS system (14, 28, 70, 89). The *rhl* system has been shown to regulate production of rhamnolipids (*rhlA, rhlB*), pyocyanin, and lectin-binding protein (*lecA*), and to be required for maximal activation of LasA protease (*lasA*) and alkaline proteases (*apr*) (50). The *rhl* system has also been reported to regulate expression of the stationary-phase factor RpoS (*rpoS*) (37), although a conflicting report stated that *rpoS* regulates QS (93). A more recent analysis proposed a complex model of *rpoS* regulation in which *rpoS* can activate the *las* system, which activates the *rhl* system, which then activates *rpoS* transcription (69).

A third LuxR homolog in *P. aeruginosa*, with 29% identity to LasR and 32% identity to RhlR, was identified and shown to act as a repressor of quorum sensing (5). This transcriptional regulator, termed *qscR* for quorum-sensing-control repressor, has been shown to repress transcription of *lasI, rhlI, hcnAB, lasB, pqsH,* and two clusters of phenazine genes (*phzA1* and *phzA2*) (5, 40). QscR has been demonstrated to interact directly with both LasR and RhlR to form heterodimers in the absence of autoinducer molecules and can associate with 3-oxo-C_{12}-HSL and C_4-HSL (40). Further study is needed to define the role of QscR in the QS-regulon.

A third signal molecule, 2-heptyl-3-hydroxy-4-quinolone or the *Pseudomonas* quinolone signal (PQS), has also been described (54). Interestingly, the chemical structure of PQS is not an acylated homoserine

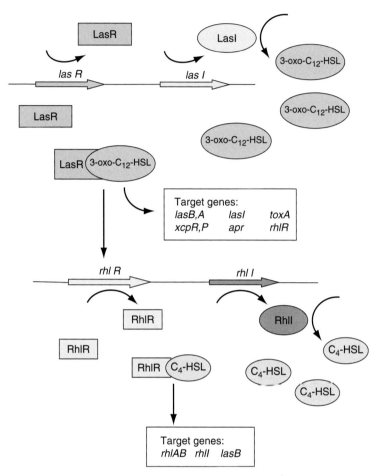

Figure 1.1. The *las* and *rhl* QS systems in *P. aeruginosa*. Both the *las* and *rhl* systems are composed of a transcriptional regulator protein (LasR, RhlR) and an autoinducer synthase (LasI, RhlI). The LasI synthase produces $3\text{-oxo-}C_{12}\text{-HSL}$, and the RhlI synthase produces $C_4\text{-HSL}$. The autoinducer then binds to its respective cognate protein. The LasR:$3\text{-oxo-}C_{12}\text{-HSL}$ complex activates transcription of genes involved in the production of several virulence factors, including *lasB*, *aprA*, and *toxA*, as well as upregulating transcription of *lasI* and of *rhlR*. The RhlR:$C_4\text{-HSL}$ complex activates transcription of several genes involved in virulence, including *rhlAB*. (see also color plate section)

lactone molecule but rather a quinolone molecule. This signal has been proposed to act as a link between the *las* and *rhl* systems (45, 54). A recent study identified several genes, including *pqsABCD* and *pqsH*, in *P. aeruginosa* PAO1 that are required for PQS biosynthesis (22). LasR is required for

PQS synthesis whereas RhlR is required for PQS bioactivity, although the precise mechanism is not clear (22). Interestingly, the *rhl* system has been shown to repress synthesis of PQS (44). The balance of 3-oxo-C_{12}-HSL and C_4-HSL concentrations appears to play a key role in the production of PQS (44). A more thorough discussion of PQS and its place in the QS system is presented elsewhere in this volume (see Chapter 2).

EXTENDING THE *P. AERUGINOSA* QS REALM

Recent efforts to map the QS regulon have elucidated a large number of genes that are regulated by the *las* and *rhl* QS systems. A recent study used transposon mutagenesis to identify QS regulated genes (92). A promoterless-*lacZ* cassette was inserted randomly into the chromosome of a *lasIrhlI* *P. aeruginosa* PAO1 mutant. The mutant library was screened for *lacZ* induction upon exposure to 3-oxo-C_{12}-HSL alone, C_4-HSL alone, or both autoinducer molecules. Of 7,000 mutants screened, 270 mutants produced a greater than 2-fold stimulation of β-galactosidase activity in response to exogenous autoinducer(s). Forty-seven of these mutants that reproducibly exhibited a greater than 5-fold β-galactosidase activity were mapped to determine the locus of disruption. A total of 39 unique genes were identified. These included some previously known QS-regulated genes, such as *rhlB* and *rhlI*, and other genes involved in known QS-regulated processes, such as phenazine synthesis, cyanide synthesis, and pyoveridine synthesis. Seven putative operons were discovered; 14 of the 39 genes identified as QS-regulated possessed putative upstream *las*-box sequences. Interestingly, these mutations could be classified into four distinct classes based upon the timing of activation and requirement for 3-oxo-C_{12}-HSL alone, C_4-HSL alone, or both autoinducer molecules for maximal induction. The authors hypothesized that based upon their results approximately 3%–4% of *P. aeruginosa* genes are controlled by quorum sensing.

The sequencing of the *P. aeruginosa* PAO1 genome allowed the development of *P. aeruginosa* high-density oligonucleotide microarrays (80). The advent of these arrays led to subsequent global analyses of QS regulation in *P. aeruginosa* PAO1. The ability to probe the expression of all known annotated genes under various experimental conditions significantly expanded the QS regulon. One microarray experiment mapped the *las* and/or *rhl* regulon by using RNA isolated from a *lasIrhlI* mutant grown in the absence or presence of 3-oxo-C_{12}-HSL and C_4-HSL (89). In this study, 616 genes were identified as QS-regulated ($p \leq 0.05$ based on three biological replicates per condition), with 394 genes being QS-activated and 222 genes being QS-repressed (89). These included 32 of the 52 previously known QS-regulated genes. Many of the 616 genes (34%) encoded proteins of unclassified or unknown function. The remainders were grouped into 24 of the 26 functional categories used to annotate *P. aeruginosa* gene products (www.pseudomonas.com), with genes encoding membrane proteins or putative enzymes representing the next two largest functional categories of QS-regulated genes. Not surprisingly, more than 50 of the 616 genes encoded known or putative virulence factors involved in attachment, colonization, dissemination, and destruction of host tissues. Of note, the microarrays revealed a link between QS and expression of type III secretion genes. Only 7% (43 genes) possessed upstream *las* boxes, suggesting that many of the genes identified are indirectly QS-regulated. Interestingly, 37 of the 616 QS-regulated genes encode known or putative transcriptional regulators. Therefore these genes may regulate the expression of many QS-regulated genes that do not possess upstream *las* boxes. A recent analysis has further characterized the recognition sites of LasR:3-oxo-C_{12}-HSL binding and found that this complex is able to recognize and activate genes that do not possess a putative *las* box and is unable to activate other genes with putative *las* boxes (71). For example, although PA1897 possesses a putative *las* box and has previously been shown to require the *las* system for activation, the *las*-dependence activation of PA1897 is apparently due to its direct regulation by *qscR* (PA1898), which is regulated by LasR. A more complex picture of the QS regulon, consisting of multiple layers of regulation, emerged (Figure 1.2).

Concurrently, another group also performed a transcriptome analysis to identify QS-regulated genes by using both a *lasIrhlI* signal mutant and a *lasRrhlR* receptor mutant (70). Transcripts were deemed to be QS-regulated if a fold-change difference of greater than or equal to 2.5, derived from comparison of transcript levels by using a *lasIrhlI* mutant grown in the absence or presence of exogenous autoinducer and the *lasRrhlR* mutant as compared with the wild-type, was observed. Transcripts identified as QS-regulated from both sets of experiments were compared; those transcripts that overlapped and had consistent regulatory expression patterns above background level were used to define a list of QS-regulated genes. By this criterion, 315 genes were identified as QS-activated and 38 genes were identified as QS-repressed. More than 87 possible operons were discovered. Importantly, a kinetic analysis of QS regulation suggested that the concentration of LasR was critical in the timing of activation or repression of QS-regulated genes.

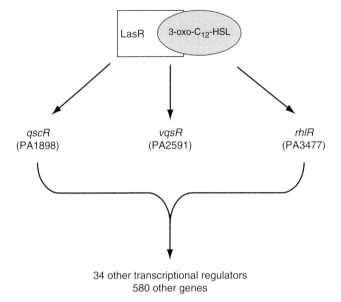

Figure 1.2. Increasing complexity of the QS regulon. Microarray analysis has revealed that conservatively more than 600 genes are part of the QS regulon. Of these genes, three transcriptional regulators, *rhlR*, *vqsR*, and *qscR*, possess putative *las* boxes that suggest these genes are directly activated by the LasR:3-oxo-C_{12}-HSL complex. These regulators may regulate expression of other QS-regulated genes that encode transcriptional regulators and genes that encode proteins representing other functional categories. The QS regulon is hypothesized to have multiple layers via several transcriptional regulatory circuits.

Although both microarray studies revealed that many genes (*c.*6%–10% of the genome) representing diverse functional groups in *P. aeruginosa* were QS-regulated, there was only approximately 50% agreement between the data sets as to the identity of these genes. Interestingly, researchers discovered that by altering the medium formulation (i.e. by using more minimal media or rich media) and oxygen availability used to cultivate *P. aeruginosa* PAO1, the detection of QS-regulated transcripts obtained by microarray analysis varied (89). For example, the level of expression of *lasR* and *rhlR* was lower in the absence of oxygen; this decrease in expression may have contributed to the inability to detect transcripts for certain QS-regulated genes under anaerobic conditions (89). Therefore, the discrepancies between both studies may reflect not only different microarray analysis tools and criteria to define QS-regulated transcripts but also the different experimental conditions used by each group.

In a third microarray analysis of the *las* and *rhl* QS regulon, the authors reported yet another set of QS-regulated genes (28). This study explored the QS regulon in cells grown planktonically, as the previous studies did, and also in cells grown as a biofilm. Taking advantage of the previous array results, the authors compared data sets to form a cohesive list of genes defined as the "general QS regulon." The authors noted that all of these genes, with the exception of one gene (PA0144), were expressed in *P. aeruginosa* biofilms. This is not surprising: it has been demonstrated that *lasI* is critical for mature biofilm formation in *P. aeruginosa* (13). Another study found that during *P. aeruginosa* biofilm development in a once-flow-through biofilm system, transcripts for 56 out of 72 previously identified QS-regulated virulence genes were detected by using microarray analysis (90). Of these, 32 genes (57%) respond to 3-oxo-C_{12}-HSL alone. Interestingly, some of these genes were detected only at day 1 (immature biofilm) or only at day 4 (mature biofilm). These data suggest that different components of the QS system are important during specific phases of biofilm development and maintenance. Interestingly, several genes that have been identified as QS-regulated, including *rhlA*, *rpoS*, *pslB* (PA2232), and PA5057–5059, have recently been shown to be important in biofilm development and maintenance (12, 30, 32, 55).

Recently, a fourth LuxR-transcriptional regulator that has been identified as QS-regulated by using microarrays (28, 70, 89) was found to be required for autoinducer synthesis and extracellular virulence factor production, as well as full pathogenicity in a *Caenorhabditis elegans* infection model (34). The gene, termed *vqsR*, was discovered by using transposon mutagenesis screening in the *P. aeruginosa* CF clinical isolate TB. *vqsR* possesses a putative *las* box upstream of its annotated translational start site, suggesting that it is directly regulated by the LasR–3-oxo-C_{12}-HSL complex (34, 89). Microarray analysis of the *vqsR* mutant compared with the wild type revealed that several genes that have been previously identified as QS-regulated, including *rhlA* and genes involved in the denitrification pathway, are also regulated by *vqsR* when grown in the presence of H_2O_2 or human serum. Interestingly, *vqsR* regulated expression of several genes that have been previously determined to be iron-regulated (8, 34). Iron plays a key role in *P. aeruginosa* pathogenesis (88). These data suggest that *vqsR* may be the link between QS and iron regulation that had been previously put forward by other studies (2, 92).

Several genes have been identified that affect activation of the QS regulon. These include the response regulator *gacA*, the CRP-homolog *vfr*, the transcriptional regulator *mvaT*, the sigma factor *rpoN*, the stationary phase sigma factor *rpoS*, and *relA*, which is involved in the stringent response (1, 17, 64, 69, 87, 93). Polyphosphate kinase (PPK) has been shown to be required for maximal 3-oxo-C_{12}-HSL and C_4-HSL synthesis and elastase and rhamnolipid production (63). The regulatory protein *RsmA*, has also been demonstrated to influence homoserine lactone auto-inducer production, rhamnolipid production, and swarming (29). Undoubtedly, the QS regulon is subject to global regulation at several levels; this regulation may reflect the importance of tight regulation of QS in response to environmental cues as well as to the metabolic state of the cell. Further study is needed to elucidate the influence of these factors on QS regulation. However, it is clear that more than just cell density (a "quorum") regulates the QS regulon in *P. aeruginosa*.

INFLUENCE OF QS ON *P. AERUGINOSA* VIRULENCE IN PLANT AND ANIMAL MODELS

To determine the role of QS in pathogenesis, several models of infection have been developed to identify virulence genes in *P. aeruginosa* in addition to the use of mammals. These include *Arabidopsis thaliana*, *Caenorhabditis elegans*, *Galleria melonella*, *Drosophila melanogaster*, and *Dictyostelium discoideum* (10, 19, 26, 33, 42, 43, 46, 56, 59, 60, 61, 62, 81). These latter systems are attractive because they allow for rapid screening of many mutants without subjecting large numbers of mammals to infection and subsequent death. Importantly, data from these experiments have shown that a number of QS-regulated *P. aeruginosa* virulence factors appear to be conserved across host species. In the *A. thaliana* model, *gacA*, *toxA*, *plcS*, and *dsbA*, as well as several genes of unknown function, were found to be important in plant pathogenesis. The *C. elegans* model identified genes that were important in two modes of killing, slow and fast killing. These two modes were found to be dependent on the strain of *P. aeruginosa* used as well as on media conditions. From studies of *C. elegans*, numerous virulence genes were identified including *gacA*, *lasR*, *rhlR*, *vqsR*, *pqsR*, and *np20*, as well as genes with no previously described functions. A paralytic mode of killing was determined to be due to production of hydrogen cyanide, a toxic virulence factor that is QS-regulated (21). By using the *D. discoideum* model, components of the *rhl* QS (*rhlR*, *rhlI*, and *rhlA*) in addition to *nfxC* were determined to be required for virulence (9). In a *G. melonella* model, genes involved in type III secretion, including *exoU* and *exoT*, were found to be important in killing as well as *gacA*, *lasR*, and *pscD*. A *Drosophila melanogaster* model revealed that genes required for twitching

Table 1.1. *Genes in the QS regulon found to be important in pathogenesis of P. aeruginosa in various host infection models*

Host species	Genes	References
Arabidopsis thaliana	*gacA*, toxA*	62
Caenorhabditis elegans	*hcnA, pqsCDEH, pqsR, phnA, lasR*, rhlR, rhlI*, gacA*, gacS, phzB*, mexA, vqsR, PA0745, PA3032*	10, 22, 34, 42
Dictyostelium discoideum	*rhlR, lasRrhlR, rhlI*, lasIrhlI*, rhlA, pscJ*	9, 59
Drosophila melanogaster	*relA, pilGHIJKL*	11, 19
Galleria mellonella	*phzB*, gacA*, lasR**	46, 60
Medicago sativa seedling	*rhlR*	72

*Genes also required for virulence in a thermal injury mouse model. Note that not all genes have been tested in the thermal injury mouse model.

motility (*pilGHIJKL, chpABCDE*), amino acid, nucleotide, and central metabolism (*pyrF, pgm, cca*), and a gene of unknown function (PA5441), were important in killing (11, 57). Both *rhlR* and *algT* have been identified as important in an alfalfa seedling (*Medicago sativa*) model of infection (72). Many of the genes found to influence pathogenesis are part of the QS regulon, either as a global regulator of QS activation (e.g. *gacAS*), QS transcriptional regulators (e.g. *lasR, rhlR, vqsR, pqsR*), or genes involved in QS autoinducer synthesis (e.g. *lasI, rhlI, pqsCDEH*) or production of extracellular secreted factors (e.g. *hcnA, rhlA*) (Table 1.1).

Research using animal models of both acute and chronic infection has supported the premise that QS significantly contributes to *P. aeruginosa* pathogenesis (15, 58, 66, 83) (Table 1.2). In a mouse model of acute pulmonary infection, *lasR, lasI, rhlI,* and *lasIrhlI* mutants were significantly attenuated in virulence (52, 83). Analysis of the *lasI* and *rhlI* mutants revealed that a *rhlI* mutant caused pneumonia in 15% of mice, as opposed to a *lasI* mutant, which caused pneumonia in 30% of mice (52). This suggests that, whereas both *lasI* and *rhlI* contribute to infection, *rhlI* regulates specific factor(s) that stimulate airway inflammation and resultant pneumonia. In addition, pili were demonstrated to contribute to pathogenesis in the same model of acute infection (82). Several QS genes have also been demonstrated to be important in the rat model of acute pneumonia. A *lasR* mutant was avirulent in the rat model (41). When a *lasIrhlI* mutant was

Table 1.2. *Genes in the QS regulon that contribute to virulence in animal models*

Host	Genes	References
mouse model of acute pneumonia	*lasR, lasI, rhlI, lasI–rhlI, pilA*	52, 82, 83
rat chronic lung infection model	*lasR, lasI–rhlI, tatC, metE, narK2, modA, PA1874*	41, 47, 57, 95
thermal injury model	*lasR, lasI, rhlI, lasI–rhlI, gacA, phzB*	42, 65, 66

tested in this model, the mutant produced milder lung pathology, induced a stronger serum antibody response, and increased interferon gamma concentrations when compared with the wild-type *P. aeruginosa* (95). In addition, *tatC*, a gene involved in a novel secretory system, the twin-argi-nine-translocation (TAT) pathway, and that was previously reported to be QS-repressed, has been shown to be important in a rat lung model (47). An ambitious high-throughput screening of 7,968 *P. aeruginosa* mutants in a rat model of chronic infection identified 214 mutants, representing 148 annotated open reading frames (ORFs), as attenuated in virulence (57). Among these genes were several previously identified QS-regulated genes, including *metE* (PA1927), *narK2* (PA3876), *modA* (PA1863), and a gene encoding a product of unknown function (PA1874). In a thermal injury mouse model, *lasR, lasI*, or *rhlI* mutants were all less lethal and exhibited a diminished ability to disseminate from the burn site (65, 66). However, a double *lasIrhlI* deficient mutant was the most attenuated in virulence (66). Several other genes, including *gacA* and *phzB*, have also been shown to be important in the burn model of infection (Table 1.2).

In addition, QS-regulated factors such as elastase and the autoinducer 3-oxo-C_{12}-HSL have been shown to have immunomodulatory effects. In a mouse thermal injury model, a *lasIrhlI* mutant failed to stimulate produc-tion of proinflammatory cytokines including interleukin-6 (IL-6), tumor necrosis factor alpha (TNF-α), and transforming growth factor beta (TGF-β), suggesting that *P. aeruginosa* QS plays a role in triggering the produc-tion of these cytokines in vivo (67). In vitro cultures of human bronchial epithelial cells exposed to a *lasR P. aeruginosa* mutant produced lower amounts of interleukin-8 (IL-8), a potent chemotactic factor, than did cells exposed to wild-type *P. aeruginosa* (76, 77). Subsequent studies showed that purified 3-oxo-C_{12}-HSL could stimulate significant IL-8 pro-duction in human bronchial epithelial cells, whereas other autoinducers such as C_6-HSL and C_4-HSL did not, and that the stimulation of IL-8 production by 3-oxo-C_{12}-HSL was due to induction of NFκB through activation of extracellular-signal-regulated kinases (ERKs) (76, 77). Purified 3-oxo-C_{12}-HSL was also found to induce cyclooxygenase-2 (Cox-2) via the NFκB pathway in vitro in human lung fibroblasts (78). Cox-2 was shown to subsequently upregulate production of prostaglandin E_2 (PGE$_2$), a potent inducer of mucus secretion, vasodilation, and edema (78). This link between 3-oxo-C_{12}-HSL and PGE$_2$ is especially intriguing because increased concentrations of PGE$_2$ have been found in *P. aeruginosa* infections in CF lungs and burn wounds (76). 3-oxo-C_{12}-HSL can also modulate immune cell responses, such as to activate interferon-gamma (IFN-γ) in T-cells, to inhibit interleukin-12 (IL-12) and TNF-α production in lipopolysaccharide (LPS)-stimulated macrophages, to promote immunoglobulin E (IgE) in interleukin-4 (IL-4)-stimulated peripheral blood mononuclear cells, and to cause apoptosis in macrophages and neutrophils (4, 76, 77, 84, 85). The immunomodulatory effect of 3-oxo-C_{12}-HSL has led to the hypothesis that QS can control T-cell responses away from a protective T-helper-1 (Th1) host response, thus promoting *P. aeruginosa* survival *in vivo* (85). Interestingly, a recent report demonstrated that both 3-oxo-C_{12}-HSL and C_4-HSL can penetrate mammalian cells and activate their respective transcriptional regulators, LasR and RhlR (94). Using LasR and RhlR chimeric transcription factors, this study showed that autoinducers are required for synthesis of functional LasR and RhlR proteins in monkey kidney COS-1 cells. Owing to the resemblance of the autoinducer binding domain of the LuxR-family protein, TraR, to a GAF or PAS domain present in a number of mammalian signaling proteins and transcription factors, the authors speculated that eukaryotic receptor proteins exist that are able to interact with autoinducers.

QS PLAYS AN ACTIVE ROLE IN VIVO

Recent evidence has suggested that *P. aeruginosa* QS is active in vivo. Several studies have investigated the presence of autoinducer molecules (3-oxo-C_{12}-HSL, C_4-HSL, PQS) and detection of transcripts for known QS regulators and QS-regulated genes in CF patient samples. Transcripts for *lasR* correlated well with the detection of transcripts for *lasA*, *lasB*, and *toxA* in sputum samples that were obtained from infected CF patients (79). The highest correlation was found between *lasR* and *lasA* detection, followed by *lasR* and *lasB*, then *lasR* and *toxA*. These results suggested that *lasR* may be coordinately regulating expression of these factors in vivo. Further analysis also revealed a relation between *lasI* expression and *lasR*, *lasA*, *lasB*, and *toxA* (18). Because the LasR-3-oxo-C_{12}-HSL complex is known to directly activate transcription of *lasI* in vitro, it was not surprising that the strongest correlation was between *lasR* and *lasI*. Interestingly, a weak but statistically significant correlation between *lasR* and *algD* transcription was found that suggested that *lasR* might to some extent regulate *algD* or be activated by a common environmental trigger. AlgD catalyzes the first step in alginate biosynthesis, resulting in the mucoid phenotype often observed in clinical isolates from *P. aeruginosa*-infected CF patients (16). In a mouse thermal injury model, *algD* was required for full virulence in a *mucD* PA14 background (97). Recently, *algR2* (*algQ*), a gene that activates *algD* expression in mucoid *P. aeruginosa* strains, was shown to negatively regulate the expression of *lasR*, *rhlR*, and *lasB* at the level of transcription in a mucoid clinical isolate, suggesting a link between alginate production and the QS regulon (39).

CF sputa have also been reported to contain 3-oxo-C_{12}-HSL, C_4-HSL, and PQS. Both 3-oxo-C_{12}-HSL and C_4-HSL could be detected at relatively low concentrations in 78% and 26% of CF sputa samples, respectively (18). Interestingly, there was no correlation between the accumulation of *lasR*, *lasI*, *lasA*, and *lasB* and the amount of autoinducer detected. The ratio of 3-oxo-C_{12}-HSL to C_4-HSL produced *in situ* in biological CF sputum samples was found to mimic those ratios found in *P. aeruginosa* grown as a biofilm under laboratory conditions (73). The authors proposed that this result suggested the organisms were growing as biofilms in CF sputa. Moreover, the knowledge that certain autoinducer ratios may indicate the growth state of *P. aeruginosa in vivo* could be used to identify therapeutic agents that interfere with *P. aeruginosa* infections. PQS has been detected at physiologically relevant amounts in the sputum, bronchoalveolar lavage fluid, and mucopurulent fluid from distal airways of endstage lungs removed at transplant in CF patients (7). In *P. aeruginosa* isolates from infant CF patients, PQS production was elevated compared with laboratory strains, suggesting that PQS is important early in the establishment of chronic infections (25).

QS AS A THERAPEUTIC TARGET

The relationship between QS and pathogenesis suggests that interference of QS in *P. aeruginosa* represents an attractive target for treatment of *P. aeruginosa* infections (also see Chapter 4). This is especially important as QS interference does not seek to kill or inhibit microbial growth, which may lead to less chance of development of antibiotic resistance. Several

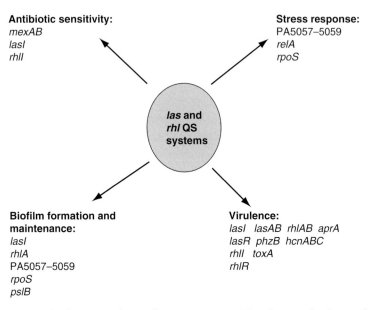

Figure 1.3. Role of QS in regulation of *P. aeruginosa*. Examples of QS-regulated genes that have been found to be implicated in regulation of biofilm formation and maintenance, stress response, antibiotic sensitivity, and virulence are depicted.

means of accomplishing this goal include inhibition of LasR and RhlR function, disruption of *lasR/lasI* and/or *rhlR/rhlI* activation, and interruption of autoinducer synthesis and activity (77). A naturally occurring compound, a halogenated furanone synthesized by *Delisea pulchra*, and several derivatives of this molecule, have been shown to inhibit QS systems in *P. aeruginosa* (27, 28, 96). One of these derivatives, C-30, was effective in decreasing bacterial load in a mouse model of chronic *P. aeruginosa* lung infection (96). Microarray analysis revealed that this compound affected transcription of approximately 80% of genes identified as QS-regulated (28). A high-throughput study to identify molecules that inhibit QS has found several candidates that are able to compete with 3-oxo-C_{12}-HSL activation of LasR and have been shown to decrease elastase production in *P. aeruginosa* (74, 75). Other mechanisms of QS interference include the use of autoinducer-specific antibodies to prevent entry into the cell, the use of lactonases that degrade the homoserine lactone ring of 3-oxo-C_{12}-HSL and C_4-HSL, the inactivation of genes that are global activators of QS, such as *gacA* or *vfr*, and the use of antisense oligonucleotides that inhibit translation of *lasR*, *lasI*, *rhlR*, and *rhlI* transcripts (77). Interestingly, human airway epithelial cells have been shown to inactivate 3-oxo-C$_{12}$-HSL (6). Determination of the identity of this cell-associated factor may prove useful in designing novel therapies.

P. aeruginosa QS systems have been demonstrated to be important in the establishment and maintenance of *P. aeruginosa* infections. Recent studies have enhanced our current understanding of the *las* and *rhl* QS systems and revealed a more complex picture than previously imagined. QS appears to play a pivotal role in *P. aeruginosa* stress response, biofilm development and maintenance, antibiotic resistance, and virulence (Figure 1.3). Further investigation of QS regulation in *P. aeruginosa* will elucidate the nuances of these systems. This is especially imperative as it becomes clear that these systems represent a novel means of therapeutic intervention.

REFERENCES

1   Albus, A., E. Pesci, L. Runyen-Janecky, S. West and B. Iglewski 1997. Vfr controls quorum sensing in *Pseudomonas aeruginosa*. *J. Bacteriol.* **179**(12): 3928–35.

2   Arevalo-Ferro, C., M. Hentzer, G. Reil *et al.* 2003. Identification of quorum-sensing regulated proteins in the opportunistic pathogen *Pseudomonas aeruginosa* by proteomics. *Environ. Microbiol.* **5**(12): 1350–69.

3   Chapon-Hervē, V., M. Akrim, A. Latifi *et al.* 1997. Regulation of the *xcp* secretion pathway by multiple quorum-sensing modulons in *Pseudomonas aeruginosa*. *Molec. Microbiol.* **24**(6): 1169–78.

4   Chhabra, S. R., C. Harty, D. S. Hooi *et al.* 2003. Synthetic analogues of the bacterial signal (quorum sensing) molecule N-(3-oxododecanoyl)-L-homoserine lactone as immune modulators. *J. Med. Chem.* **46**(1): 97–104.

5   Chugani, S. A., M. Whiteley, K. M. Lee *et al.* 2001. QscR, a modulator of quorum-sensing signal synthesis and virulence in *Pseudomonas aeruginosa*. *Proc. Natn. Acad. Sci. USA* **98**(5): 2752–7.

6   Chun, C. K., E. A. Ozer, M. J. Welsh, J. Zabner and E. P. Greenberg 2004. Inactivation of a *Pseudomonas aeruginosa* quorum-sensing signal by human airway epithelia. *Proc. Natn. Acad. Sci. USA* **101**(10): 3587–90.

7   Collier, D. N., L. Anderson, S. L. McKnight *et al.* 2002. A bacterial cell to cell signal in the lungs of cystic fibrosis patients. *FEMS Microbiol. Lett.* **215**(1): 41–6.

8   Cornelis, P. and S. Aendekerk 2004. A new regulator linking quorum sensing and iron uptake in *Pseudomonas aeruginosa*. *Microbiology* **150**(4): 752–6.

9   Cosson, P., L. Zulianello, O. Join-Lambert *et al.* 2002. *Pseudomonas aeruginosa* virulence analyzed in a Dictyostelium discoideum host system. *J. Bacteriol.* **184**(11): 3027–33.

10 Darby, C., C. L. Cosma, J. H. Thomas and C. Manoil 1999. Lethal paralysis of *Caenorhabditis elegans* by *Pseudomonas aeruginosa*. *Proc. Natn. Acad. Sci. USA* **96**(26): 15202–7.

11 D'Argenio, D. A., L. A. Gallagher, C. A. Berg and C. Manoil 2001. *Drosophila* as a model host for *Pseudomonas aeruginosa* infection. *J. Bacteriol.* **183**(4): 1466–71.

12 Davey, M. E., N. C. Caiazza and G. A. O'Toole 2003. Rhamnolipid surfactant production affects biofilm architecture in *Pseudomonas aeruginosa* PAO1. *J. Bacteriol.* **185**(3): 1027–36.

13 Davies, D. G., M. R. Parsek, J. P. Pearson *et al.* 1998. The involvement of cell-to-cell signals in the development of a bacterial biofilm. *Science* **280**(5361): 295–8.

14 de Kievit, T., P. C. Seed, J. Nezezon, L. Passador and B. H. Iglewski 1999. RsaL, a novel repressor of virulence gene expression in *Pseudomonas aeruginosa*. *J. Bacteriol.* **181**(7): 2175–84.

15 de Kievit, T. R. and B. H. Iglewski 2000. Bacterial quorum sensing in pathogenic relationships. *Infect Immun* **68**(9): 4839–49.

16 Deretic, V., J. F. Gill and A. M. Chakrabarty 1987. Gene *algD* coding for GDPmannose dehydrogenase is transcriptionally activated in mucoid *Pseudomonas aeruginosa*. *J. Bacteriol.* **169**(1): 351–8.

17 Diggle, S. P., K. Winzer, A. Lazdunski, P. Williams and M. Camara 2002. Advancing the quorum in *Pseudomonas aeruginosa*: MvaT and the regulation of N-acylhomoserine lactone production and virulence gene expression. *J. Bacteriol.* **184**(10): 2576–86.

18 Erickson, D. L., R. Endersby, A. Kirkham *et al.* 2002. *Pseudomonas aeruginosa* quorum-sensing systems may control virulence factor expression in the lungs of patients with cystic fibrosis. *Infect. Immun.* **70**(4): 1783–90.

19 Erickson, D. L., J. L. Lines, E. C. Pesci, V. Venturi and D. G. Storey 2004. *Pseudomonas aeruginosa relA* contributes to virulence in *Drosophila melanogaster*. *Infect. Immun.* **72**(10): 5638–45.

20 Fuqua, W., S. Winans and E. Greenberg 1994. Quorum sensing in bacteria: the LuxR\LuxI family of cell density-responsive transcriptional regulators. *J. Bacteriol.* **176**: 269–75.

21 Gallagher, L. A. and C. Manoil 2001. *Pseudomonas aeruginosa* PAO1 kills *Caenorhabditis elegans* by cyanide poisoning. *J. Bacteriol.* **183**(21): 6207–14.

22 Gallagher, L. A., S. L. McKnight, M. S. Kuznetsova, E. C. Pesci and C. Manoil 2002. Functions required for extracellular quinolone signaling by *Pseudomonas aeruginosa*. *J. Bacteriol.* **184**(23): 6472–80.

23 Gambello, M., S. Kaye and B. Iglewski 1993. LasR of *Pseudomonas aeruginosa* is a transcriptional activator of the alkaline protease gene (*apr*) and an enhancer of exotoxin A expression. *Infect. Immun.* **61**: 1180–4.

24 Gambello, M. J. and B. H. Iglewski 1991. Cloning and characterization of the *Pseudomonas aeruginosa lasR* gene, a transcriptional activator of elastase expression. *J. Bacterial.* **173**(9): 3000–9.

25 Guina, T., S. O. Purvine, E. C. Yi *et al.* 2003. Quantitative proteomic analysis indicates increased synthesis of a quinolone by *Pseudomonas aeruginosa* isolates from cystic fibrosis airways. *Proc. Natn. Acad. Sci. USA* **100**(5): 2771–6.

26 Hendrickson, E. L., J. Plotnikova, S. Mahajan-Miklos, L. G. Rahme and F. M. Ausubel 2001. Differential roles of the *Pseudomonas aeruginosa* PA14 *rpoN* gene in pathogenicity in plants, nematodes, insects, and mice. *J. Bacteriol.* **183**(24): 7126–34.

27 Hentzer, M., K. Riedel, T. B. Rasmussen *et al.* 2002. Inhibition of quorum sensing in *Pseudomonas aeruginosa* biofilm bacteria by a halogenated furanone compound. *Microbiology* **148**(1): 87–102.

28 Hentzer, M., H. Wu, J. B. Andersen *et al.* 2003. Attenuation of *Pseudomonas aeruginosa* virulence by quorum sensing inhibitors. *EMBO J.* **22**(15): 3803–15.

29 Heurlier, K., F. Williams, S. Heeb *et al.* 2004. Positive control of swarming, rhamnolipid synthesis, and lipase production by the posttranscriptional RsmA/RsmZ system in *Pseudomonas aeruginosa* PAO1. *J. Bacteriol.* **186**(10): 2936–45.

30 Heydorn, A., B. Ersboll, J. Kato *et al.* 2002. Statistical analysis of *Pseudomonas aeruginosa* biofilm development: impact of mutations in genes involved in twitching motility, cell-to-cell signaling, and stationary-phase sigma factor expression. *Appl. Environ. Microbiol.* **68**(4): 2008–17.

31 Hoang, T. T., S. A. Sullivan, J. K. Cusick and H. P. Schweizer 2002. Beta-ketoacyl acyl carrier protein reductase (FabG) activity of the fatty acid biosynthetic pathway is a determining factor of 3-oxo-homoserine lactone acyl chain lengths. *Microbiology* **148**(12): 3849–56.

32 Jackson, K. D., M. Starkey, S. Kremer, M. R. Parsek and D. J. Wozniak 2004. Identification of *psl*, a locus encoding a potential exopolysaccharide that is essential for *Pseudomonas aeruginosa* PAO1 biofilm formation. *J. Bacteriol.* **186**(14): 4466–75.

33 Jander, G., L. G. Rahme and F. M. Ausubel 2000. Positive correlation between virulence of *Pseudomonas aeruginosa* mutants in mice and insects. *J. Bacteriol.* **182**(13): 3843–5.

34 Juhas, M., L. Wiehlmann, B. Huber *et al.* 2004. Global regulation of quorum sensing and virulence by VqsR in *Pseudomonas aeruginosa*. *Microbiology* **150**(4): 831–41.

35 Kiratisin, P., K. D. Tucker and L. Passador 2002. LasR, a transcriptional activator of *Pseudomonas aeruginosa* virulence genes, functions as a multimer. *J. Bacteriol.* **184**(17): 4912–19.

36 Lamb, J. R., H. Patel, T. Montminy, V. E. Wagner and B. H. Iglewski 2003. Functional domains of the RhlR transcriptional regulator of *Pseudomonas aeruginosa*. *J. Bacteriol.* **185**(24): 7129–39.

37 Latifi, A., M. Foglino, K. Tanaka, P. Williams and A. Lazdunski 1996. A hierarchical quorum-sensing cascade in *Pseudomonas aeruginosa* links the transcriptional activators LasR and RhlR (VsmR) to expression of the stationary-phase sigma factor RpoS. *Molec. Microbiol.* **21**: 1137–46.

38 Latifi, A., M. Winson, M. Foglino *et al.* 1995. Multiple homologues of LuxR and LuxI control expression of virulence determinants and secondary metabolites through quorum sensing in *Pseudomonas aeruginosa* PAO1. *Molec. Microbiol.* **17**: 333–43.

39 Ledgham, F., C. Soscia, A. Chakrabarty, A. Lazdunski and M. Foglino 2003. Global regulation in *Pseudomonas aeruginosa*: the regulatory protein AlgR2 (AlgQ) acts as a modulator of quorum sensing. *Res. Microbiol.* **154**(3): 207–13.

40 Ledgham, F., I. Ventre, C. Soscia *et al.* 2003. Interactions of the quorum sensing regulator QscR: interaction with itself and the other regulators of *Pseudomonas aeruginosa* LasR and RhlR. *Molec. Microbiol.* **48**(1): 199–210.

41 Lesprit, P., F. Faurisson, O. Join-Lambert *et al.* 2003. Role of the quorum-sensing system in experimental pneumonia due to *Pseudomonas aeruginosa* in rats. *Am. J. Respir. Crit. Care Med.* **167**(11): 1478–82.

42 Mahajan-Miklos, S., L. G. Rahme and F. M. Ausubel 2000. Elucidating the molecular mechanisms of bacterial virulence using non-mammalian hosts. *Molec. Microbiol.* **37**(5): 981–8.

43 Mahajan-Miklos, S., M. W. Tan, L. G. Rahme and F. M. Ausubel 1999. Molecular mechanisms of bacterial virulence elucidated using a *Pseudomonas aeruginosa-Caenorhabditis elegans* pathogenesis model. *Cell* **96**(1): 47–56.

44 McGrath, S., D. S. Wade and E. C. Pesci 2004. Dueling quorum sensing systems in *Pseudomonas aeruginosa* control the production of the Pseudomonas quinolone signal (PQS). *FEMS Microbiol. Lett.* **230**(1): 27–34.

45 McKnight, S. L., B. H. Iglewski and E. C. Pesci 2000. The Pseudomonas quinolone signal regulates *rhl* quorum sensing in *Pseudomonas aeruginosa*. *J. Bacteriol.* **182**(10): 2702–8.

46 Miyata, S., M. Casey, D. W. Frank, F. M. Ausubel and E. Drenkard 2003. Use of the *Galleria mellonella* caterpillar as a model host to study the role of the type III secretion system in *Pseudomonas aeruginosa* pathogenesis. *Infect. Immun.* **71**(5): 2404–13.

47 Ochsner, U. A., A. Snyder, A. I. Vasil and M. L. Vasil 2002. Effects of the twin-arginine translocase on secretion of virulence factors, stress response, and pathogenesis. *Proc. Natn. Acad. Sci. USA* **99**(12): 8312–17.

48 Passador, L., J. Cook, M. Gambello, L. Rust and B. Iglewski 1993. Expression of *Pseudomonas aeruginosa* virulence genes requires cell-to-cell communication. *Science* **260**: 1127–30.

49 Passador, L., K. Tucker, K. Guertin, M. Journet, A. Kende and B. Iglewski 1996. Functional analysis of the *Pseudomonas aeruginosa* autoinducer PAI. *J. Bacteriol.* **178**, 5995–6000.

50 Pearson, J., E. Pesci and B. Iglewski 1997. Roles of *Pseudomonas aeruginosa las* and *rhl* quorum-sensing systems in control of elastase and rhamnolipid biosynthesis genes. *J. Bacteriol.* **179**(18): 5756–67.

51 Pearson, J., C. Van Delden and B. Iglewski 1999. Active efflux and diffusion are involved in transport of *Pseudomonas aeruginosa* cell-to-cell signals. *J. Bacteriol.* **181**(4): 1203–10.

52 Pearson, J. P., M. Feldman, B. H. Iglewski and A. Prince 2000. *Pseudomonas aeruginosa* cell-to-cell signaling is required for virulence in a model of acute pulmonary infection. *Infect. Immun.* **68**(7): 4331–4.

53 Pesci, E., J. Pearson, P. Seed and B. Iglewski 1997. Regulation of *las* and *rhl* quorum sensing in *Pseudomonas aeruginosa*. *J. Bacteriol.* **179**(10): 3127–32.

54 Pesci, E. C., J. B. Milbank, J. P. Pearson *et al.* 1999. Quinolone signaling in the cell-to-cell communication system of *Pseudomonas aeruginosa*. *Proc. Natn. Acad. Sci. USA* **96**(20): 11229–34.

55 Pham, T. H., J. S. Webb and B. H. Rehm 2004. The role of polyhydroxyalkanoate biosynthesis by *Pseudomonas aeruginosa* in rhamnolipid and alginate production as well as stress tolerance and biofilm formation. *Microbiology* **150**(10): 3405–13.

56 Plotnikova, J. M., L. G. Rahme and F. M. Ausubel 2000. Pathogenesis of the human opportunistic pathogen *Pseudomonas aeruginosa* PA14 in Arabidopsis. *Plant Physiol.* **124**(4): 1766–74.

57 Potvin, E., D. E. Lehoux, I. Kukavica-Ibrulj *et al.* 2003. *In vivo* functional genomics of *Pseudomonas aeruginosa* for high-throughput screening of new virulence factors and antibacterial targets. *Environ. Microbiol.* **5**(12): 1294–308.

58 Preston, M. J., P. C. Seed, D. S. Toder *et al.* 1997. Contribution of proteases and LasR to the virulence of *Pseudomonas aeruginosa* during corneal infections. *Infect. Immun.* **65**(8): 3086–90.

59 Pukatzki, S., R. H. Kessin and J. J. Mekalanos 2002. The human pathogen *Pseudomonas aeruginosa* utilizes conserved virulence pathways to infect the social amoeba *Dictyostelium discoideum*. *Proc. Natn. Acad. Sci. USA* **99**(5): 3159–64.

60 Rahme, L. G., F. M. Ausubel, H. Cao *et al.* 2000. Plants and animals share functionally common bacterial virulence factors. *Proc. Natn. Acad. Sci. USA* **97**(16): 8815–21.

61 Rahme, L. G., E. J. Stevens, S. F. Wolfort *et al.* 1995. Common virulence factors for bacterial pathogenicity in plants and animals. *Science* **268**(5219): 1899–902.

62 Rahme, L. G., M. W. Tan, L. Le *et al.* 1997. Use of model plant hosts to identify *Pseudomonas aeruginosa* virulence factors. *Proc. Natn. Acad. Sci. USA* **94**(24): 13245–50.

63 Rashid, M. H., K. Rumbaugh, L. Passador *et al.* 2000. Polyphosphate kinase is essential for biofilm development, quorum sensing, and virulence of *Pseudomonas aeruginosa*. *Proc. Natn. Acad. Sci. USA* **97**(17): 9636–41.

64 Reimmann, C., M. Beyeler, A. Latifi *et al.* 1997. The global activator GacA of *Pseudomonas aeruginosa* PAO positively controls the production of the auto-inducer N-butyryl-homoserine lactone and the formation of the virulence factors pyocyanin, cyanide, and lipase. *Molec. Microbiol.* **24**(2): 309–19.

65 Rumbaugh, K. P., J. A. Griswold and A. N. Hamood 1999. Contribution of the regulatory gene *lasR* to the pathogenesis of *Pseudomonas aeruginosa* infection of burned mice. *J. Burn Care Rehabil.* **20**(1): 42–9.

66 Rumbaugh, K. P., J. A. Griswold, B. H. Iglewski and A. N. Hamood 1999. Contribution of quorum sensing to the virulence of *Pseudomonas aeruginosa* in burn wound infections. *Infect. Immun.* **67**(11): 5854–62.

67 Rumbaugh, K. P., A. N. Hamood and J. A. Griswold 2004. Cytokine induction by the *P. aeruginosa* quorum sensing system during thermal injury. *J. Surg. Res.* **116**(1): 137–44.

68 Rust, L., E. C. Pesci and B. H. Iglewski 1996. Analysis of the *Pseudomonas aeruginosa* elastase (lasB) regulatory region. *J. Bacteriol.* **178**(4): 1134–40.

69 Schuster, M., A. C. Hawkins, C. S. Harwood and E. P. Greenberg 2004. The *Pseudomonas aeruginosa* RpoS regulon and its relationship to quorum sensing. *Molec. Microbiol.* **51**(4): 973–85.

70 Schuster, M., C. P. Lostroh, T. Ogi and E. P. Greenberg 2003. Identification, timing, and signal specificity of *Pseudomonas aeruginosa* quorum-controlled genes: a transcriptome analysis. *J. Bacteriol.* **185**(7): 2066–79.

71 Schuster, M., M. L. Urbanowski and E. P. Greenberg 2004. Promoter specificity in *Pseudomonas aeruginosa* quorum sensing revealed by DNA binding of purified LasR. *Proc. Natn. Acad. Sci. USA* **101**: 15833–9.

72 Silo-Suh, L., S. J. Suh, P. A. Sokol and D. E. Ohman 2002. A simple alfalfa seedling infection model for *Pseudomonas aeruginosa* strains associated with cystic fibrosis shows AlgT (sigma-22) and RhlR contribute to pathogenesis. *Proc. Natn. Acad. Sci. USA* **99**(24): 15699–704.

73 Singh, P. K., A. L. Schaefer, M. R. Parsek *et al.* 2000. Quorum-sensing signals indicate that cystic fibrosis lungs are infected with bacterial biofilms. *Nature* **407**(6805): 762–4.

74 Smith, K. M., Y. Bu and H. Suga 2003. Induction and inhibition of *Pseudomonas aeruginosa* quorum sensing by synthetic autoinducer analogs. *Chem. Biol.* **10**(1): 81–9.

75 Smith, K. M., Y. Bu and H. Suga 2003. Library screening for synthetic agonists and antagonists of a *Pseudomonas aeruginosa* autoinducer. *Chem. Biol.* **10**(6): 563–71.

76 Smith, R. S. and B. H. Iglewski 2003. *P. aeruginosa* quorum-sensing systems and virulence. *Curr. Opin. Microbiol.* **6**(1): 56–60.

77 Smith, R. S. and B. H. Iglewski 2003. *Pseudomonas aeruginosa* quorum sensing as a potential antimicrobial target. *J. Clin. Invest.* **112**(10): 1460–5.

78 Smith, R. S., R. Kelly, B. H. Iglewski and R. P. Phipps 2002. The *Pseudomonas* autoinducer N-(3-oxododecanoyl) homoserine lactone induces cyclooxygenase-2 and prostaglandin E2 production in human lung fibroblasts: implications for inflammation. *J. Immunol.* **169**(5): 2636–42.

79 Storey, D., E. Ujack, H. Rabin and I. Mitchell 1998. *Pseudomonas aeruginosa* lasR transcription correlates with the transcription of lasA, lasB, and toxA in chronic lung infections associated with cystic fibrosis. *Infect. Immun.* **66**(6): 2521–8.

80 Stover, C. K., X. Q. Pham, A. L. Erwin *et al.* 2000. Complete genome sequence of *Pseudomonas aeruginosa* PAo1, an opportunistic pathogen. *Nature* **406**(6799): 959–64.

81 Tan, M. W., L. G. Rahme, J. A. Sternberg, R. G. Tompkins and F. M. Ausubel 1999. *Pseudomonas aeruginosa* killing of *Caenorhabditis elegans* used to identify *P. aeruginosa* virulence factors. *Proc. Natn. Acad. Sci. USA* **96**(5): 2408–13.

82 Tang, H., M. Kays and A. Prince 1995. Role of *Pseudomonas aeruginosa* pili in acute pulmonary infection. *Infect. Immun.* **63**(4): 1278–85.

83 Tang, H. B., E. DiMango, R. Bryan *et al.* 1996. Contribution of specific *Pseudomonas aeruginosa* virulence factors to pathogenesis of pneumonia in a neonatal mouse model of infection. *Infect. Immun.* **64**(1): 37–43.

84 Tateda, K., Y. Ishii, M. Horikawa *et al.* 2003. The *Pseudomonas aeruginosa* auto-inducer N-3-oxododecanoyl homoserine lactone accelerates apoptosis in macro-phages and neutrophils. *Infect. Immun.* **71**(10): 5785–93.

85 Telford, G., D. Wheeler, P. Williams *et al.* 1998. The *Pseudomonas aeruginosa* quorum-sensing signal molecule N-(3-oxododecanoyl)-L-homoserine lactone has immunomodulatory activity. *Infect. Immun.* **66**(1): 36–42.

86 Toder, D., M. Gambello and B. Iglewski 1991. *Pseudomonas aeruginosa* LasA; a second elastase gene under transcriptional control of *lasR. Molec. Microbiol.* **5**: 2003–10.

87 van Delden, C., R. Comte and A. M. Bally 2001. Stringent response activates quorum sensing and modulates cell density-dependent gene expression in *Pseudomonas aeruginosa. J. Bacteriol.* **183**(18): 5376–84.

88 Vasil, M. L. and U. A. Ochsner 1999. The response of *Pseudomonas aeruginosa* to iron: genetics, biochemistry and virulence. *Molec. Microbiol.* **34**(3): 399–413.

89 Wagner, V. E., D. Bushnell, L. Passador, A. I. Brooks and B. H. Iglewski 2003. Microarray analysis of *Pseudomonas aeruginosa* quorum-sensing regulons: effects of growth phase and environment. *J. Bacteriol.* **185**(7): 2080–95.

90 Wagner, V. E., R. J. Gillis and B. H. Iglewski 2004. Transcriptome analysis of quorum-sensing regulation and virulence factor expression in *Pseudomonas aeruginosa. Vaccine* **22**(suppl. 1): S15–20.

91 Whiteley, M. and E. P. Greenberg 2001. Promoter specificity elements in *Pseudomonas aeruginosa* quorum-sensing-controlled genes. *J. Bacteriol.* **183**(19): 5529–34.

92 Whiteley, M., K. M. Lee and E. P. Greenberg 1999. Identification of genes controlled by quorum sensing in *Pseudomonas aeruginosa. Proc. Natn. Acad. Sci. USA* **96**(24): 13904–9.

93 Whiteley, M., M. R. Parsek and E. P. Greenberg 2000. Regulation of quorum sensing by RpoS in *Pseudomonas aeruginosa. J. Bacteriol.* **182**(15): 4356–60.

94 Williams, S. C., E. K. Patterson, N. L. Carty *et al.* 2004. *Pseudomonas aeruginosa* autoinducer enters and functions in mammalian cells. *J. Bacteriol.* **186**(8): 2281–7.

95 Wu, H., Z. Song, M. Givskov *et al.* 2001. *Pseudomonas aeruginosa* mutations in *lasI* and *rhlI* quorum sensing systems result in milder chronic lung infection. *Microbiology* **147**(5): 1105–13.

96 Wu, H., Z. Song, M. Hentzer *et al.* 2004. Synthetic furanones inhibit quorum-sensing and enhance bacterial clearance in *Pseudomonas aeruginosa* lung infection in mice. *J. Antimicrob. Chemother.* **53**(6): 1054–61.

97 Yorgey, P., L. G. Rahme, M. W. Tan and F. M. Ausubel 2001. The roles of *mucD* and alginate in the virulence of *Pseudomonas aeruginosa* in plants, nematodes and mice. *Molec. Microbiol.* **41**(5): 1063–76.

V. E. WAGNER AND B. H. IGLEWSKI

CHAPTER 2

The *Pseudomonas aeruginosa* quinolone signal

Everett C. Pesci

*The Brody School of Medicine at East Carolina University,
Greenville, NC, USA*

INTRODUCTION

Pseudomonas aeruginosa is a ubiquitous Gram-negative bacterium that is a major source of acute infections for immunocompromised individuals. This opportunistic pathogen also infects the lungs of most cystic fibrosis (CF) patients, causing a chronic infection that produces progressive lung damage throughout the life of the patient. *P. aeruginosa*'s ability to survive in almost any surroundings is augmented by an intricate cell-to-cell signaling scheme that controls a large number of cell functions. Through our ongoing attempts to eavesdrop on *P. aeruginosa*, we have learned that communities of this organism appear to be constantly chattering among themselves as they adapt to their environment. The *las* and *rhl* quorum sensing systems of *P. aeruginosa* are acyl-homoserine lactone-based signal systems that have been well characterized and are nicely reviewed in Chapter 1 of this book. The focus of this chapter will be a different type of signal, which has only recently been identified. The signal is 2-heptyl-3-hydroxy-4-quinolone and is referred to as the *Pseudomonas* quinolone signal (PQS). This signal is unique in that it is the only known quinolone molecule used as a cell-to-cell signal and *P. aeruginosa* is the only organism known to produce it.

DISCOVERY OF PQS

PQS was discovered while studying the effects of the *rhl* quorum sensing system on *lasB* induction. The *lasB* gene encodes LasB elastase, a protease considered to be a major *P. aeruginosa* virulence factor (1, 33). The regulation of this gene is complex, as it is directly controlled in a

Bacterial Cell-to-Cell Communication: Role in Virulence and Pathogenesis, ed. D. R. Demuth and R. J. Lamont. Published by Cambridge University Press. © Cambridge University Press 2005.

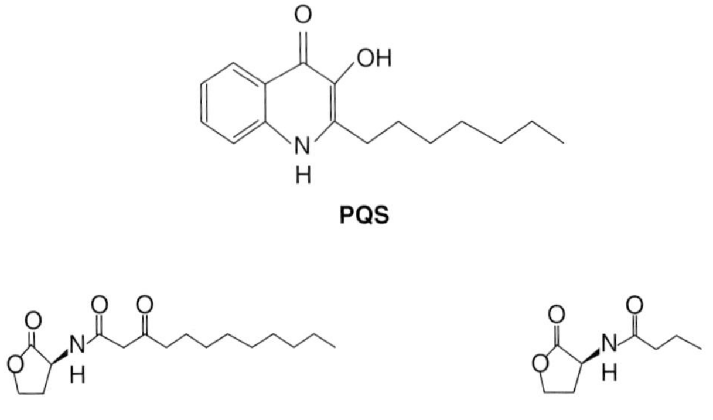

PQS

3-oxo-C$_{12}$-HSL

C$_4$-HSL

Figure 2.1. Structures of the three *Pseudomonas aeruginosa* cell-to-cell signals.

positive manner by both the *las* quorum sensing system (LasR and N-3-oxododecanoyl-L-homoserine lactone (3-oxo-C$_{12}$-HSL)) and the *rhl* quorum sensing system [RhlR and *N*-butyryl-L-homoserine lactone (C$_4$-HSL)] (2, 14, 29) (see Figure 2.1 for signal structures). In a *P. aeruginosa* *lasR* mutant, *lasB* is not induced because the entire quorum sensing cascade is interrupted (note that the *las* quorum sensing system induces the *rhl* quorum sensing system) (17, 31). The addition of C$_4$-HSL to a *lasR* mutant strain caused a minor induction of *lasB*; exogenous 3-oxo-C$_{12}$-HSL had no effect (28). However, when an organic extract of a spent culture supernatant from wild-type *P. aeruginosa* was added to a growing culture of the *lasR* mutant, the *lasB* gene was significantly induced (30). Because this induction of *lasB* could not be complemented with exogenous 3-oxo-C$_{12}$-HSL or C$_4$-HSL, it was obvious that a third *P. aeruginosa* cell-to-cell signal was involved. The signal was purified from *P. aeruginosa*; chemical analysis showed that it was 2-heptyl-3-hydroxy-4-quinolone (PQS) (30) (Figure 2.1). This structure was confirmed by acquiring synthetic PQS and showing that its chemical properties were identical to those of natural PQS (30). The synthetic compound also induced *lasB* in a manner similar to that of natural PQS (30). Thus, a new bacterial cell-to-cell signal had been identified.

RELATIONSHIP OF PQS TO THE *P. AERUGINOSA* QUORUM SENSING CIRCUITRY

The first clues as to the relationship between PQS and the *las* and *rhl* quorum sensing systems surfaced during the initial purification of PQS.

PQS was not produced at a detectable level by a *lasR* mutant, indicating that the *las* quorum sensing system directly or indirectly controlled a factor necessary for PQS production (30). It was also noted that PQS could not induce *lasB* in a strain that did not produce RhlR (30). Although this made it appear as though RhlR and PQS could be an effector protein and signal team, this has not been proven. An attempt to develop an *Escherichia coli* bioassay in which RhlR was expressed in the presence of a *lasB'–lacZ* reporter fusion and exogenous PQS failed (30). In addition, radiolabelled PQS did not specifically associate with cells expressing RhlR or LasR (E. Pesci, unpublished data) as is the case for C_4-HSL and 3-oxo-C_{12}-HSL, respectively (26, 29). Therefore, the specific target for PQS has remained elusive.

The link between quorum sensing systems and PQS grew stronger with the finding that PQS could induce *rhlI*, the gene that encodes the C_4-HSL synthase (25). PQS had a smaller positive effect on *lasR* and *rhlR* expression and no effect on *lasI* expression (25). Together, PQS and C_4-HSL were able to restore the expression of *rhlI* in a *lasR* mutant to a level similar to that seen in the wild-type strain (25). This indicated that PQS and C_4-HSL have an additive effect on *rhlI* expression. It also suggests that PQS could be acting as a linker between the *las* and *rhl* quorum sensing systems. Overall, the available data showed that the *las* quorum sensing system must induce a gene (or genes) required for PQS synthesis, and that PQS has a positive effect on the *rhl* quorum sensing system. The link implied from this is complicated because it has been shown that *rhlI* is expressed at a normal level in a mutant that does not produce PQS (13). This leads to the speculation that *rhlI* is controlled in multiple ways to ensure that this gene is induced when needed.

THE GENETICS OF PQS SYNTHESIS

Additional evidence for quorum sensing control of PQS production began to accumulate with the identification of genes required for PQS synthesis. Because PQS production required an active *las* quorum sensing system, a logical place to search for PQS biosynthetic genes was in a pool of mutants harboring insertions in quorum-sensing-controlled genes (34). This search proved fruitful with the discovery of a 3-oxo-C_{12}-HSL-controlled monooxygenase homolog, encoded by gene PA2587 (*pqsH*), that was required for PQS production (B. Bullman, W. Calfee, and E. Pesci, unpublished data). The function of PqsH has not been determined, but it is most likely adding the hydroxy group to 2-heptyl-4-quinolone in the final step of PQS synthesis (see Figure 2.2). Some evidence for this role was gathered by two

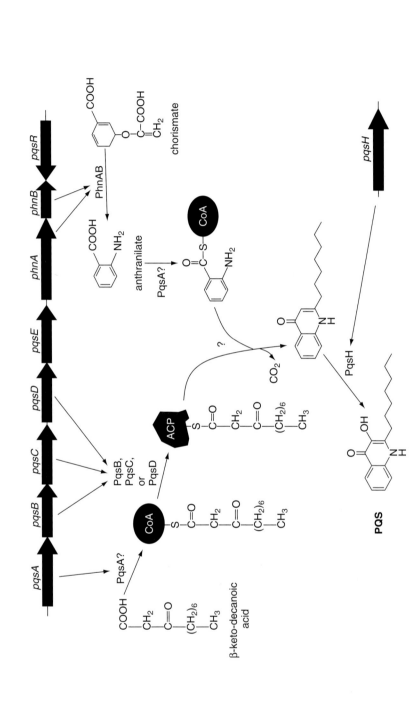

Figure 2.2. Proposed synthesis scheme for PQS. Enzymatic activities for proteins encoded by the PQS synthetic region have not been proven, but a possible synthesis scheme can be proposed based on homologies to known enzymes. In this model, the putative coenzyme A (CoA) ligase, PqsA, links CoA to either a ten-carbon acyl chain or anthranilate. The question marks by PqsA indicate the possibility of the enzyme's acting in one reaction or the other. The acyl chain is then transferred to an acyl carrier protein (ACP) through the action of one of the three putative acyl carrier protein synthases encoded by *pqsB*, *pqsC*, and *pqsD*. These different putative ACP synthases could be responsible for the great variability seen in 2-alkyl-4-quinolones. While this is happening, chorismate is converted to anthranilate through the action of the putative anthranilate synthase PhnAB. Anthranilate is then charged with CoA, possibly by PqsA as mentioned above. Anthranilate-CoA would then be condensed with the β-keto-decanoic acid attached to the ACP and the direct PQS precursor would result. The question mark by this reaction indicates that the condensation occurs via the action of an unidentified enzyme. The putative monooxygenase encoded by *pqsH* would then add the hydroxy group to the third carbon of the precursor to produce PQS.

separate lines of research. First, it was determined through an agar plate cross-feeding assay that a *pqsH* mutant can feed a *pqsC* or *pqsD* mutant with a compound that allows the production of PQS (13). Because these strains are most likely affected at an early step of PQS production, the *pqsH* mutant must be able to make a compound that is close to, or the direct precursor of, PQS. Second, radiolabeled 2-heptyl-4-quinolone is converted to PQS by a wild-type strain (9). This result suggested that this compound is the immediate precursor of PQS and implied that a monooxygenase should be responsible for the conversion of the precursor into PQS.

The main PQS biosynthetic gene cluster was then discovered because of the involvement of anthranilate in PQS synthesis. The basic synthetic pathway for quinolone compounds has been elucidated (6). This pathway suggested that anthranilate would be a precursor for PQS synthesis and this was shown to be the case when radiolabeled anthranilate was converted into PQS (3). *P. aeruginosa* encodes at least four anthranilate synthases, one of which is PhnAB. The *phnA* and *phnB* genes putatively encode the large and small subunit, respectively, of anthranilate synthase (11). PhnAB was initially believed to be directly involved in the synthesis of the blue phenazine compound pyocyanin (11). This role was made less likely with the discovery that *P. aeruginosa* encodes two separate, duplicate phenazine synthetic gene clusters, each with its own anthranilate synthase homolog (23). A survey of mutants that contained insertions in and around the *phnAB* region showed that this location was required for PQS synthesis (13). These mutants were originally pooled because they did not produce pyocyanin, indicating that PQS controls pyocyanin production (13). This also implies that the regulation of pyocyanin by *phnAB* happens indirectly through PQS. The genes of this region that were identified as being required for PQS production were: PA0998 and PA0999, both of which are β-keto-acyl–acyl carrier protein synthase homologs; PA1001 (*phnA*); and PA1003, which is homologous to members of the LysR-type regulator family (13). This genetic region is summarized in Figure 2.2. This study also showed that the monooxygenase homolog encoded by *pqsH* (gene PA2587) was required for PQS synthesis (13). At the same time, a parallel study led to the identification of this region as being important for the autolytic phenotype exhibited by older, plate-grown cultures of *P. aeruginosa* (7). It was shown that the phenotype of a mutant that produced colonies with concentric zones of lysis could be suppressed by the mutation of genes within the PQS synthetic region (7). Included in the genes that affected autolysis were: gene PA0996, which encodes a hydroxybenzoate CoA ligase homolog; gene PA0997, which encodes a beta-keto-acyl–acyl-carrier protein synthase homolog; PA0998, PA0999, PA1003, and PA2587 (the latter four genes were identified above) (7). All of these genes were found to be necessary for PQS production (7). To simplify the nomenclature of these genes, genes PA0996 through PA1000 were named *pqsABCDE* (7, 13). Gene PA1003 was previously published as *mvfR* (4), and was renamed *pqsR* because of its location in the PQS synthetic region and its role in regulating PQS production (discussed below) (7, 13). Gene PA2587 was subsequently named *pqsH* (7, 13). In addition, gene PA4444, which was responsible for the autolytic phenotype discussed above and encodes a homolog of a monooxygenase (like *pqsH*), was named *pqsL* (7). The loss of this gene actually causes PQS production to be increased 4-fold (7) and it has been shown that *pqsH* and *pqsL* may be competing to convert 2-heptyl-4-quinolone to PQS or an alternate compound (21).

Overall, this group of genes appears to encode much of what is required for the synthesis of PQS from basic precursor molecules. For the sake of caution, it must be pointed out that this statement is based only on homology, and true enzymatic roles for the encoded proteins have not been determined. Our proposed synthesis scheme is included in Figure 2.2.

THE REGULATION OF PQS SYNTHESIS

Mutant complementation studies and reverse transcriptase PCR experiments showed that *pqsABCDE* and *phnAB* are transcribed as two separate polycistronic operons (13, 24). As mentioned above, insertions in *pqsA*, *pqsB*, *pqsC*, *pqsD*, or *phnA* all lead to the loss of PQS (7, 13). The necessity of the *phnB* gene for PQS synthesis has not been determined; a mutant with an insertion in *pqsE*, which encodes a hypothetical protein, still makes a normal amount of PQS (13). Most interestingly, the *pqsE* mutant was still defective in pyocyanin production (13). This mutant was also found to be unable to respond to exogenous PQS, indicating that PqsE is important for the cell to respond to PQS (13). Further studies are necessary to determine the role of *pqsE* in the cellular response to PQS.

Preliminary studies on the expression of the *pqsABCDE* operon have shown that it is regulated in a very complex manner. A *pqsA'–lacZ* fusion was not induced in either a *lasI* or a *pqsR* mutant, indicating that *pqsABCDE* expression is positively controlled by the *las* quorum sensing system and PqsR (24). The production of PQS was greatly reduced in these mutant strains, thereby confirming the reporter fusion data (24). The reliance of *pqsA* on a functional *pqsR* gene was also confirmed in a separate study (9). Most interestingly, when the *pqsA'–lacZ* reporter was placed in a *rhlI* mutant, it was induced to a level that was greater than 7-fold higher than that seen in the wild type strain (24). PQS production was also increased 6-fold in this strain (24). These results indicated that *pqsABCDE* is regulated in a negative manner by the *rhl* quorum sensing system. Whether these regulation events are the result of direct or indirect interactions of LasR, RhlR, and/or PqsR with the promoter region of *pqsA* is not known. It is worthy of note that the *pqsA* promoter region has two sequences that are similar to quorum sensing operator sequences, which might allow for dual or alternate binding of both LasR and RhlR (24). Although the importance of these sequences has not been directly investigated, there are data that suggest that some type of competition is occurring at the *pqsA* promoter. The addition of exogenous C_4-HSL was able to inhibit the induction of *pqsABCDE* by exogenous 3-oxo-C_{12}-HSL in a mutant that produces neither signal (24). In addition, the production of PQS, which was induced by 3-oxo-C_{12}-HSL, was inhibited by C_4-HSL (24). Taken together, these data imply that a competition between C_4-HSL and 3-oxo-C_{12}-HSL occurs during the induction of *pqsABCDE*. A model for the proposed regulation scheme for PQS is presented in Figure 2.3.

The second operon of the PQS synthetic region is also tightly controlled. The expression of *phnAB* has been shown to be directly regulated by PqsR. A *pqsR* mutant does not express *phnAB* and PqsR was shown to bind specifically to the *phnAB* promoter region (4). PqsR also appeared to have no effect on the expression of *lasR* or *rhlR* (4). Overall, the available data suggest that the main role of PqsR is to act as a master regulator for the two operons of the PQS synthetic region to which it is adjacent.

Finally, the regulation of PqsR is also quite interesting. It has been found that this regulator is inactivated through cleavage. This regulatory event was discovered when cells expressing a PqsR–GST fusion protein were exposed for a short time to sterile, spent supernatant from a stationary-phase culture. The spent supernatant caused PqsR to be cleaved at amino acid 147 (total length is 332 amino acids) (4). This effect was found to be dependent on the bacterial growth phase: spent supernatant from log-phase cultures did not cause PqsR cleavage (4). This is apparently an autoregulation event, because spent supernatant from a *pqsR* mutant had no effect on cleavage (4). These data indicate that PqsR is negatively auto-regulated by a stationary-phase factor responsible for site-specific cleavage of the regulator. Overall, the regulation of PQS production is both interesting and complex; further studies are necessary to fully understand PQS production.

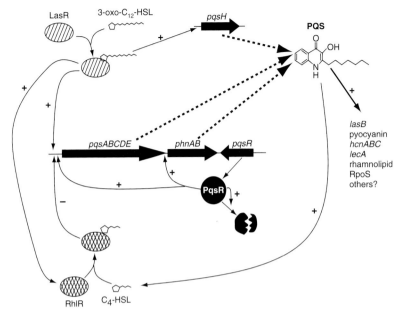

Figure 2.3. Proposed model for the regulation of PQS production. The *las* and *rhl* quorum sensing systems have a positive and negative effect, respectively, on the *pqsABCDE* operon. Whether these are direct or indirect effects is not clear. The *las* quorum sensing system also positively affects the transcription of *pqsH*. PqsR is required for the production of *pqsABCDE*; this regulator also directly interacts with the *phnAB* promoter to induce transcription. It is possible that the effects of the *las* and *rhl* quorum sensing systems are being directed through *pqsR*. This is currently being determined. PqsR is autoregulated either directly or indirectly through site-specific cleavage as indicated by the split protein. Together, *pqsH*, *pqsABCDE*, and *phnAB* produce enzymes that are responsible for the synthesis of PQS. PQS has positive effects on several genes or products including the *rhl* quorum sensing system. The mechanism of action for PQS is presumably through an unidentified regulatory protein.

THE TIMING OF PQS PRODUCTION

One of the unique features of PQS, compared with other quorum sensing signals, is the timing of its production. Unlike 3-oxo-C_{12}-HSL and C_4-HSL, the major production of which is started during the mid to late log phase of growth (27, 28), PQS production occurs at a later stage of growth. McKnight *et al.* (25) showed that PQS production began during the end of log-phase growth, and greatly increased during the stationary phase of growth. The production was at a maximal level when cultures had grown for 30–42 h and then greatly decreased after 48 h of growth (25). These cultures were grown in a rich medium, PTSB, and PQS was indirectly detected by using a *P. aeruginosa* bioassay. Most interestingly, PQS was produced in this manner despite the use of a *P. aeruginosa* strain that contained a constitutively expressed, mutated LasR protein which induced *las* quorum-sensing-controlled genes in the absence of 3-oxo-C_{12}-HSL (30). This indicated that PQS production required a stationary-phase factor that was not regulated by the *las* quorum sensing system (25). Not surprisingly, this expression pattern is similar to that of *phnAB*, which is also maximally expressed during stationary phase with a large decrease in expression thereafter (4). This decrease in *phnAB* expression, and thus in PQS production, is most likely due to the stationary-phase-dependent cleavage of PqsR (4). In addition to these data, two other studies (10, 20) have shown that PQS production starts during the late log phase of growth, becomes maximal during the stationary phase of growth, and decreases after reaching a peak concentration. Although PQS production was found to be slightly earlier in the studies of Lepine *et al.* (20), their more sensitive liquid chromatography/mass spectrometry detection assay and the use of LB media still showed a stationary-phase peak concentration of PQS that then declined. Most interestingly, this study found that a similar *P. aeruginosa* secondary metabolite, 2-heptyl-4-hydroxyquinoline *N*-oxide, was produced in a similar time frame but its concentration did not decrease after reaching a maximal level (20). This suggests that a specific PQS alteration may occur. Finally, Diggle *et al.* (10) used a thin layer chromatography detection assay to confirm that PQS production in LB media begins in the late log phase of growth and becomes maximal later during the stationary phase of growth. Their growth-curve study was stopped before PQS concentrations would have begun to decrease (10). The studies of Diggle *et al.* (10) also showed that the addition of exogenous PQS would induce *lecA* expression by *P. aeruginosa*. However, unlike other acyl homoserine lactone signals, PQS did not advance the timing of a gene that it controlled (*lecA*) (10). (Note: *lecA* encodes for the virulence determinant PA-IL lectin.) Once again, it is apparent that PQS activity relies on a stationary-phase factor that has not yet been identified.

Overall, the later timing of PQS production compared with other cell-to-cell signals suggests a role in responding to stationary-phase growth. *P. aeruginosa* has devoted numerous control mechanisms to ensure that cell-to-cell signaling occurs at the proper time (see (10) for a summary of quorum sensing timing), thereby leading one to conclude that, when it comes to cell-to-cell signaling, "timing is everything."

PQS AND *P. AERUGINOSA* VIRULENCE

Many cell-to-cell signals have been shown to be important for virulence (see (8) for review); PQS is not an exception. Strains that contain mutations within genes required for PQS production have been identified in many virulence gene identification experiments. PQS mutants are greatly reduced in the ability to produce paralytic killing in the nematode *Caenorhabditis elegans* (12, 13, 22). This was determined to be caused by a lack of hydrogen cyanide production, owing to the positive control that PQS exerts over the *hcnABC* operon (12, 13). A *pqsR* mutant was also unable to produce rotting in a lettuce leaf virulence assay and was greatly reduced in virulence for the burned mouse model of *P. aeruginosa* infection (32). In addition, this strain was found to be less virulent when tested in both an acute and a chronic model of mouse lung infection (18).

The importance of PQS exhibited in the virulence assays described above apparently stems from its ability to control the production of several virulence factors. In addition to hydrogen cyanide, which was mentioned previously, PQS controls the production of LasB elastase (3, 10, 25, 30) and pyocyanin (4, 12, 13, 32). Pyocyanin is important in *P. aeruginosa* pathogenesis (18) in that, when introduced into mouse lungs, it caused a large influx of neutrophils. However, the virulence of a *pqsR* mutant could not be complemented by the addition of exogenous pyocyanin, indicating that PQS regulates multiple important virulence factors (18). To this end, PQS has also been shown to control the production of hemolytic activity (32), PA-IL lectin (10), rhamnolipid (10, 16), Rhl R (10), Rpo S (10), and C_4-HSL (10, 25). It has been proposed that PQS acts directly or indirectly through RhlR (30) and controls primarily *rhl* quorum-sensing-regulated genes (10), but this theory has not yet been tested in a microarray experiment.

At the bacterial community level, PQS is known to have an effect on the growth of *P. aeruginosa* on stainless steel coupons (10); the overproduction of PQS causes autolysis, possibly through the activation of lysogenic phage or bacteriocin (7). These effects imply that PQS is likely to be involved in the formation of *P. aeruginosa* communities such as biofilms.

With regard to human infections, PQS has been found in sputum samples from cystic fibrosis patients infected with *P. aeruginosa* (5). The amount of PQS in sputum correlated nicely with the density of *P. aeruginosa* in the sample (5). PQS was also seen in bronchial alveolar lavage fluid from infected CF patients and in mucopurulent airway fluid from a lung resected from a CF transplant patient (5). Not surprisingly, the majority (9 out of 10) of *P. aeruginosa* isolates from CF patients produced PQS in laboratory cultures (5). One study has also shown that 4 out of 5 *P. aeruginosa* CF strains actually produced more PQS than the laboratory strain PAO1 (15). Furthermore, these strains produced PQS earlier (log phase) than the laboratory strain (stationary phase). This same study found that growing *P. aeruginosa* in low-magnesium media, which are thought to simulate the CF lung, caused PQS production to be increased 5-fold (15). Taken together, these data can allow one to theorize that PQS production increases, perhaps permanently, once a *P. aeruginosa* strain enters the lung of a CF patient.

Overall, numerous studies have shown that PQS is important for the virulence of *P. aeruginosa*. How this signal functions during infections is not known, but it is apparently providing some type advantage as *P. aeruginosa* adapts to its environment *in vivo*.

THE POTENTIAL OF PQS AS A DRUG TARGET

One of the most exciting aspects of cell-to-cell signaling is that the signals and their synthetic pathways are attractive targets for drug development. Agents that target these pathways would most likely not be bactericidal or bacteriostatic, but they hopefully will decrease virulence and augment a known antibiotic and/or allow the immune system to better clear the infecting organism. With this in mind, preliminary experiments have shown that the PQS synthetic pathway is a viable target. Calfee *et al.* (3) showed that an anthranilate analog, methyl anthranilate, inhibited the synthesis of PQS (3). This presumably would occur by the analog entering the PQS synthetic pathway before the condensation of anthranilate and the fatty acid chain (see Figure 2.2). Methyl anthranilate also caused a major decrease in the production of elastase activity by *P. aeruginosa*, suggesting that the analog interfered with the signaling pathway and thereby inhibited the induction of a PQS-controlled gene (*lasB*) (3). The effect of methyl anthranilate was also confirmed when Diggle *et al.* (10) showed this analog inhibited *lecA* induction and pyocyanin production, both of which are PQS-controlled virulence factors. These studies have been expanded upon to show that multiple anthranilate analogs can inhibit PQS signaling (S. McKnight and E. Pesci, unpublished data). This promising line of research will require a better understanding of the functions of different enzymes in the PQS synthetic pathway.

OTHER PQS-LIKE MOLECULES SYNTHESIZED BY *P. AERUGINOSA*

The 4-quinolones ("pyo compounds") produced by *P. aeruginosa* were first discovered because of the antibiotic activity they exhibited (6, 19). PQS was overlooked because it apparently does not have antibiotic activity, and this was at least partly proven by showing that it has no effect on *E. coli* or *Staphylococcus aureus* (30). However, PQS is one of at least 56 quinolone-type compounds produced by *P. aeruginosa* (21). These compounds all appear to fall under the control of *pqsR* and are most likely synthesized by the genes of the PQS synthetic region (9). The roles of these different secondary metabolites must be assessed on an individual basis. At this point, many of the compounds are known to have antibiotic activity; except for PQS, none has been shown to act directly as a cell-to-cell signal.

CONCLUDING REMARKS

Only five years have passed since the *Pseudomonas* quinolone signal was identified (30). In that time, a great deal has been learned about the synthesis and regulation of PQS. The signal has also been found to play an important role in the virulence of *P. aeruginosa*. Despite what has been learned, there is obviously a great deal that is not yet known about this fascinating *P. aeruginosa* cell-to-cell signal. The mechanism by which it causes gene induction, its role during infections, and the exploitation of its synthetic pathway for drug development are all future directions that will provide a wealth of knowledge about how and why *P. aeruginosa* uses PQS to help ensure survival. As we continue to keep our collective ears turned toward the *P. aeruginosa* culture plate, it is likely that this bacterium will once again prove to be a fascinating conversationalist.

ACKNOWLEDGEMENTS

E. C. Pesci is supported by a research grant from the National Institutes of Allergy and Infectious Disease (grant R01-AI46682). A thank-you is extended to T. R. de Kievit, E. A. Ling, D. S. Wade, J. P. Coleman, and A. J. Pesci for help in chapter preparation and thoughtful insight.

REFERENCES

1   Blackwood, L. L., R. M. Stone, B. H. Iglewski and J. E. Pennington 1983. Evaluation of *Pseudomonas aeruginosa* exotoxin A and elastase as virulence factors in acute lung infection. *Infect. Immun.* **39**: 198–201.

2 Brint, J. M. and D. E. Ohman 1995. Synthesis of multiple exoproducts in *Pseudomonas aeruginosa* is under the control of RhlR-RhlI, another set of regulators in strain PAO1 with homology to the autoinducer-responsive LuxR-LuxI family. *J. Bacteriol.* **177**: 7155–63.

3 Calfee, M. W., J. P. Coleman and E. C. Pesci 2001. Interference with *Pseudomonas* quinolone signal synthesis inhibits virulence factor expression by *Pseudomonas aeruginosa*. *Proc. Natn. Acad. Sci. USA* **98**: 11633–7.

4 Cao, H., G. Krishnan, B. Goumnerov *et al.* 2001. A quorum sensing-associated virulence gene of encodes a LysR-like transcription regulator with a unique self-regulatory mechanism. *Proc. Natn. Acad. Sci. USA* **98**: 14613–18.

5 Collier, D. N., L. Anderson, S. L. McKnight *et al.* 2002. A bacterial cell to cell signal in the lungs of cystic fibrosis patients. *FEMS Microbiol. Lett.* **215**: 41–6.

6 Cornforth, J. W. and A. T. James 1956. Structure of a naturally occurring antagonist of dihydrostreptomycin. *Biochem. J.* **63**: 124–30.

7 D'Argenio, D. A., M. W. Calfee, P. B. Rainey and E. C. Pesci 2002. Autolysis and autoaggregation in *Pseudomonas aeruginosa* colony morphology mutants. *J. Bacteriol.* **184**: 6481–9.

8 de Kievit, T. R. and B. H. Iglewski 2000. Bacterial quorum sensing in pathogenic relationships. *Infect. Immun.* **68**: 4839–49.

9 Deziel, E., F. Lepine, S. Milot, *et al.* 2004. Analysis of *Pseudomonas aeruginosa* 4-hydroxy-2-alkylquinolines (HAQs) reveals a role for 4-hydroxy-2-heptylquinoline in cell-to-cell communication. *Proc. Natn. Acad. Sci. USA* **101**: 1339–44.

10 Diggle, S. P., K. Winzer, S. R. Chhabra *et al.* 2003. The *Pseudomonas aeruginosa* quinolone signal molecule overcomes the cell density-dependency of the quorum sensing hierarchy, regulates rhl-dependent genes at the onset of stationary phase and can be produced in the absence of LasR. *Molec. Microbiol.* **50**: 29–43.

11 Essar, D. W., L. Eberly, A. Hadero and I. P. Crawford 1990. Identification and characterization of genes for a second anthranilate synthase in *Pseudomonas aeruginosa*: interchangeability of the two anthranilate synthases and evolutionary implications. *J. Bacteriol.* **172**: 884–900.

12 Gallagher, L. A. and C. Manoil 2001. *Pseudomonas aeruginosa* PAO1 kills *Caenorhabditis elegans* by cyanide poisoning. *J. Bacteriol.* **183**: 6207–14.

13 Gallagher, L. A., S. L. McKnight, M. S. Kuznetsova, E. C. Pesci and C. Manoil 2002. Functions required for extracellular quinolone signaling by *Pseudomonas aeruginosa*. *J. Bacteriol.* **184**: 6472–80.

14 Gambello, M. J. and B. H. Iglewski 1991. Cloning and characterization of the *Pseudomonas aeruginosa* lasR gene, a transcriptional activator of elastase expression. *J. Bacteriol.* **173**: 3000–9.

15 Guina, T., S. O. Purvine, E. C. Yi *et al.* 2003. Quantitative proteomic analysis indicates increased synthesis of a quinolone by *Pseudomonas aeruginosa* isolates from cystic fibrosis airways. *Proc. Natn. Acad. Sci. USA* **100**: 2771–6.

16 Kohler, T., C. van Delden, L. K. Curty, M. M. Hamzehpour and J. C. Pechere 2001. Overexpression of the MexEF-OprN multidrug efflux system affects cell-to-cell signaling in *Pseudomonas aeruginosa*. *J. Bacteriol.* **183**: 5213–22.

17 Latifi, A., M. Foglino, K. Tanaka, P. Williams and A. Lazdunski 1996. A hierarchical quorum-sensing cascade in *Pseudomonas aeruginosa* links the transcriptional activators LasR and RhlR (VsmR) to expression of the stationary-phase sigma factor RpoS. *Molec. Microbiol.* **21**: 1137–46.

18 Lau, G. W., H. Ran, F. Kong, D. J. Hassett and D. Mavrodi 2004. *Pseudomonas aeruginosa* pyocyanin is critical for lung infection in mice. *Infect. Immun.* **72**: 4275–8.

19 Leisinger, T. and R. Margraff. 1979. Secondary metabolites of the fluorescent pseudomonads. *Microbiol. Rev.* **43**: 422–42.

20 Lepine, F., E. Deziel, S. Milot and L. G. Rahme 2003. A stable isotope dilution assay for the quantification of the *Pseudomonas* quinolone signal in *Pseudomonas aeruginosa* cultures. *Biochim. Biophys. Acta* **1622**: 36–41.

21 Lepine, F., S. Milot, E. Deziel, J. He and L. G. Rahme 2004. Electrospray/mass spectrometric identification and analysis of 4-hydroxy-2-alkylquinolines (HAQs) produced by *Pseudomonas aeruginosa*. *J. Am. Soc. Mass Spectrom.* **15**: 862–9.

22 Mahajan-Miklos, S., M. W. Tan, L. G. Rahme and F. M. Ausubel 1999. Molecular mechanisms of bacterial virulence elucidated using a *Pseudomonas aeruginosa-Caenorhabditis elegans* pathogenesis model. *Cell* **96**: 47–56.

23 Mavrodi, D. V., R. F. Bonsall, S. M. Delaney *et al.* 2001. Functional analysis of genes for biosynthesis of pyocyanin and phenazine-1-carboxamide from *Pseudomonas aeruginosa* PAO1. *J. Bacteriol.* **183**: 6454–65.

24 McGrath, S., D. S. Wade and E. C. Pesci 2004. Dueling quorum sensing systems in *Pseudomonas aeruginosa* control the production of the *Pseudomonas* quinolone signal (PQS). *FEMS Microbiol. Lett.* **230**: 27–34.

25 McKnight, S. L., B. H. Iglewski and E. C. Pesci 2000. The *Pseudomonas* quinolone signal regulates *rhl* quorum sensing in *Pseudomonas aeruginosa*. *J. Bacteriol.* **182**: 2702–8.

26 Passador, L., K. D. Tucker, K. R. Guertin *et al.* 1996. Functional analysis of the *Pseudomonas aeruginosa* autoinducer PAI. *J. Bacteriol.* **178**: 5995–6000.

27 Pearson, J. P., K. M. Gray, L. Passador *et al.* 1994. Structure of the autoinducer required for expression of *Pseudomonas aeruginosa* virulence genes. *Proc. Natn. Acad. Sci. USA* **91**: 197–201.

28 Pearson, J. P., L. Passador, B. H. Iglewski and E. P. Greenberg 1995. A second N-acylhomoserine lactone signal produced by *Pseudomonas aeruginosa*. *Proc. Natn. Acad. Sci. USA* **92**: 1490–4.

29 Pearson, J. P., E. C. Pesci and B. H. Iglewski 1997. Roles of *Pseudomonas aeruginosa las* and *rhl* quorum-sensing systems in control of elastase and rhamnolipid biosynthesis genes. *J. Bacteriol.* **179**: 5756–67.

30 Pesci, E. C., J. B. Milbank, J. P. Pearson *et al.* 1999. Quinolone signaling in the cell-to-cell communication system of *Pseudomonas aeruginosa*. *Proc. Natn. Acad. Sci. USA* **96**: 11229–34.

31 Pesci, E. C., J. P. Pearson, P. C. Seed and B. H. Iglewski 1997. Regulation of *las* and *rhl* quorum sensing in *Pseudomonas aeruginosa*. *J. Bacteriol.* **179**: 3127–32.

32 Rahme, L. G., M. W. Tan, L. Le *et al.* 1997. Use of model plant hosts to identify *Pseudomonas aeruginosa* virulence factors. *Proc. Natn. Acad. Sci. USA* **94**: 13245–50.

33 Tamura, Y., S. Suzuki and T. Sawada 1992. Role of elastase as a virulence factor in experimental *Pseudomonas aeruginosa* infection in mice. *Microb. Pathogen.* **12**: 237–44.

34 Whiteley, M., K. M. Lee and E. P. Greenberg 1999. Identification of genes controlled by quorum sensing in *Pseudomonas aeruginosa*. *Proc. Natn. Acad. Sci. USA* **96**: 13904–9.

CHAPTER 3

Quorum-sensing-mediated regulation of plant–bacteria interactions and *Agrobacterium tumefaciens* virulence

Catharine E. White and Stephen C. Winans

Department of Microbiology,
Cornell University, Ithaca, NY, USA

INTRODUCTION

Plant-associated bacteria have a wide range of interactions with their hosts, from non-specific associations to more dedicated symbiotic or pathogenic interactions. Many complex interactions take place between plant roots and associated bacteria, fungi, and protozoa in a highly diverse and dense community within the rhizosphere. Bacterial cell-to-cell communication systems in this ecological niche appear to affect biofilm formation, pathogenesis, and production of siderophores and antibiotics. These activities are no doubt important in root colonization as well as in symbiosis and pathogenesis. Exciting developments and current studies in understanding the many complex interactions in the rhizosphere include both the characterization of the microbial communities involved and the responses of the plant hosts to these communities. Cell-to-cell signaling between members of the community is no doubt critical for these interactions to sense population densities and diffusion barriers in the rhizosphere. Such studies are beyond the scope of this chapter, but we refer the reader to recent reviews of this field (43, 65, 82).

Perhaps the best-characterized group of soil bacteria that serves as the model for understanding plant–bacteria associations is the Rhizobiaceae. This family, in the alpha subgroup of the Proteobacteria, includes members of the genera *Rhizobium*, *Sinorhizobium*, *Mesorhizobium*, *Azorhizobium*, and *Bradyrhizobium* (collectively referred to here as rhizobia), which form symbiotic relationships with host plants, and several pathogenic species of the genus *Agrobacterium* (including *A. tumefaciens*, *A. rhizogenes*, *A. vitis*, and *A. rubi*, here referred to as agrobacteria). Rhizobia form symbiotic relationships with specific host plants by inducing formation of nodules

Bacterial Cell-to-Cell Communication: Role in Virulence and Pathogenesis, ed. D. R. Demuth and R. J. Lamont. Published by Cambridge University Press. © Cambridge University Press 2005.

on the plant roots. Colonizing bacteria are restricted to these nodules, and benefit the host by fixing atmospheric nitrogen. In contrast, pathogenic agrobacteria cause tumors or other neoplasias on a wide variety of host plants. Although rhizobia are beneficial to their plant hosts and agrobacteria are plant pathogens, their initial interactions with the host plants are very similar. The colonizing bacteria must respond to host-released signals during the initial stages of infection and attach to the plant surface. In both cases, these bacteria synthesize and respond to diffusible chemical signals called acyl-homoserine lactones (AHLs). AHL-mediated cell-to-cell communication in rhizobia is reviewed elsewhere (30, 65, 84). For the remainder of this chapter we will focus on *A. tumefaciens*.

THE Ti PLASMIDS OF *A. TUMEFACIENS* AND THE DISCOVERY OF AHLs AS CONJUGAL PHEROMONES

A. tumefaciens are primarily soil-dwelling bacteria, but many isolates carry large (200 kb) plasmids that are called tumor-inducing (Ti) plasmids, which are required for the formation of crown gall tumors on infected plants. There is significant variation between strains of *A. tumefaciens*; isolates are most often classified according to the opines that they can catabolize (despite the fact that Ti plasmids can generally catabolize more than one type of opine) (94). The best-characterized *A. tumefaciens* strains contain either so-called octopine-type Ti plasmids (including strains A6, B6, Ach5, 15966, and R10) or nopaline-type Ti plasmids (including strains C58 and T37). The full genome sequence of strain C58 has been completed by two different groups (32, 87). Its circular chromosome shows striking synteny with that of *Sinorhizobium meliloti*, whereas its linear chromosome and two plasmids appear unique. A composite sequence of the octopine-type Ti plasmid has also been published (89).

Almost all of the genes that are required for tumor formation are located on these Ti plasmids. Tumor formation requires the products of approximately 25 *vir* genes (71, 85), which process and transfer oncogenic fragments of DNA (T-DNA) from the Ti plasmid into the host cell, where they are integrated into the host genomic DNA. Sequence analysis indicates that the T-DNA transfer apparatus evolved from a bacterial conjugation system (42, 48, 49, 68). In the host cell, the T-DNA-encoded genes direct the overproduction of phytohormones that cause the cell proliferation that is the hallmark of this disease (56). The T-DNA also carries a number of genes whose products synthesize opines in the infected plant cells. These compounds are released from the plant cells, taken up by the agrobacteria via dedicated ABC-type uptake systems, and used as sources of energy, carbon, and, in some cases, nitrogen or phosphorus (17). The genes required for opine uptake and catabolism are also located on the Ti plasmid and are positively regulated by the cognate opine (17). Although there is a remarkable diversity in the types of opine used by *A. tumefaciens*, there is always a perfect match between the types of opine encoded on the T-DNA and the opine-catabolic genes of a Ti plasmid.

Ti plasmids have a second complete set of DNA transfer genes. These *tra* and *trb* genes direct the interbacterial conjugal transfer of the entire Ti plasmid (24) and are organized into three operons (Figure 3.1) (1). The *traAFBH* and the *traCDG* operons are divergently transcribed with the origin of transfer (*oriT*) lying within the intergenic region between these operons. The third operon includes the *traI* and *trbBCDEJKL* genes, most of which are expected to be involved in mating pair formation and synthesis of the conjugal pili. Some of these genes share a common ancestry with a number of the *vir* genes, although sequence similarities between the *tra* and *vir* genes are generally rather weak. Ti plasmid conjugation is reviewed in more detail elsewhere (24, 86).

Ti plasmid conjugation played a critical role in research on the mechanism of *A. tumefaciens* virulence. In early experiments, it was discovered that the formation of crown gall tumors on plants by *A. tumefaciens* required a genetic element that could be horizontally transferred by cell–cell contact, suggesting that this trait was plasmid-encoded (44, 45). This plasmid was subsequently visualized by gel electrophoresis (79). Later, a portion of this plasmid (the T-DNA) was found in infected plant cells in the bacteria-induced tumors (57).

In these early studies, conjugal transfer of the Ti plasmid was detectable only in crown gall tumors (44, 45, 79), suggesting that these tumors produce some factor that is essential for conjugation. Following the discovery that crown gall tumors produce opines, Kerr and colleagues soon demonstrated that conjugation could occur *ex planta* in the presence of these compounds (31, 46). In these experiments, conjugation of the octopine-type Ti plasmids (such as pTiR10) was stimulated by the opine octopine, whereas conjugation of nopaline-type Ti plasmids (such as pTiC58) required agrocinopine A or B. Opines that stimulate conjugation are referred to as conjugal opines.

In 1991, Kerr and colleagues discovered a second compound that stimulated conjugal transfer of Ti plasmids; however, unlike opines, this compound was produced by the bacteria themselves (91). This diffusible compound, originally called conjugation factor (CF), stimulated conjugation in

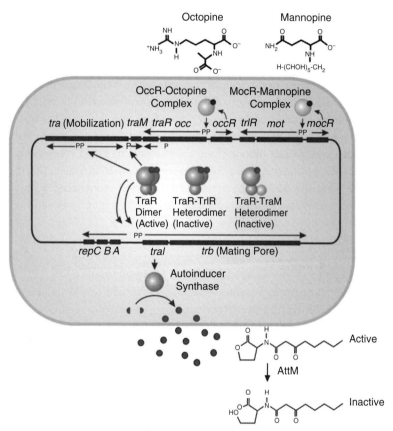

Figure 3.1. The TraR–TraI quorum sensing system of *Agrobacterium tumefaciens*. TraI synthesizes OOHL, which diffuses across the cellular envelope. At high population density, OOHL accumulates in the cell and binds to TraR. TraR activates expression of the two *tra* promoters, the *trb* operon which includes *traI*, *traM*, and *repABC*. OccR activates transcription of *traR* in the presence of octopine, and MocR activates *trlR* in the presence of mannopine. TraM and TrlR are anti-activators of TraR. Finally, at the stationary phase of growth, *attM* is transcribed, resulting in inactivation of OOHL.

A. tumefaciens strains carrying octopine-type plasmids but only when they were cultured in the presence of octopine. Conjugation of Ti plasmids that otherwise transferred with low efficiency was dramatically stimulated by CF, but only when it was provided to the conjugal donors. Not long after the discovery that CF stimulates conjugation, it was shown that this compound could specifically increase the transcription of a *tra–lacZ* fusion (66). By using this fusion as a reporter, four bioactive compounds were identified, including 3-oxo-C8-homoserine lactone (OOHL), 3-oxo-C6-homoserine lactone (OHHL), and two chemical derivatives of those compounds that may have been produced during purification (90). Following the identification of these compounds as the CF of octopine-type plasmids, it was shown that nopaline-type plasmids produce OOHL as the only active compound (38).

N-acyl-homoserine lactones (AHLs) have also been discovered in many other proteobacteria, beginning with *Vibrio fischeri* (21). These compounds are released and detected by populations of bacteria and are thought to permit cell-to-cell communication within a population, leading to coordination of a wide variety of behaviors. These signaling systems are thought to allow cells to estimate their population densities and to carry out various types of behavior only at high population densities, a phenomenon often referred to as autoinduction or quorum sensing.

IDENTIFICATION OF A Ti-PLASMID-ENCODED AHL SYNTHASE AND AHL RECEPTOR

The genes required for the synthesis and detection of the AHLs in the nopaline- and octopine-type plasmids were discovered concurrently by two different groups. In a study using random Tn5 mutagenesis, Farrand and colleagues isolated an insertion that eliminated the requirement of inducing opines for conjugation (3). Sequence analysis indicated that this transposon had inserted upstream of a gene called *traR* and was probably being over-expressed from an outward-reading promoter on the transposon (66). The role of *traR* in *tra* gene expression was verified by expressing this gene from a constitutive promoter along with a *tra–lacZ* fusion in a Ti-plasmid-less strain of *A. tumefaciens*. β-Galactosidase expression was induced only in the presence of OOHL or OHHL, although the former inducer was far more potent. This experiment suggested that TraR was indeed a receptor for OOHL, and that it activated expression of conjugation genes.

Bioassays using a *tra–lacZ* reporter strain indicated that *A. tumefaciens* synthesizes AHLs, suggesting that it encoded a protein homologous to LuxI of *V. fischeri*. To search for such a gene, Farrand and colleagues tested random Tn3 insertions in the nopaline-type Ti plasmid for production of OOHL (39). All of the insertions that disrupted OOHL synthesis mapped to one gene, *traI*, which was indeed a *luxI* homolog. TraR upregulated the expression of *traI* in the presence of exogenous autoinducer, strongly increasing OOHL production. The *traI* gene is the first gene in the *trb* operon, which encodes a type IV secretion system that is required for conjugation.

The *traR* and *traI* genes of the octopine-type Ti plasmid were identified at about the same time. The *traR* gene was discovered in a random Tn5*gus* mutagenesis study designed to identify genes that were inducible by octo-pine (9–12). As octopine was known to stimulate Ti plasmid conjugation, all octopine-inducible fusions were screened for defects in conjugation. Although none of them caused a decrease in conjugation, one of the fusions unexpectedly showed enhanced conjugation (29). This transposon inser-tion had disrupted a gene called *traM*, which is adjacent to *traR*. From this fusion, *traR* was eventually identified and found to stimulate conjugation when overexpressed. The reason that a *traM* null mutation elevated conjugation was unraveled in later studies (see below).

The *luxI* homolog *traI* on the octopine-type Ti plasmid was identified by screening a cosmid library for the synthesis of OOHL (29). Like *traI* of nopaline-type Ti plasmids, this *traI* gene appeared to be the first gene in the *trb* operon. Both OOHL production and conjugation were disrupted by a Tn5 insertion that mapped to the *traI* gene. However, when *traI* was expressed *in trans* in the same strain, OOHL production was restored but conjugation was not. These results indicated that *traI* was indeed the first gene of the *trb* operon.

In an effort to identify the reaction mechanism for OOHL synthesis, TraI was purified as a His6-tagged fusion and found to produce OOHL in vitro when added to a complex extract of bacterial proteins and small molecules (55). TraI also synthesized OOHL in a defined buffer supple-mented with 3-oxooctanoyl-ACP and S-adenosylmethionine (SAM). This was the first time that the activity of any LuxI-type protein had been reconstituted in vitro and the substrates identified. 3-Oxooctanoyl-CoA was inactive as a fatty acid group donor. Similarly, a variety of possible precursors for the homoserine lactone ring were tested, but only SAM was active. In the same study, purified TraI could transfer a radiolabeled car-boxyl carbon of SAM to the OOHL product, providing further evidence that SAM is the precursor for the homoserine lactone ring. A similar reaction mechanism was later described for other LuxI-type proteins (64, 78). Recently, the crystal structures of two proteins from this family were solved, EsaI of *Pantoea stewartii* and LasI of *Pseudomonas aeruginosa* (33, 83). As EsaI and LasI synthesize different AHLs, 3-oxo-C6- and 3-oxo-C12-homoserine lactone, respectively, comparisons of these two structures are useful for understanding not only the reaction mechanism but also the determinants of acyl-ACP specificity.

REGULATION OF *traR* EXPRESSION

The discovery of the TraR–TraI quorum sensing system of the Ti plasmids helped explain previous observations that conjugation of these plasmids requires opines. For octopine-type plasmids such as pTiR10, the *traR* gene is at the distal end of the *occ* operon, which is activated by octopine via OccR (Figure 3.1) (28). When *traR* is expressed from a constitutive promoter, conjugation does not require octopine. Therefore, regulation of *traR* transcription by OccR fully explains the requirement of octopine for conjugation. In the nopaline-type plasmids, *traR* expression is also directly regulated by opines but through a different mechanism. In pTiC58, the conjugal opines agrocinopine A and B induce expression of the *acc* genes and the divergently transcribed *arc* operon, which includes *traR* (2). In early studies of this system it was found that a null mutation of *accR* causes constitutive expression of these genes, and *traR* overexpression also results in constitutive conjugation but has no effect on the *acc* genes. More recent evidence from experiments with purified AccR and promoter DNA indicates that the expression of the *arc* and *acc* operons is directly repressed by AccR, and that this repression is relieved by agrocinopines (67).

The regulation of *traR* by opines appears to have evolved independently several times. OccR is an activator of transcription when it binds to the conjugal opine octopine, whereas AccR acts as a repressor in the absence of agrocinopines. OccR and AccR are not related proteins: OccR is in the LysR family of transcriptional regulators but AccR is related to the Lac repressor (2, 28). Finally, these *traR* genes are located within completely dissimilar operons.

Regulation of *traR* expression by opines has since been described for a number of other plasmids in *A. tumefaciens*, and in general these follow the models described above for octopine- and nopaline-type plasmids (Figure 3.2). On pTiChry5, *traR* is in a two gene operon called *arc* (58). The expression of this operon is induced by the agrocinopines C and D, which are thought to relieve repression by an AccR homolog that is encoded nearby (58). Agrocinopines C and D are also known to induce the transfer of the agropine-type Ti plasmid pTiBo542 (23). Another nopaline-type plasmid is pATK84b from *A. radiobacter* isolate K84 (15, 34, 35). This *Agrobacterium* strain is unable to induce the formation of crown gall tumors, and in fact pATK84b does not carry the *vir* genes or T-DNA of the Ti plasmids. However, pATK84b does allow this strain to compete for opines as a source of nutrients, as it has functional copies of nopaline and agrocinopine A and B catabolic genes (15, 16). Interestingly, pATK84b also has two copies of *traR* (59). One copy, $traR_{noc}$, is the last gene of the *nox* operon and is induced by nopaline, whereas $traR_{acc}$ is induced by agrocinopines A and B. Each of these two *traR* genes is required for induction of transfer by the cognate opine (59).

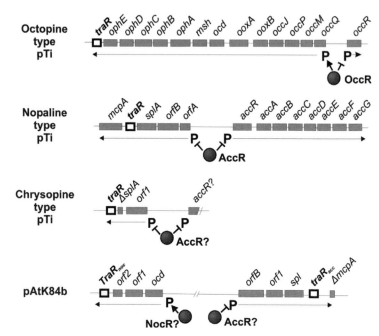

Figure 3.2. Regulation of *traR* on an octopine-type Ti plasmid and two nopaline-type Ti plasmids. Also shown is pAtK84b, which does not have *vir* genes or T-DNA but does carry opine catabolic genes. On octopine-type Ti plasmids, *traR* is activated by OccR in response to octopine. On the nopaline-type Ti plasmids, *traR* is transcribed when repression by AccR is relieved by agrocinopines. pAtK84b has two *traR* genes: one is thought to be activated by NocR in reponse to nopaline, and the other is repressed by AccR.

POST-TRANSCRIPTIONAL REGULATION OF TraR ACTIVITY

The TraR–TraI regulatory system is inhibited by a surprising number of proteins, one of which is TraM. A null mutation in *traM* elevates *tra* gene transcription and conjugation, whereas overexpression of *traM* abolishes *tra* gene expression in a strain expressing wild-type levels of TraR (26, 38). Providing high levels of OOHL did not stimulate *tra* gene expression. However, *tra* gene transcription was restored by overexpressing TraR, suggesting that TraM and TraR may interact stoichiometrically.

TraR was demonstrated by yeast two-hybrid assays and far western immunoblots to interact directly with TraM (40). Deletion and point mutations showed that binding is mediated by the C-termini of both proteins (40). TraM could both prevent the interaction of TraR with its DNA binding site and bind to TraR in protein–DNA complexes to cause release of TraR from DNA (51). Analysis by surface plasmon resonance showed that TraR–TraM binding is stable and has an affinity in the nanomolar range (75). TraR–TraM complexes eluted from a gel filtration column at a molecular mass of approximately 60 kDa in one study (75) or 157 kDa in a second study (80). The former is consistent with a TraR monomer binding one or two TraM monomers; the latter is consistent with two TraR dimers binding two TraM dimers. This discrepancy must be due to differences in experimental conditions. Several TraM point mutants were isolated that still bound TraR but did not inhibit TraR activity (46, 70), suggesting that binding and inactivation may occur in two sequential steps. The TraM structure was solved by two groups using X-ray crystallography (75, 8). TraM crystallized as a dimer, with each monomer consisting largely of two antiparallel α-helices. The monomers were intimately associated in the dimer, and significant hydrophobic surface was buried upon dimerization.

It is not clear how TraM–TraR interactions disrupt TraR–DNA interactions; however, two different mechanisms have been suggested. Chen and colleagues proposed that the TraM and TraR homodimers may dissociate upon binding to each other so that a TraR–TraM anti-activation complex can form, perhaps interacting through the surface of TraM that is otherwise buried in the homodimer interface (8). Evidence for this model is that TraM dimers, when their subunits are covalently cross-linked to each other, can bind to TraR but are unable to inactivate the protein (8). Mutational analyses of TraM also suggest that residues involved in initial binding to TraR are different from those required for TraR inactivation (51, 75). A second study reached different conclusions, proposing that two dimers of TraM are "clamped" between two dimers of TraR, with part of each of the four TraR DNA-binding domains bound into one of two narrow grooves on the surface of each TraM dimer (80). This complex would physically inhibit the TraR dimers from binding to DNA. Structural analysis of TraR–TraM complexes will help to resolve the intriguing question of how TraM mediates inactivation of TraR, although the static structure will not provide a complete picture, inasmuch as TraR binding and inactivation are thought to occur in several sequential steps.

The adaptive significance of TraM is not understood. On both octopine- and nopaline-type plasmids, expression of *traM* is positively regulated by TraR and OOHL (25, 36), indicating that this quorum-sensing regulon induces its own antagonist and suggesting that TraM may be acting as a governor to limit expression of this regulon (Figure 3.1). TraM is a highly conserved member of TraR–TraI quorum-sensing systems, and is even associated with these systems on the symbiosis megaplasmids of *Rhizobium* spp. (36).

Another TraR inhibitor, designated TrlR, is encoded only on octopine-type Ti plasmids. These Ti plasmids incite tumors that synthesize several opines, including mannopine. In the course of analyzing the *mot* operon, which is required for the utilization of mannopine, a gene, designated *trlR*, was identified. This gene is virtually identical to *traR* except for a frameshift mutation between the N-terminal pheromone binding domain and the C-terminal DNA binding domain (60). As might be expected, TrlR is inactive in *tra* gene expression, whereas correction of the frameshift mutation by site-directed mutagenesis resulted in a *trlR* mutant that encoded a fully functional protein (95). Identical genes were subsequently identified on a number of other octopine-type Ti plasmids, indicating that the frameshift within *trlR* is not a laboratory artefact but in fact is disseminated in nature (95). TrlR is not only non-functional but also inhibits TraR activity by forming inactive heterodimers (7). Similar truncated alleles of LuxR (made in the laboratory) have similar properties in that they block the function of native LuxR, probably by forming inactive heterodimers (14).

Expression of *trlR* is induced in response to mannopine, probably via the MocR protein (Figure 3.1) (60, 95). Mannopine decreases conjugation specifically due to *trlR*, as expressing this gene from a constitutive promoter also blocked conjugation (7). In the same study it was found that expression of *trlR* was strongly repressed by favored catabolites, including succinate, glutamine, and tryptone. As these nutrients could restore TraR activity by blocking *trlR* gene expression, it was speculated that TrlR functions as an inhibitor of the energetically expensive process of conjugation when nutrients are limiting.

TraR activity is regulated by a third gene, *attM*, although in this case the mechanism of inhibition is more indirect than described above. This gene is a homologue of *aiiA* from *Bacillus cereus* (32, 87); both genes direct the inactivation of AHLs by hydrolyzing their lactone ring, forming the corresponding N-acyl homoserine (Figure 3.1) (18, 19, 89). The *attM* gene lies within the *attKLM* operon, where *attK* and *attL* encode a predicted semi-aldehyde dehydrogenase and alcohol dehydrogenase, respectively (88). If these enzymes form a pathway, the predicted final product of OOHL metabolism would likely be N-(3-oxooctanoyl)aspartate.

The *attKLM* operon is repressed by the product of the divergent *attJ* gene; expression is induced at stationary phase in response to either carbon or nitrogen limitation (83). This provides suggestive evidence that *attKLM* is a catabolic operon that is induced only if its natural substrates are present and favored carbon sources are absent. It is likely that the natural substrates interact with AttJ to derepress expression of this operon. However, AHLs are not necessarily the natural substrates for these enzymes, as expression of the *attKLM* operon is not induced by any tested AHLs (89) (Y. Chai and S. C. Winans, unpublished data). Even if this pathway is not dedicated to inactivation and catabolism of OOHL, it is possible that *A. tumefaciens* can recycle these signal molecules. AHL recycling as an energy and nitrogen source has been demonstrated in *Variovorax paradoxus* and in *Pseudomonas aeruginosa* (37, 47).

THE ROLE OF OOHL IN TraR MATURATION

TraR activity is checked at yet another level, in that the protein is rapidly degraded by cytoplasmic proteases in the absence of OOHL. When TraR is strongly overexpressed in *E. coli* or in *A. tumefaciens*, it accumulates in the insoluble fraction of cell lysates. This led one group to suggest that, in the absence of OOHL, TraR binds to the cytoplasmic membrane and that it is released by OOHL to the cytoplasm, where it dimerizes and binds DNA (70). However, another group had a different interpretation, in which this insoluble protein accumulated in inclusion bodies that formed owing to TraR overproduction and were biologically irrelevant. When *traR* was mildly overexpressed, the protein accumulated only when OOHL was present, and was otherwise degraded by cytoplasmic proteases. This finding was first made by western immunoblots and confirmed by pulse–chase experiments (97). These data led us to conclude that OOHL rescues TraR from proteolysis during or directly after translation, and may act as a scaffold for TraR folding. In addition, purified apo-TraR was rapidly degraded by trypsin to oligopeptides, whereas TraR–OOHL complexes were more resistant to the proteases and were cleaved only at the accessible interdomain linkers of the protein (97). Again, this indicates that TraR absolutely requires OOHL to fold into its native and stable conformation.

Contrary to the well-established lock-and-key model of protein–ligand recognition, it is becoming increasingly evident that ligand binding induces conformational changes in many other protein receptors, from ordering of short loops to disorder-to-order transitions of the entire polypeptide chain (20). In many cases these proteins do not accumulate in the absence of a regulatory signal. The term "intrinsically unstructured" was coined by Dyson and Wright (20) to signify the importance of these proteins in cellular processes and as models for studying protein-folding mechanisms. Like TraR, many proteins that are intrinsically unstructured are involved in highly time-dependent processes. These include cell-cycle regulators required for inducing apoptosis, regulatory RNA binding proteins, proteins involved at various points in signal transduction pathways, and many other transcriptional regulators involved in activation or repression (20). It has also been suggested that one function of a disorder-to-order transition upon binding is to increase the binding rates for a faster response to the stimulus by increasing the likelihood that initial long-range interactions will occur (72). Incorporation of OOHL into the folding process may also optimize the specificity of the protein–ligand interaction, as has been suggested for some other intrinsically unstructured proteins (20).

STRUCTURE AND FUNCTION STUDIES OF TraR

Knowing that TraR requires OOHL for solubility enabled purification of amounts sufficient for X-ray crystallography. There are now two crystal structures of TraR available: both are ternary complexes of TraR, OOHL, and the DNA binding site (called a *tra* box) for TraR (81, 92). These structures support previous biochemical data but also have some unexpected features (Figure 3.3). They confirm earlier findings that TraR binds OOHL via its N-terminal domain in a 1:1 mole ratio, and that it binds DNA as a dimer via its C-terminal domain (70, 97). The N-terminal domain has an α–β–α structure, with one molecule of OOHL embedded between a layer of α-helix and the β-sheet (81, 92). OOHL is therefore deeply buried within the N-terminal pheromone-binding domain of each TraR subunit (Figure 3.3). The hydrophobic atoms of OOHL pack with hydrophobic residues in the core of the domain; the polar oxo-groups and amine form hydrogen bonds with nearby residues. When bound to TraR, the pheromone appears to be completely protected from solvent (81, 92). This engulfment of the ligand in the core of the N-terminal domain supports our previous model that apo-TraR is to at least some degree unstructured and that OOHL participates in the protein-folding process.

The main dimerization interface of TraR is composed of a long α-helix in the N-terminal domain of each subunit that is parallel with the same helix of the opposite subunit. Contributions of residues buried in this interface to dimer formation were confirmed by mutational analysis (52). Although the N-terminal domains are sufficient for dimer formation, the C-terminal domains also dimerize via two parallel helices, although the interface is not as extensive (81, 92). The structure of the C-terminal domains, each a four-helix bundle containing a helix–turn–helix DNA binding motif, is highly conserved and places the LuxR family members

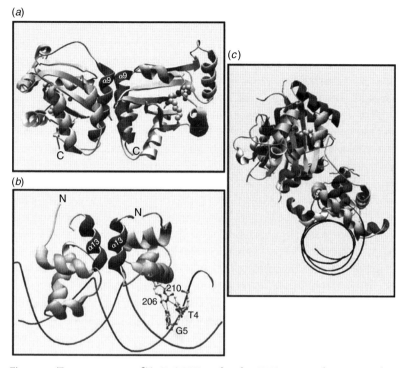

(a)

(c)

(b)

Figure 3.3. Ternary structure of TraR, OOHL, and *tra* box DNA. (*a*) Just the N-terminal domains of a dimer. The dimerization interface is highlighted, and is along the length of α-helix 9 of each monomer. The OOHL and water molecule bound in each N-terminal domain are shown in space-fill and can be seen most clearly in the right-most monomer. The linker between the N-terminal domains and the C-terminal domains is marked with a C. (*b*) Just the C-terminal domains of the TraR dimer bound to DNA. Hydrogen bonds between the side-chains of arginine 206 and 210 of the recognition helix and bases T4 and G5 of the binding site are shown. The dimerization interface is also highlighted, along α-helix 13 of each monomer, and the linker between the NTD and CTD of each monomer is marked with an N. (*c*) A view of a full dimer down the long axis of the DNA, showing the asymmetry of the crystallized protein due to the flexible linkers. Coordinates for these models are from references 81 and 92. (See also color plate section)

in the larger NarL–FixJ family of transcriptional regulators (53). The DNA is in B-form and the consensus *tra* box, an 18 bp sequence with perfect dyad symmetry, has a 30° bend toward the sides of the dimer (81, 92). Although there is an extensive interface between the protein and the DNA, the main contributors to specificity and affinity of binding are two arginine residues in the recognition helix that have hydrogen bonds with two bases in the major groove of each half site (81, 92) (C. E. White and S. C. Winans, unpublished data). One of these residues had been previously shown as critical for DNA binding affinity (50). TraR had also been shown to bind to target DNA sequences with high specificity (96). In fact there is little variation in most of the native *tra* boxes of the Ti plasmid, and the two bases that make direct contacts with two arginines of TraR as described above are absolutely conserved (see Figure 3.4*b*).

The N-terminal pheromone-binding domain and the C-terminal DNA-binding domain of each monomer are connected by a 12-residue unstructured linker (81, 92). This flexibility may be the reason for a pronounced asymmetry of the crystallized protein. The C-terminal domains of each dimer have a two-fold axis of symmetry, and the N-terminal domains also have a two-fold axis of symmetry, but these axes lie at a 90° angle to each other (81, 92). The presence of the flexible linkers connecting the C-terminal domains of the dimer to the N-terminal domains suggests that there is a high degree of flexibility between the two domains *in vivo*. This has been proposed to play a role at divergently transcribed promoters, as discussed in the next section.

There is significant interest in understanding how AHLs convert their receptors from inactive to active forms, and how these proteins can discriminate between their cognate ligand and similar AHLs found in nature. When TraR is expressed at native levels in the cell it binds to its cognate ligand, OOHL, with high specificity (93). There are four hydrogen bonds between residues in the binding pocket of TraR and the polar groups of OOHL, in addition to numerous hydrophobic and packing interactions surrounding the ligand (81, 92). Mutational analyses confirm that the polar interactions between residues of TraR and the OOHL are critical for binding and stability of TraR (6, 52). One of these polar interactions is of considerable interest as it involves a variable group on the OOHL. The 3-oxo group of OOHL is not common to all AHLs and therefore could be an important determinant of specificity of binding. In the binding pocket, this group makes a water-mediated hydrogen bond to a threonine residue. In an attempt to alter binding specificity for a 3-unsubstituted AHL (C8-homoserine lactone), point mutations were made of this residue to increase hydrophobicity and displace the water molecule from its binding site (6). These mutations resulted in a broadened specificity of binding rather than in altered specificity. In the same study, point mutations were also constructed to change the preference of binding to AHLs with shorter fatty acid tails, by increasing hydrophobic bulk in the binding pocket. Some of the mutations did alter specificity of binding but the stability of these mutants was also decreased. These data suggest that the residues in the binding pocket that contact the OOHL play dual roles in folding or stability of the

N-terminal domain and ligand recognition. As these residues are buried in the hydrophobic core of the protein, they make many contributions to packing interactions with surrounding residues in addition to OOHL. Therefore, incorporating OOHL into the protein-folding process may be important not only for regulation of TraR activity but also for specificity of the protein–ligand interaction.

TraR AS AN ACTIVATOR OF TRANSCRIPTION

Seven TraR-dependent promoters have been identified on the octopine-type Ti plasmid (Figure 3.4). These include all of the genes required for conjugal transfer, arranged in three operons: *traAFBH* and *traCDG-yci*, both required for conjugal DNA processing, and the *traI–trb* operon (27). The TraR anti-activator *traM* is also activated by TraR (C. Fuqua, personal communication). More recently, three TraR-dependent promoters of the *repABC* operon have been described (61). This operon is required for both vegetative replication of the plasmid and partitioning into daughter cells. Promoter structure at this operon is quite complex, as there is at least one additional promoter for *rep* that is not TraR-dependent, but is repressed by RepA (62). This ensures plasmid maintenance and partitioning when TraR is inactive. Activation of *repABC* by TraR increases Ti plasmid copy number about seven-fold, thereby enhancing the expression of all Ti-plasmid-encoded genes (61). This could lead to enhanced uptake and catabolism of opines as well as enhanced conjugation and possibly enhanced T-DNA transfer.

The transcription start sites at all of these TraR-dependent promoters have been mapped; each has an identifiable *tra* box (27, 61) (C. Fuqua, personal communication). The *traA* and *traC* promoters are divergently transcribed from *tra* box I (the consensus sequence); the *traI* and two *rep* promoters (*repAP1* and *repAP2*) are divergently transcribed from *tra* box II (Figure 3.4). The *tra* boxes of P*traA*, P*traC*, P*traI*, P*repA1*, and P*repA3* are centered at approximately −45 nucleotides upstream from the transcription start site and overlap the −35 element of each promoter. In contrast, the *tra* boxes of the P*traM* and P*repA2* promoters are centered at −66 nucleotides from the transcription start site (Figure 3.5). These promoters are reminiscent of the class I and class II promoters first described for CRP (5). At class II promoters the activator overlaps the −35 element and can make multiple contacts with RNA polymerase. These contacts recruit RNA polymerase to the promoter and possibly affect later steps in transcription initiation. At class I promoters the activator binds farther

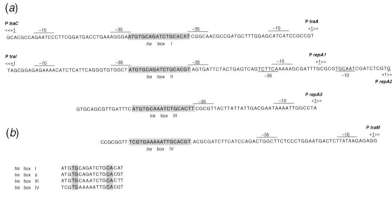

Figure 3.4. (*a*) DNA sequences and *tra* boxes of TraR-dependent promoters. The *tra* box for each promoter is highlighted, transcription start sites are marked with a + 1, and the first gene of the operon is marked above each start site (translation starts are not shown). Each of the three TraR-dependent *rep* promoters (*repA1*, *repA2*, and *repA3*) is indicated. (*b*) Alignment of the four known *tra* boxes of pTiR10. The two positions of each half-site that make sequence-specific contacts with Arg206 and Arg210 of TraR are highlighted.

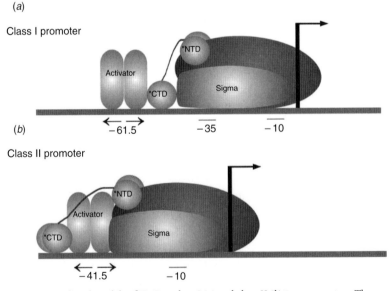

Figure 3.5. Predicted models of TraR at class I (*a*) and class II (*b*) type promoters. These models are based on studies of other activators (5).

upstream on the same face of the DNA as polymerase and can contact only the α C-terminal domain, which is connected to the rest of the polymerase by a flexible linker (5).

As TraR can activate transcription at both class I- and class II-like promoters, the activator is predicted to contact at least the α C-terminal domain of RNA polymerase. At class II promoters TraR is thought to make additional contacts with σ70. This prediction is based on the observation that LuxR, which also activates transcription from a class II promoter, interacts with both the α C-terminal domain and σ70 of RNA polymerase (25, 41, 73). A potential RNA polymerase contact site, or activating region, has been identified on the surface of the TraR C-terminal domain and overlaps with a patch on the LuxR surface that is also critical for activation (22) (C. E. White and S. C. Winans, unpublished data). Mutations of residues in this position do not disrupt protein stability and DNA binding but are required for activation. However, this patch on LuxR is predicted to interact with σ70 (76), whereas the same region of TraR is predicted to contact the α C-terminal domain of RNA polymerase (C. E. White and S. C. Winans, unpublished data). Two additional residues have been identified on the TraR N-terminal domain that are thought to contact the α C-terminal domain (50, 69). Owing to the high degree of flexibility between the TraR domains, these residues could lie very close to the patch of residues that we identified on the C-terminal domain (81, 92). Both sets of residues could together form a contact site for the α C-terminal domain. This potentially novel activating region, formed from two different sets of residues from different parts of the TraR protein, may be critical in the activity of TraR at divergent promoters. However, further studies are required to fully understand the interactions of TraR with RNA polymerase at TraR-dependent promoters.

PERSPECTIVES AND FUTURE STUDIES

In this chapter we have discussed the details of how the cell-to-cell signaling system of *A. tumefaciens* is regulated, what genes are activated by TraR, and how TraR functions as a quorum-sensing transcriptional regulator. However, many of the biochemical details of this system remain to be solved. One major question involves the inactivation of TraR by its anti-activator TraM. Although the mutagenesis studies of TraM suggest that there are two steps in its activity, binding to TraR and then deactivation, the mechanism of inhibition remains elusive (51, 75). These studies along with the structural studies of TraM led to the model that, following complex formation, the TraM dimer dissociates to inactivate TraR (8). Further structural studies of TraM–TraR complexes are critical to understanding this mechanism.

Although the mechanisms by which TrlR acts are fairly well resolved, we really have very little idea of why TrlR is adaptive. Why does one opine, octopine, stimulate Ti plasmid conjugation and replication whereas another, mannopine, inhibits both activities? The third gene we described that inhibits TraR activity, *attM*, offers even more puzzles. Did this gene evolve to degrade AHLs specifically, or some other substrate? If AttM catalyzes the first step in OOHL degradation and AttK and AttL are also involved, what is the enzymatic pathway? Can OOHL be mineralized and used by *A. tumefaciens* as a carbon and nitrogen source, or is the function of AttM only to regulate the cellular concentration of OOHL?

We also do not fully understand how OOHL protects TraR against proteolysis. Even if OOHL does not act as a scaffold in the folding process, major rearrangements must occur in the N-terminal domain of TraR upon binding to the ligand to sequester the molecule in the core of the domain and help the protein to evade proteolysis. Future biophysical studies of the interaction of apo-TraR with OOHL will be very exciting and reveal not only how OOHL contributes to TraR stability but also the specificity of the interaction. The LuxR family members have evolved to be highly selective in AHL binding; structural studies of other receptors and comparisons of these to TraR are critical to understanding binding specificity and how these proteins recognize their targets. A number of other LuxR homologs, including LuxR and RhlR, have been purified; unlike TraR, they do not bind to their cognate AHLs irreversibly (54, 77).

Although a possible RNA polymerase contact site has been identified on the C-terminal domain of TraR, it is very likely that there is at least one more patch critical to activation at class II-like promoters that has not yet been identified. The details of how TraR interacts with RNA polymerase at TraR-dependent promoters, particularly the divergent *traA–traC* and *traI–repABC* promoters, remain to be determined. Although we often use models of transcription activation that are based on CRP, it is already apparent that there are major differences in how these two proteins interact with RNA polymerase. The predicted contact site with the α C-terminal domain of RNA polymerase on TraR is between the dimerization interface and the edge of the protein on the C-terminal domain, whereas on CRP this same activating region is on the edge of the dimer, contacting the α-subunit as it lies adjacent to CRP in the minor groove of the DNA (4). Moreover, the DNA bend in the activator binding site is also likely to influence how it interacts with polymerase. CRP induces an 80° bend in the DNA, resulting in kinks in each half-site, whereas the angle is only 30° in TraR–DNA complexes (63, 81, 92). Finally, although CRP can activate transcription at divergent promoters, the complex promoter architecture at *tra* box II is unique (5, 27, 61), with two class II-type and one class I-type promoter all being activated from one TraR-binding site! There are also a number of differences in the mechanism of activation between LuxR and TraR. The LuxR C-terminal domain is sufficient for activation, whereas two residues have been identified on the TraR N-terminal domain that are predicted to contact RNA polymerase (13, 50, 74). TraR requires a supercoiled template for transcription activation, but LuxR does not (77, 96).

In this chapter we have focused on the biochemistry of this system, but there are many intriguing questions of how the quorum-sensing system of *A. tumefaciens* relates to its role as a plant pathogen. As the quorum-sensing system is induced only in the presence of opines, it is active only on transformed plants and only after infection. This in itself seems to be an anomaly, as most signaling systems in plant pathogens are thought to be important in measuring a population of sufficient size before inducing virulence so that pathogenesis might be successful. Perhaps increase in Ti plasmid copy number by conjugation and replication in a plant-associated population benefits *A. tumefaciens* by increasing gene dose of the virulence and opine import and catabolic genes. As the *traI* and *traR* genes are both on the Ti plasmid, then all conjugal donors in the population both release and detect the OOHL signal. The obvious result of this is that donors are counting other donors instead of possible recipients that do not carry the Ti plasmid, so perhaps donors conjugate with each other to increase overall Ti plasmid copy number in the population. In support of this, sequence examination suggests that the Ti plasmid may not have a potent entry-exclusion system. Conjugation between donor cells would also serve to increase copy number of all Ti plasmid genes in bacteria colonizing a tumor, including all opine catabolic operons, and conjugation operons. This would also increase the gene dosage of T-DNA and *vir* genes, which could stimulate further rounds of tumorigenesis on infected plants.

ACKNOWLEDGEMENTS

The authors thank the members of their laboratory for helpful discussions and review of the paper. Research in the authors' laboratory is supported by grants from NIH (GM42893) and NSF (MCB-9904917).

REFERENCES

1 Alt-Morbe, J., J. L. Stryker, C. Fuqua *et al.* 1996. The conjugal transfer system of *Agrobacterium tumefaciens* octopine-type Ti plasmids is closely related to the transfer system of an IncP plasmid and distantly related to Ti plasmid vir genes. *J. Bacteriol.* **178**: 4248–57.

2 Beck von Bodman, S., G. T. Hayman and S. K. Farrand 1992. Opine catabolism and conjugal transfer of the nopaline Ti plasmid pTiC58 are coordinately regulated by a single repressor. *Proc. Natn. Acad. Sci. USA* **89**: 643–7.

3 Beck von Bodman, S., J. E. McCutchan and S. K. Farrand 1989. Characterization of conjugal transfer functions of *Agrobacterium tumefaciens* Ti plasmid pTiC58. *J. Bacteriol.* **171**: 5281–9.

4 Benoff, B., H. Yang, C. L. Lawson *et al.* 2002. Structural basis of transcription activation: the CAP-alpha CTD-DNA complex. *Science* **297**: 1562–6.

5 Busby, S. and R. H. Ebright 1999. Transcription activation by catabolite activator protein (CAP). *J. Molec. Biol.* **293**: 199–213.

6 Chai, Y. and S. C. Winans 2004. Site-directed mutagenesis of a LuxR-type quorum-sensing transcription factor: alteration of autoinducer specificity. *Molec. Microbiol.* **51**: 765–76.

7 Chai, Y., J. Zhu and S. C. Winans 2001. TrlR, a defective TraR-like protein of *Agrobacterium tumefaciens*, blocks TraR function in vitro by forming inactive TrlR:TraR dimers. *Molec. Microbiol.* **40**: 414–21.

8 Chen, G., J. W. Malenkos, M. R. Cha, C. Fuqua and L. Chen 2004. Quorum-sensing antiactivator TraM forms a dimer that dissociates to inhibit TraR. *Molec. Microbiol.* **52**: 1641–51.

9 Cho, K., C. Fuqua, B. S. Martin and S. C. Winans 1996. Identification of *Agrobacterium tumefaciens* genes that direct the complete catabolism of octopine. *J. Bacteriol.* **178**: 1872–80.

10 Cho, K., C. Fuqua and S. C. Winans 1997. Transcriptional regulation and locations of *Agrobacterium tumefaciens* genes required for complete catabolism of octopine. *J. Bacteriol.* **179**: 1–8.

11 Cho, K. and S. C. Winans 1993. Altered-function mutations in the *Agrobacterium tumefaciens* OccR protein and in an OccR-regulated promoter. *J. Bacteriol.* **175**: 7715–19.

12 Cho, K. and S. C. Winans 1996. The putA gene of *Agrobacterium tumefaciens* is transcriptionally activated in response to proline by an Lrp-like protein and is not autoregulated. *Molec. Microbiol.* **22**: 1025–33.

13 Choi, S. H. and E. P. Greenberg 1991. The C-terminal region of the *Vibrio fischeri* LuxR protein contains an inducer-independent lux gene activating domain. *Proc. Natn. Acad. Sci. USA* **88**: 11115–19.

14 Choi, S. H. and E. P. Greenberg 1992. Genetic evidence for multimerization of LuxR, the transcriptional activator of *Vibrio fischeri* luminescence. *Molec. Mar. Biol. Biotechnol.* **1**: 408–13.

15 Clare, B. G., A. Kerr and D. A. Jones 1990. Characteristics of the nopaline catabolic plasmid in *Agrobacterium* strains K84 and K1026 used for biological control of crown gall disease. *Plasmid* **23**: 126–37.

16 Dessaux, Y., A. Petit, S. K. Farrand and P. J. Murphy 1998. Opines and opine-like molecules involved in plant/*Rhizobiaceae* interactions. In H. P. Spaink, A. Kondorosi and P. J. Hooykaas (eds), *The Rhizobiaceae*, pp. 173–97. Dordrecht, The Netherlands: Kluwer Academic Publishers.

17 Dessaux, Y., A. Petit and J. Tempe 1992. Opines in *Agrobacterium* biology. In D. P. S. Verma (ed.), *Molecular Signals in Plant-Microbe Communications*, pp. 109–36. Ann Arbor, MI: CRC Press.

18 Dong, Y. H., A. R. Gusti, Q. Zhang, J. L. Xu and L. H. Zhang 2002. Identification of quorum-quenching N-acyl homoserine lactonases from *Bacillus* species. *Appl. Environ. Microbiol.* **68**: 1754–9.

19 Dong, Y. H., L. H. Wang, J. L. Xu *et al.* 2001. Quenching quorum-sensing-dependent bacterial infection by an N-acyl homoserine lactonase. *Nature* **411**: 813–17.

20 Dyson, H. J. and P. E. Wright 2002. Coupling of folding and binding for unstructured proteins. *Curr. Opin. Struct. Biol.* **12**: 54–60.

21 Eberhard, A., A. L. Burlingame, C. Eberhard *et al.* 1981. Structural identification of autoinducer of *Photobacterium fischeri* luciferase. *Biochemistry* **20**: 2444–9.

22 Egland, K. A. and E. P. Greenberg 2001. Quorum sensing in *Vibrio fischeri*: analysis of the LuxR DNA binding region by alanine-scanning mutagenesis. *J. Bacteriol.* **183**: 382–6.

23 Ellis, J. G., A. Kerr, A. Petit and J. Tempe 1982. Conjugal transfer of nopaline and agropine Ti-plasmids: the role of agrocinopines. *Molec. Gen. Genet.* **186**: 269–73.

24 Farrand, S. K. 1998. Conjugal plasmids and their transfer. In H. P. Spaink, A. Kondorosi and P. J. J. Hooykaas (eds), *The Rhizobiaceae: Molecular Biology of Model Plant-associated Bacteria*, pp. 199–233. Dordrecht, The Netherlands: Kluwer Academic Publishers.

25 Finney, A. H., R. J. Blick, K. Murakami, A. Ishihama and A. M. Stevens 2002. Role of the C-terminal domain of the alpha subunit of RNA polymerase in LuxR-dependent transcriptional activation of the lux operon during quorum sensing. *J. Bacteriol.* **184**: 4520–8.

26 Fuqua, C., M. Burbea and S. C. Winans 1995. Activity of the Agrobacterium Ti plasmid conjugal transfer regulator TraR is inhibited by the product of the traM gene. *J. Bacteriol.* **177**: 1367–73.

27 Fuqua, C. and S. C. Winans 1996. Conserved *cis*-acting promoter elements are required for density-dependent transcription of *Agrobacterium tumefaciens* conjugal transfer genes. *J. Bacteriol.* **178**: 435–40.

28 Fuqua, C. and S. C. Winans 1996. Localization of OccR-activated and TraR-activated promoters that express two ABC-type permeases and the *traR* gene of Ti plasmid pTiR10. *Molec. Microbiol.* **20**: 1199–210.

29 Fuqua, W. C. and S. C. Winans 1994. A LuxR-LuxI type regulatory system activates *Agrobacterium* Ti plasmid conjugal transfer in the presence of a plant tumor metabolite. *J. Bacteriol.* **176**: 2796–806.

30 Gage, D. J. 2004. Infection and invasion of roots by symbiotic, nitrogen-fixing rhizobia during nodulation of temperate legumes. *Microbiol. Molec. Biol. Rev.* **68**: 280–300.

31 Genetello, C., N. Van Larebeke, M. Holsters *et al.* 1977. Ti plasmids of *Agrobacterium* as conjugative plasmids. *Nature* **265**: 561–3.

32 Goodner, B., G. Hinkle, S. Gattung *et al.* 2001. Genome sequence of the plant pathogen and biotechnology agent *Agrobacterium tumefaciens* C58. *Science* **294**: 2323–8.

33 Gould, T. A., H. P. Schweizer and M. E. Churchill 2004. Structure of the *Pseudomonas aeruginosa* acyl-homoserinelactone synthase LasI. *Molec. Microbiol.* **53**: 1135–46.

34 Hayman, G. T. and S. K. Farrand 1990. *Agrobacterium* plasmids encode structurally and functionally different loci for catabolism of agrocinopine-type opines. *Molec. Gen. Genet.* **223**: 465–73.

35 Hayman, G. T. and S. K. Farrand 1988. Characterization and mapping of the agrocinopine-agrocin 84 locus on the nopaline Ti plasmid pTiC58. *J. Bacteriol.* **170**: 1759–67.

36 He, X., W. Chang, D. L. Pierce, L. O. Seib, J. Wagner and C. Fuqua 2003. Quorum sensing in *Rhizobium* sp. strain NGR234 regulates conjugal transfer (*tra*) gene expression and influences growth rate. *J. Bacteriol.* **185**: 809–22.

37 Huang, J. J., J. I. Han, L. H. Zhang and J. R. Leadbetter 2003. Utilization of acyl-homoserine lactone quorum signals for growth by a soil pseudomonad and *Pseudomonas aeruginosa* PAO1. *Appl. Environ. Microbiol.* **69**: 5941–9.

38 Hwang, I., D. M. Cook and S. K. Farrand 1995. A new regulatory element modulates homoserine lactone-mediated autoinduction of Ti plasmid conjugal transfer. *J. Bacteriol.* **177**: 449–58.

39 Hwang, I., P. L. Li, L. Zhang *et al.* 1994. TraI, a LuxI homologue, is responsible for production of conjugation factor, the Ti plasmid N-acylhomoserine lactone autoinducer. *Proc. Natn. Acad. Sci. USA* **91**: 4639–43.

40 Hwang, I., A. J. Smyth, Z. Q. Luo and S. K. Farrand 1999. Modulating quorum sensing by antiactivation: TraM interacts with TraR to inhibit activation of Ti plasmid conjugal transfer genes. *Molec. Microbiol.* **34**: 282–94.

41 Johnson, D. C., A. Ishihama and A. M. Stevens 2003. Involvement of region 4 of the sigma 70 subunit of RNA polymerase in transcriptional activation of the *lux* operon during quorum sensing. *FEMS Microbiol. Lett.* **228**: 193–201.

42 Kado, C. I. 1994. Promiscuous DNA transfer system of *Agrobacterium tumefaciens*: role of the virB operon in sex pilus assembly and synthesis. *Molec. Microbiol.* **12**: 17–22.

43 Kent, A. D. and E. W. Triplett 2002. Microbial communities and their interactions in soil and rhizosphere ecosystems. *A. Rev. Microbiol.* **56**: 211–36.

44 Kerr, A. 1971. Acquisition of virulence by non-pathogenic isolates of *Agrobacterium radiobacter. Physiol. Plant Pathol.* **1**: 241–6.

45 Kerr, A. 1969. Transfer of virulence between isolates of *Agrobacterium. Nature* **223**: 1175–6.

46 Kerr, A., P. Manigault and J. Tempe 1977. Transfer of virulence *in vivo* and *in vitro* in *Agrobacterium. Nature* **265**: 560–1.

47 Leadbetter, J. R. and E. P. Greenberg 2000. Metabolism of acyl-homoserine lactone quorum-sensing signals by *Variovorax paradoxus. J. Bacteriol.* **182**: 6921–6.

48 Lessl, M., D. Balzer, W. Pansegrau and E. Lanka 1992. Sequence similarities between the RP4 Tra2 and the Ti VirB region strongly support the conjugation model for T-DNA transfer. *J. Biol. Chem.* **267**: 20471–80.

49 Lessl, M. and E. Lanka 1994. Common mechanisms in bacterial conjugation and Ti-mediated T-DNA transfer to plant cells. *Cell* **77**: 321–4.

50 Luo, Z. Q. and S. K. Farrand 1999. Signal-dependent DNA binding and functional domains of the quorum-sensing activator TraR as identified by repressor activity. *Proc. Natn. Acad. Sci. USA* **96**: 9009–14.

51 Luo, Z. Q., Y. Qin and S. K. Farrand 2000. The antiactivator TraM interferes with the autoinducer-dependent binding of TraR to DNA by interacting with the C-terminal region of the quorum-sensing activator. *J. Biol. Chem.* **275**: 7713–22.

52 Luo, Z. Q., A. J. Smyth, P. Gao, Y. Qin and S. K. Farrand 2003. Mutational analysis of TraR. Correlating function with molecular structure of a quorum-sensing transcriptional activator. *J. Biol. Chem.* **278**: 13173–82.

53 Maris, A. E., M. R. Sawaya, M. Kaczor-Grzeskowiak *et al.* 2002. Dimerization allows DNA target site recognition by the NarL response regulator. *Nat. Struct. Biol.* **9**: 771–8.

54 Medina, G., K. Juarez, B. Valderrama and G. Soberon-Chavez 2003. Mechanism of *Pseudomonas aeruginosa* RhlR transcriptional regulation of the *rhlAB* promoter. *J. Bacteriol.* **185**: 5976–83.

55 Moré, M. I., L. D. Finger, J. L. Stryker *et al.* 1996. Enzymatic synthesis of a quorum-sensing autoinducer through use of defined substrates. *Science* **272**: 1655–8.

56 Morris, R. O. 1990. Genes specifying auxin and cytokinin biosynthesis in prokaryotes. In P. J. Davies (ed.), *Plant Hormones and Their Role in Plant Growth and Development*, pp. 636–55. Dordrecht, The Netherlands: Kluwer Academic Publishers.

57 Nester, E. W., D. J. Merlo, M. H. Drummond *et al.* 1977. The incorporation and expression of *Agrobacterium* plasmid genes in crown gall tumors. *Basic Life Sci.* **9**: 181–96.

58 Oger, P. and S. K. Farrand 2001. Co-evolution of the agrocinopine opines and the agrocinopine-mediated control of TraR, the quorum-sensing activator of the Ti plasmid conjugation system. *Molec. Microbiol.* **41**: 1173–85.

59 Oger, P. and S. K. Farrand 2002. Two opines control conjugal transfer of an *Agrobacterium* plasmid by regulating expression of separate copies of the quorum-sensing activator gene traR. *J. Bacteriol.* **184**: 1121–31.

60 Oger, P., K. S. Kim, R. L. Sackett, K. R. Piper and S. K. Farrand 1998. Octopine-type Ti plasmids code for a mannopine-inducible dominant-negative allele of traR, the quorum-sensing activator that regulates Ti plasmid conjugal transfer. *Molec. Microbiol.* **27**: 277–88.

61 Pappas, K. M. and S. C. Winans 2003. A LuxR-type regulator from *Agrobacterium tumefaciens* elevates Ti plasmid copy number by activating transcription of plasmid replication genes. *Molec. Microbiol.* **48**: 1059–73.

62 Pappas, K. M. and S. C. Winans 2003. The RepA and RepB autorepressors and TraR play opposing roles in the regulation of a Ti plasmid *repABC* operon. *Molec. Microbiol.* **49**: 441–55.

63 Parkinson, G., C. Wilson, A. Gunasekera *et al.* 1996. Structure of the CAP-DNA complex at 2.5 angstroms resolution: a complete picture of the protein-DNA interface. *J. Molec. Biol.* **260**: 395–408.

64 Parsek, M. R., D. L. Val, B. L. Hanzelka, J. E. Cronan, Jr. and E. P. Greenberg 1999. Acyl homoserine-lactone quorum-sensing signal generation. *Proc. Natn. Acad. Sci. USA* **96**: 4360–5.

65 Pierson, L. S. I., D. W. Wood and S. B. von Bodman 1999. Quorum sensing in plant-associated bacteria. In G. M. Dunny and S. C. Winans (eds), *Cell-Cell Signaling in Bacteria*, pp. 101–15. Washington, DC: ASM Press.

66 Piper, K. R., S. Beck von Bodman and S. K. Farrand 1993. Conjugation factor of *Agrobacterium tumefaciens* regulates Ti plasmid transfer by autoinduction. *Nature* **362**: 448–50.

67 Piper, K. R., S. Beck Von Bodman, I. Hwang and S. K. Farrand 1999. Hierarchical gene regulatory systems arising from fortuitous gene associations: controlling quorum sensing by the opine regulon in *Agrobacterium*. *Molec. Microbiol.* **32**: 1077–89.

68 Pohlman, R. F., H. D. Genetti and S. C. Winans 1994. Common ancestry between IncN conjugal transfer genes and macromolecular export systems of plant and animal pathogens. *Molec. Microbiol.* **14**: 655–68.

69 Qin, Y., Z. Q. Luo and S. K. Farrand 2004. Domains formed within the N-terminal region of the quorum-sensing activator TraR are required for transcriptional activation and direct interaction with RpoA from agrobacterium. *J. Biol. Chem.* **279**: 40844–51.

70 Qin, Y., Z. Q. Luo, A. J. Smyth *et al.* 2000. Quorum-sensing signal binding results in dimerization of TraR and its release from membranes into the cytoplasm. *EMBO J.* **19**: 5212–21.

71 Sheng, J. and V. Citovsky 1996. *Agrobacterium*-plant cell DNA transport: have virulence proteins, will travel. *Plant Cell* **8**: 1699–710.

72 Shoemaker, B. A., J. J. Portman and P. G. Wolynes 2000. Speeding molecular recognition by using the folding funnel: the fly-casting mechanism. *Proc. Natn. Acad. Sci. USA* **97**: 8868–73.

73 Stevens, A. M., N. Fujita, A. Ishihama and E. P. Greenberg 1999. Involvement of the RNA polymerase alpha-subunit C-terminal domain in LuxR-dependent activation of the *Vibrio fischeri* luminescence genes. *J. Bacteriol.* **181**: 4704–7.

74 Stevens, A. M. and E. P. Greenberg 1997. Quorum sensing in *Vibrio fischeri*: essential elements for activation of the luminescence genes. *J. Bacteriol.* **179**: 557–62.

75 Swiderska, A., A. K. Berndtson, M. R. Cha *et al.* 2001. Inhibition of the *Agrobacterium tumefaciens* TraR quorum-sensing regulator. Interactions with the TraM anti-activator. *J. Biol. Chem.* **276**: 49449–58.

76 Trott, A. E. and A. M. Stevens 2001. Amino acid residues in LuxR critical for its mechanism of transcriptional activation during quorum sensing in *Vibrio fischeri*. *J. Bacteriol.* **183**: 387–92.

77 Urbanowski, M. L., C. P. Lostroh and E. P. Greenberg 2004. Reversible acyl-homoserine lactone binding to purified *Vibrio fischeri* LuxR protein. *J. Bacteriol.* **186**: 631–7.

78 Val, D. L. and J. E. Cronan, Jr 1998. In vivo evidence that S-adenosylmethionine and fatty acid synthesis intermediates are the substrates for the LuxI family of autoinducer synthases. *J. Bacteriol.* **180**: 2644–51.

79 Van Larebeke, N., G. Engler, M. Holsters *et al.* 1974. Large plasmid in *Agrobacterium tumefaciens* essential for crown gall-inducing ability. *Nature* **252**: 169–70.

80 Vannini, A., C. Volpari and S. Di Marco 2004. Crystal structure of the quorum-sensing protein TraM and its interaction with the transcriptional regulator TraR. *J. Biol. Chem.* **279**: 24291–6.

81 Vannini, A., C. Volpari, C. Gargioli *et al.* 2002. The crystal structure of the quorum sensing protein TraR bound to its autoinducer and target DNA. *EMBO J.* **21**: 4393–401.

82 Walker, T. S., H. P. Bais, E. Grotewold and J. M. Vivanco 2003. Root exudation and rhizosphere biology. *Plant Physiol.* **132**: 44–51.

83 Watson, W. T., T. D. Minogue, D. L. Val, S. Beck von Bodman and M. E. Churchill 2002. Structural basis and specificity of acyl-homoserine lactone signal production in bacterial quorum sensing. *Molec. Cell* **9**: 685–94.

84 Whitehead, N. A., A. M. Barnard, H. Slater, N. J. Simpson and G. P. Salmond 2001. Quorum-sensing in Gram-negative bacteria. *FEMS Microbiol. Rev.* **25**: 365–404.

85 Winans, S. C. 1992. Two-way chemical signaling in Agrobacterium-plant interactions. *Microbiol. Rev.* **56**: 12–31.

86 Winans, S. C., J. Zhu and M. I. Moré 1999. Cell density-dependent gene expression by *Agrobacterium tumefaciens* during colonization of crown gall tumors. In G. M. Dunny and S. C. Winans (eds), *Cell-Cell Signaling in Bacteria*, pp. 117–28. Washington, DC: ASM Press.

87 Wood, D. W., J. C. Setubal, R. Kaul *et al.* 2001. The genome of the natural genetic engineer *Agrobacterium tumefaciens* C58. *Science* **294**: 2317–23.

88 Zhang, H. B., C. Wang and L. H. Zhang 2004. The quormone degradation system of *Agrobacterium tumefaciens* is regulated by starvation signal and stress alarmone (p)ppGpp. *Molec. Microbiol.* **52**: 1389–401.

89 Zhang, H. B., L. H. Wang and L. H. Zhang 2002. Genetic control of quorum-sensing signal turnover in *Agrobacterium tumefaciens. Proc. Natn. Acad. Sci. USA* **99**: 4638–43.

90 Zhang, L., P. J. Murphy, A. Kerr and M. E. Tate 1993. Agrobacterium conjugation and gene regulation by N-acyl-L-homoserine lactones. *Nature* **362**: 446–8.

91 Zhang, L. H. and A. Kerr 1991. A diffusible compound can enhance conjugal transfer of the Ti plasmid in *Agrobacterium tumefaciens. J. Bacteriol.* **173**: 1867–72.

92 Zhang, R. G., T. Pappas, J. L. Brace *et al.* 2002. Structure of a bacterial quorum-sensing transcription factor complexed with pheromone and DNA. *Nature* **417**: 971–4.

93 Zhu, J., J. W. Beaber, M. I. Moré *et al.* 1998. Analogs of the autoinducer 3-oxooctanoyl-homoserine lactone strongly inhibit activity of the TraR protein of *Agrobacterium tumefaciens. J. Bacteriol.* **180**: 5398–405.

94 Zhu, J., P. M. Oger, B. Schrammeijer *et al.* 2000. The bases of crown gall tumorigenesis. *J. Bacteriol.* **182**: 3885–95.

95 Zhu, J. and S. C. Winans 1998. Activity of the quorum-sensing regulator TraR of *Agrobacterium tumefaciens* is inhibited by a truncated, dominant defective TraR-like protein. *Molec. Microbiol.* **27**: 289–97.

96 Zhu, J. and S. C. Winans. 1999. Autoinducer binding by the quorum-sensing regulator TraR increases affinity for target promoters in vitro and decreases TraR turnover rates in whole cells. *Proc. Natn. Acad. Sci. USA* **96**: 4832–7.

97 Zhu, J. and S. C. Winans 2001. The quorum-sensing transcriptional regulator TraR requires its cognate signaling ligand for protein folding, protease resistance, and dimerization. *Proc. Natn. Acad. Sci. USA* **98**: 1507–12.

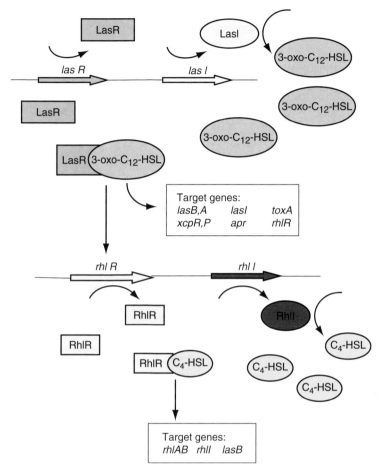

Figure 1.1. The *las* and *rhl* QS systems in *P. aeruginosa*. Both the *las* and *rhl* systems are composed of a transcriptional regulator protein (LasR, RhlR) and an autoinducer synthase (LasI, RhlI). The LasI synthase produces $3\text{-oxo-}C_{12}\text{-HSL}$, and the RhlI synthase produces $C_4\text{-HSL}$. The autoinducer then binds to its respective cognate protein. The LasR:$3\text{-oxo-}C_{12}\text{-HSL}$ complex activates transcription of genes involved in the production of several virulence factors, including *lasB*, *aprA*, and *toxA*, as well as upregulating transcription of *lasI* and of *rhlR*. The RhlR:$C_4\text{-HSL}$ complex activates transcription of several genes involved in virulence, including *rhlAB*.

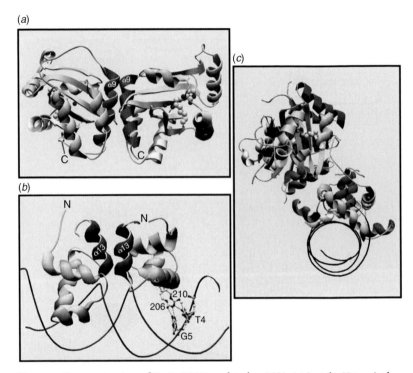

Figure 3.3. Ternary structure of TraR, OOHL, and *tra* box DNA. (*a*) Just the N-terminal domains of a dimer. The dimerization interface is highlighted, and is along the length of α-helix 9 of each monomer. The OOHL and water molecule bound in each N-terminal domain are shown in space-fill and can be seen most clearly in the right-most monomer. The linker between the N-terminal domains and the C-terminal domains is marked with a C. (*b*) Just the C-terminal domains of the TraR dimer bound to DNA. Hydrogen bonds between the side-chains of arginine 206 and 210 of the recognition helix and bases T4 and G5 of the binding site are shown. The dimerization interface is also highlighted, along α-helix 13 of each monomer, and the linker between the NTD and CTD of each monomer is marked with an N. (*c*) A view of a full dimer down the long axis of the DNA, showing the asymmetry of the crystallized protein due to the flexible linkers. Coordinates for these models are from references 81 and 92.

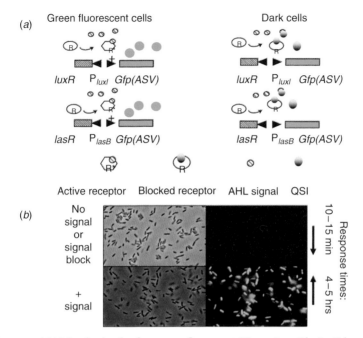

Figure 4.2. (*a*) Molecular details of two green fluorescent QS monitors. The LuxR-based monitor responds to a variety of AHL signals except for BHL, whereas the LasR-based monitor primarily responds to OdDHL. (*b*) Microscopic inspection of *Escherichia coli* cells carrying the LuxR-based monitor (right panels show fluorescent microscopy). The cells respond to the addition of 10 nM OHHL by producing detectable amounts of the unstable GFP protein. If the OHHL is removed by washing, or if a QSI compound is added which blocks GFP synthesis, growing cells lose the GFP signal in 4–5 hours.

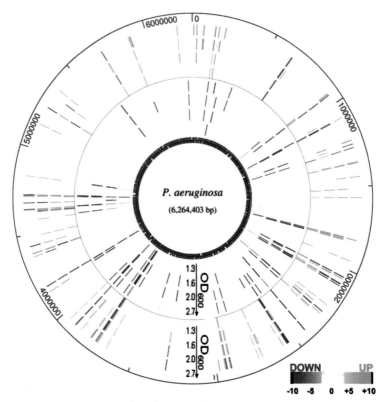

Figure 4.4. Gene expression atlas of QS-controlled genes and QSI target genes in *Pseudomonas aeruginosa* PAO1. Upregulated genes are shown in green; downregulated genes are red. QS-controlled gene expression is shown in the outer half of the atlas; furanone C-30 target genes are indicated in the inner half. QS-controlled genes were identified by growing a *P. aeruginosa* PAO1 Δ*lasIrhlI* mutant with or without exogenous AHL signals (2 μM OdDHL and 5 μM BHL) and retrieving samples for microarray analysis at several culture densities. C-30 target genes were identified by growing *P. aeruginosa* PAO1 with or without 10 μM furanone C-30. *P. aeruginosa* PAO1 ORFs are shown in red (sense strand) and blue (antisense strand) in the innermost circle. The outermost scale gives the gene localization (in base pairs) in the genome. The color scale bar gives the correlation between color coding and fold-change in gene expression.

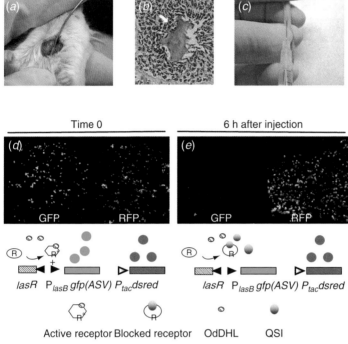

Figure 4.5. The pulmonary dose–response model. (*a*) Mice are challenged intratracheally with alginate beads containing *P. aeruginosa*. (*b*) Photomicrographs of mouse lung tissue infected with *P. aeruginosa*. The arrow points at an alginate bead surrounded by numerous PMNs. Adapted from reference (126). (*c*) QSI drugs can be injected intravenously in the tail vein. (*d, e*) Mouse lung tissue infected with *P. aeruginosa* carrying the LasR-based monitor *PlasB–gfp* for detection of cell-to-cell signaling (green fluorescence) and a tag for simple identification in tissue samples (red fluorescence) examined by SCLM. (*d*) Mice were administered C-30 via intravenous injection at time zero. (*e*) Infected animals were sacrificed in groups of three at the time point indicated and the lung tissue samples were examined by SCLM according to Hentzer *et al.* (43). Parts (*d*) and (*e*) are adapted from (43).

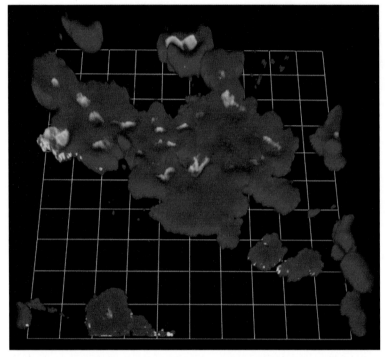

Figure 9.3. Three-dimensional reconstruction of a *Staphylococcus aureus* biofilm. Cells expressing a quorum-controlled green fluorescent protein reporter are green; the remaining biofilm is red from staining with propidium iodide. Each side of the grid represents about 600 μm.

CHAPTER 4

Jamming bacterial communications: new strategies to combat bacterial infections and the development of biofilms

Michael Givskov

Center for Biomedical Microbiology, BioCentrum Technical University of Denmark

Morten Hentzer

Carlsberg Biosector, Carlsberg A/S, Valby, Denmark

INTRODUCTION

The growth and activity of microorganisms affect our lives in both positive and negative ways. We have, since early times, tried to combat unwanted microbes and utilize those expressing useful traits. Microorganisms can cause diseases and chronic infections in humans, animals, and plants. In medicine, agriculture and fish farming, treatment scenarios are based on antimicrobial compounds such as antibiotics, with toxic and growth-inhibitory properties. Control of growth by eradication of bacteria is one of the most important scientific achievements. Unfortunately, bacteria have become gradually more and more resistant to antibiotics, and infections caused by resistant bacteria are on a dramatic increase. It has recently become apparent that the bacterial lifestyle also contributes significantly to this problem. The traditional way of culturing bacteria as planktonic, liquid cultures imprinted the view that bacteria live as unicellular organisms. Although it must be emphasized that such test-tube studies have led to fundamental insights into basic life processes and have unraveled complex intracellular regulatory networks, it is now clear that in nature microbial activity is mainly associated with surfaces and we as scientists must therefore turn our attention to this sessile mode of growth (33). It appears that the ability to form surface-associated, structured and cooperative consortia (referred to as biofilms) is one of the most remarkable and ubiquitous characteristics of bacteria (12). In this sessile life form, bacteria can cause various problems in industrial settings, ranging from corrosion and biofouling to food contamination. In clinical microbiology, the biofilm mode of bacterial growth has attracted particular attention. Many persistent and chronic infections (including pulmonary

Bacterial Cell-to-Cell Communication: Role in Virulence and Pathogenesis, ed. D. R. Demuth and R. J. Lamont. Published by Cambridge University Press. © Cambridge University Press 2005.

infections of cystic fibrosis (CF) patients, periodontitis, otitis media, biliary tract infection, and endocarditis) as well as the colonization of medical implants (catheters, artificial heart valves, etc.) are now believed to be intrinsically linked to the formation of bacterial biofilms (15).

Biofilm infections are becoming more and more common owing to the more frequent insertion of artificial medical implants. Taking the change in age distribution (as the proportion of elderly people increases) in the developed world into account, it will be of the utmost importance to address this major medical and public health problem in the twenty-first century. A recent public announcement from the US National Institutes of Health stated that more than 80% of all microbial infections involve biofilms (17). The capability of forming a biofilm within the human body is therefore considered to represent a pathogenic trait *per se*. Biofilm infections (often caused by opportunistic pathogenic bacteria) are particularly problematic as they give rise to chronic infections, inflammation, and tissue damage. For the clinician, the major problem is that bacteria living in biofilms can often withstand the host defense systems and are markedly tolerant to antibiotics, often exceeding the highest deliverable doses of antibiotics and thus making an efficient treatment impossible. Treatment of Biofilm infections therefore calls for new strategies.

WHY QS INHIBITORS?

The most obvious alternative to antibiotic-mediated killing or growth inhibition would be to attenuate the bacteria with respect to their pathogenicity. The ability to organize structurally and to distribute activities between the different bacteria demands a high degree of coordinated cell–cell interaction reminiscent of that seen in multicellular organisms. Such interactions involve cell-to-cell communication systems to adjust the various functions of specialized members of the population. In fact, many bacteria employ intercellular communication systems that rely on small signal molecules to monitor their own population densities in a process known as quorum sensing (QS) as described in detail in previous chapters. Signal molecules that are released by the cells modulate the activity of other cells in the vicinity, thus regulating collective activities. Cell-to-cell signaling is referred to as QS because the system enables a given bacterial species to sense when a critical (i.e. quorate) population density has been reached in the host and in response activate the expression of target genes required for succession (32).

A diverse range of bacterial metabolites is now recognized as intercellular QS signals. The list includes peptides, butyrolactones, palmitic acid methyl esters, quinones, and cyclic dipeptides. Although the principle behind signal-mediated gene expression in both Gram-positive and Gram-negative bacteria is shared, the molecular mechanisms and signal molecules differ. Peptides serve as the signal molecules in many Gram-positive bacteria; the signals are sensed via two component phosphorelay systems (54). Among Gram-negative bacteria, N-acyl-L-homoserine lactone (AHL)-dependent QS systems are particularly widespread. These systems are used to coordinate expression of traits that are fundamental to the interaction of bacteria with each other, with their environment, and particularly with higher organisms, covering a variety of functions ranging from pathogenic to symbiotic interactions. These include expression of bioluminescence in the specialized light organs of squids and fish, surface colonization, and biofilm development, as well as production of virulence factors and hydrolytic enzymes during infection of eukaryotic hosts.

Back in the mid 1990s, the strictly cell-density-dependent formation of a swarm colony observed in *Serratia liquefaciens* led to the hypothesis that a QS system triggered colony expansion once a critical size, the quorum, had been attained. The behavior of the *swrI* knockout mutant on swarm media in the presence or absence of exogenously added BHL (*N*-butanoyl-L-homoserine lactone) clearly supported this hypothesis (27). The *swr* system controls swarming but not swimming motility and was therefore the first published example of a surface-associated behavior that is controlled by means of QS. Production of several virulence factors was in fact controlled by the *swr* QS system by means of the QS-controlled *lipB* transporter (94). *lipB* is part of an operon encoding a type I secretion system, which is responsible for the secretion of extracellular lipase, metalloprotease, and S-layer protein. Accordingly, QS systems seem to link virulence with surface activity, which has opened a new perspective for controlling undesired microbial activity. Throughout the years, we and our collaborators have documented that blockade of AHL-mediated communication represents an effective approach to interfere with surface colonization and to attenuate the virulence of bacterial pathogens (34, 43, 89). We have denoted compounds capable of this as anti-pathogenic drugs. QS systems regulate (in the classical sense) non-essential phenotypes, so if the new drugs could be based on QS inhibitor (QSI) compounds the bacteria should not, at least in theory, meet the same hard selective pressure as is imposed by antibiotics. The major advantage of this creative strategy for antipathogenic therapy is that it is likely to circumvent the problem of resistance, which is intimately connected to the use of conventional antibacterial agents such as antibiotics. Although resistant mutants may arise, they are not expected to have a selective growth advantage *per se* and thus should not outcompete the parental strain. Furthermore, QSI compounds are not expected to eliminate communities of helpful and beneficial bacteria present in the host (for example, the gut flora). The antipathogenic drug approach is generic in nature and relevant to a broad spectrum of scientists such as microbiologists and medical doctors working in the field of infectious diseases and artificial implants and those engineers who are involved in maintenance of industrial facilities and water pipelines.

PSEUDOMONAS AERUGINOSA AND QS

Pseudomonas aeruginosa is an increasingly prevalent opportunistic human pathogen and is the most common Gram-negative bacterium found in nosocomial and life-threatening infections of immunocompromised patients (114). Patients with cystic fibrosis (CF) are especially disposed to *P. aeruginosa* infections and for these persons the bacterium is responsible for high rates of morbidity and mortality (46, 60). *P. aeruginosa* makes several virulence factors that contribute to its pathogenesis, as described in Chapter 1 of this volume. QS plays a key role in orchestrating the expression of many of these virulence factors, such as exoproteases, siderophores, exotoxins, and several secondary metabolites, and participates in the development of biofilms (18, 43, 81, 121). *P. aeruginosa* possesses two QS systems: the LasR–LasI and the RhlR–RhlI, with the cognate signal molecules OdDHL (*N*-[3-oxo-dodecanoyl]-L-homoserine lactone) and BHL, respectively. The two systems do not operate independently: the *las* system positively regulates expression of the *rhl* system (57, 86). Intertwined in this QS hierarchy is the quinolone signal (PQS) system, which provides a link between the *las* and *rhl* QS systems (85) (70).

In addition to mediate communication between bacteria, immunoassays *in vitro* on human leukocytes have shown that OdDHL possesses immunomodulatory properties, e.g. inhibition of lymphocyte proliferation and downregulation of tumor necrosis factor alpha and IL-12 production (110). OdDHL has been demonstrated to activate T-cells *in vivo* to produce the inflammatory cytokine γ-interferon (104), thereby potentially promoting a Th-2 dominated response leading to increased tissue damage and inflammation. OdDHL also possesses proinflammatory, immunomodulatory, and vasorelaxant properties (7). *In vivo* evidence has left little doubt that QS controlled gene expression plays an important role in the development of *P. aeruginosa* infections (29, 82, 125, 126).

THE INVOLVEMENT OF QS IN *P. AERUGINOSA* BIOFILM DEVELOPMENT

A biofilm is a structured community of bacterial cells enclosed in a self-produced polymeric matrix and adherent to an inert or living surface. Bacterial biofilms are considered ubiquitous in nature (12). In most cases biofilms form at the interface between a solid surface and an aqueous phase. According to the prevailing model, biofilm development is believed to proceed through a temporal series of stages (79). This hypothesis has gained momentum from the isolation of mutants that appear to be arrested at certain stages of this development (37, 62, 75, 87, 113). In the initial phase, bacteria attach to a surface, grow, and then proliferate to form microcolonies. These microcolonies develop into hydrated structures in which bacterial cells are enmeshed in a matrix of self-produced slime (58). For reviews see (109, 120). This slime is commonly referred to as exopolymeric substances (EPS) and may, dependent on the type of bacteria and their overall metabolic status, consist of all major classes of macro-molecules (proteins, polysaccharides, DNA, and RNA) in addition to pepti-doglycan, lipids, and phospholipids (118, 124). Mature biofilms typically consist of "towers" and "mushrooms" of cells enmeshed in copious amounts of EPS, separated by channels and interstitial voids to allow convective flow to transport nutrients to interior parts of the biofilm and remove waste products. This structural heterogeneity is commonly referred to as the biofilm architecture.

The involvement of QS in *P. aeruginosa* biofilm architecture is based on circumstantial evidence. Davies *et al.* (18) demonstrated that a *lasI* mutant formed flat and undifferentiated biofilms in contrast to the wild type, which formed the characteristic biofilm architecture. The authors also found that the flat biofilms formed by the quorum mutant were eradicated with SDS treatment, suggesting that development of the characteristic biofilm toler-ance to antimicrobial treatment would in fact require proper biofilm differ-entiation, which in turn would rely on QS-dependent gene expression. Studies by others revealed no differences between the biofilms of the wild type and those formed by signal-negative mutants (44, 108). Purevdorj *et al.* (88) reported minor structural differences between wild-type and mutant biofilms, but these differences were only apparent when particular hydro-dynamic conditions were used for growing the biofilms. This indicates that the experimental settings influence the "look" of the biofilm.

One of the key issues for biofilm structure seems to be the growth conditions. O'Toole *et al.* (74) showed the catabolite repression control (Crc) protein to be involved; Klausen *et al.* (53) demonstrated that the availability of different carbon sources affected *P. aeruginosa* biofilm structures. In general, glucose-grown biofilms are very heterogeneous, exhibiting discrete towers and mushroom structures separated by water channels and voids. On the contrary, biofilms grown under identical conditions but with citrate as the carbon source appear flat and uniform. In addition, when the medium is supplemented with nitrate as alternative electron acceptor to oxygen, compact and less differentiated biofilms are obtained (authors' unpublished results).

The involvement of QS in biofilm formation has also been demonstrated for *Burkholderia cepacia* (48, 49), *Aeromonas hydrophila* (61), *Pseudomonas putida* (105) and *S. liquefaciens* (56). AHL-negative mutants of *B. cepacia* and *A. hydrophila* showed defects in the late stages of biofilm development and thus were unable to form typical structured biofilms. Recent work on *S. liquefaciens* MG1 has demonstrated that expression of QS-controlled genes is crucial at a specific stage for the development of its characteristic filamentous biofilm (56).

The advent of cDNA microarray technology has provided great insight into differential gene expression in biofilm bacteria. Despite the striking differences between the lifestyles of planktonic and biofilm bacteria, only 1%–3% of the 5,570 predicted genes in *P. aeruginosa* PAO1 show differential expression in the two modes of growth (39, 41, 43, 119). The gene expression profile of *P. aeruginosa* biofilms exhibited greater similarity to that of planktonic cultures in stationary phase than to that of planktonic cultures in mid exponential growth or early stationary phase, hence supporting the view that biofilms are dominated by bacteria with stationary-phase physiology. The genes upregulated during biofilm growth are involved in many different cellular processes, such as transcription and translation, energy metabolism, intermediate metabolism, transport, cofactor biosynthesis, amino acid biosynthesis, and cell-wall synthesis. Notably, a large number of genes related to denitrification and anaerobic respiration were upregulated in biofilms, indicating that anaerobiosis is important in microbial biofilm physiology. A study of the transient gene expression during biofilm development showed that many genes involved in anaerobic respiration became further activated as the biofilm matured. At late stages of biofilm development, transcription of a gene cluster related to the filamentous phage Pf1 became strongly activated (more than 200-fold) (39, 41, 119). The Pf1-like genes have been shown to be involved in bacterial cell death and biofilm dispersal (117).

The involvement of QS in biofilm formation and development has been established by several research groups (18, 20, 95) but the identity of a class of specific QS-controlled genes that influence biofilm formation has remained elusive. It is noteworthy that most attempts to identify QS-controlled genes in *P. aeruginosa* have been performed on planktonic cultures and not on biofilm-growing cells, hence neglecting the possibility that the constitution of the QS regulon can be influenced by environmental factors. In our laboratory, we performed a parallel identification of QS-controlled genes in planktonic and biofilm-growing *P. aeruginosa*. We found that, indeed, the QS regulon is influenced by environmental conditions; a number of biofilm-specific QS-controlled genes could clearly be identified. These genes are of course prime suspects for mediating the role of QS in biofilm development (39, 41). Interestingly, many of the biofilm-specific quorum-sensing-controlled genes were related to responses to iron-limitation. We hypothesize that in the biofilm mode of growth either the availability of iron is reduced, owing to binding to the EPS matrix, or, alternatively, biofilm cells require elevated amounts of iron relative to planktonic cells to support growth. The increased demand for iron during biofilm growth could be due to anaerobiosis as many of the genes comprising the nitrate respiratory pathway, which we found to be differentially upregulated in biofilms, are iron-containing proteins.

NATURAL BLOCKERS: A EUKARYOTIC DEFENSE STRATEGY AGAINST BIOFILMS AND BIOFOULING

The ability of bacteria to form biofilms is a major challenge for living organisms such as humans, animals, and marine eukaryotes (55, 59). Marine plants are, in the absence of more advanced immune systems, prone to disease (11, 30). Bacteria can be highly detrimental to marine algae and other eukaryotes (59). The Australian red macroalga *Delisea pulchra* (23) produces a range of halogenated furanone compounds (for structures and numbering see Figure 4.1), which display antifouling and antimicrobial properties (21, 22, 92). This particular alga originally attracted the attention of marine biologists because it was devoid of extensive surface colonization, i.e. biofouling, unlike other plants in the same environment. Biofouling is primarily caused by marine invertebrates and plants, but bacteria are believed to be the first colonizers of submerged surfaces, providing an initial conditioning biofilm to which other marine organisms may attach (93). Therefore, the abundance and composition of the bacterial community on the surface will significantly affect the subsequent development of a biofouling community (4, 38). To cope with this, eukaryotes have developed chemical defense mechanisms (19, 22, 116)

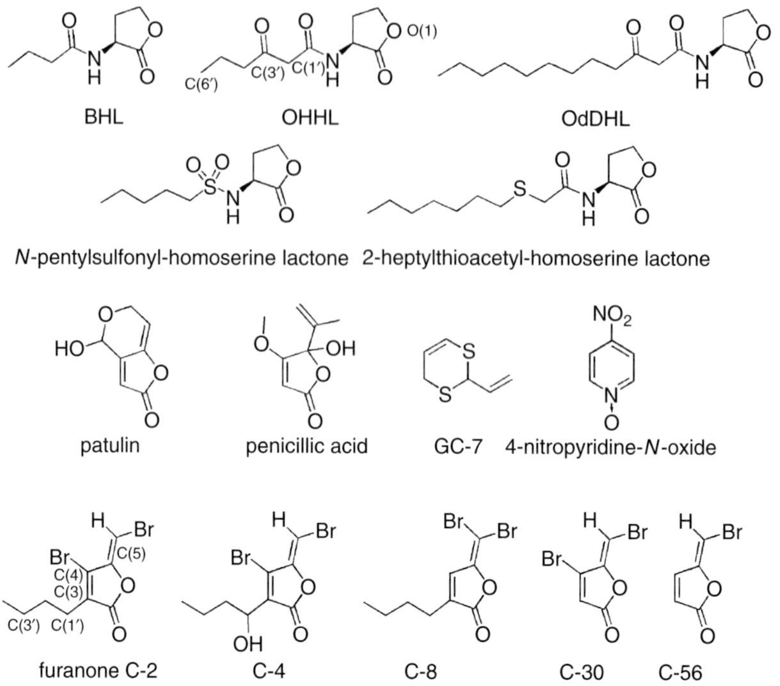

Figure 4.1. AHL signals and QSI compounds. The upper line shows the *Pseudomonas aeruginosa* signal molecules BHL and OdDHL and the *Vibrio fischeri* signal OHHL with carbon atom numbering. The rest are QSI compounds. The second line shows the two synthetic sulfur compounds developed from AHL by rational drug design. The third line shows the fungal metabolites patulin and penicillic acid, a metabolite (GC-7) isolated from garlic, and the synthetic compound 4-NPO. The bottom line shows three natural furanone compounds, C-2 (with carbon atom numbering), C-4 and C-8, isolated from *Delisea pulchra*, and two synthetic derivatives, C-30 and C-56.

which in several cases are secondary metabolites that inhibit bacterial colonization-relevant phenotypes (52, 67, 100).

The effect of furanones on bacterial colonization phenotypes is due to interference with specific cell processes rather than to general toxicity or surface modification (52, 68). The *D. pulchra* furanone compounds consist of a furan ring structure with a substituted acyl chain at the C-3 position and a bromine substitution at the C-4 position. The substitution at the C-5 position may vary in terms of side-chain structure. The natural furanones are halogenated at various positions by bromide, iodide, or chloride (23). For furanone structures, see Figure 4.1. *D. pulchra* produces a minimum of

30 different species of halogenated furanone compound, which are stored in specialized vesicles. The compounds are released at the surface of the thallus at concentrations ranging from 1 to 100 ng cm^{-2} (26, 93). Field experiments have demonstrated that the surface concentration of furanones is inversely correlated to the degree of colonization by marine bacteria (52).

Several of the furanone compounds that exhibit structural similarity to the short-chain AHL molecules inhibited swarming motility of *S. liquefaciens* MG1 (34). This, taken together with inhibition of LuxR-controlled transcriptional fusions, strongly suggested that furanone compounds competed with the cognate signals for the SwrR and the LuxR receptors. Based on the observation that biofilm formation on submerged surfaces precedes the attraction of higher fouling organisms, Givskov *et al.* (34) hypothesized that the *D. pulchra* furanones constitute a specific means of eukaryotic interference with bacterial communication, surface colonization and virulence factor expression, i.e. multicellular behavior. Extensive experimental evidence in support of this hypothesis has accumulated during the years and include the observations that furanones inhibit QS-regulated expression of *Vibrio fischeri* bioluminescence (63), virulence factor production and pathogenesis in *P. aeruginosa* (42, 43), luminescence and virulence *in vivo* of the black tiger prawn pathogen *Vibrio harveyi* (64), and finally virulence of *Erwinia carotovora* (66).

Furanone-mediated displacement of ^{3}H-labeled OHHL molecules from LuxR-overproducing *Escherichia coli* cells supported the assumption of a direct interaction between furanones and LuxR homologs (63). At the same time it was puzzling that there was no substantial affinity of a labeled furanone for *E. coli* cells overproducing LuxR (65). This apparent paradox was resolved with the finding that halogenated furanones accelerate the degradation of the LuxR protein (65). The authors discovered that the half-life of the protein is reduced up to 100-fold in the presence of halogenated furanones. This suggests that halogenated furanones modulate LuxR activity by destabilization, rather than by protecting the transcriptional activator from interaction with the cognate signal. The furanone-dependent reduction in the cellular concentration of the LuxR protein correlated with a reduction in expression of a plasmid encoded P_{luxI}-*gfp*(ASV) fusion, suggesting that the reduction in LuxR concentration is the mechanism by which furanones control QS (65).

SYNTHETIC ANALOGS

AHL-dependent QS systems may be jammed in several ways, in particular by inhibition of AHL signal synthesis, increased AHL signal degradation, and blockade of AHL signal reception. Although it is conceivable that AHL biosynthesis could be effectively obstructed by blockade of AHL synthases, no specific inhibitors for this class of enzymes have yet been derived (80). Dong *et al.* (24) isolated an enzyme, named AiiA, from *Bacillus* sp. that inactivates AHLs by hydrolyzing the lactone bond. It was shown that transgenic plants expressing AiiA exhibit enhanced resistance to *Erwinia carotovora* infections. The authors suggested that, because this bacterium employs AHL-dependent QS to control expression of plant pathogenic traits, the attenuated virulence is likely to be a direct consequence of signal degradation. Whether this strategy is applicable to the treatment of human infections remains to be seen.

Molecules capable of antagonizing binding of the cognate AHL signal to the LuxR-type receptor would block signal transduction and therefore jam the communication system. Competitive inhibitors are likely to be structurally related to the native AHL signal in order to bind to and occupy the AHL binding site but fail to activate the LuxR-type receptor. Non-competitive inhibitors may show little or no structural similarity to AHL signals, as these molecules are thought to bind differently to the receptor protein. However, the recent finding by Chun *et al.* (9) that the human airway might protect itself by producing a lactonase which targets and inactivates certain derivatives of AHLs (in particular OdDHL) might limit the value of putative drugs based on agonist design.

Synthetic AHL analogs described so far mainly fall into two categories: (i) compounds differing from the cognate inducers by the length and composition of the acyl side chain; and (ii) compounds in which the lactone ring has been substituted by other heterocycles or carbocycles. With respect to (i), Hanzelka and Greenberg (36) performed a series of detailed studies of analogs to OHHL in experimental scenarios including *E. coli* harboring a *ptac–luxR* overexpression system or the *lux* operon devoid of the *luxI* gene. From these studies, it was obvious that the length of the acyl side chain plays a critical role in binding to the LuxR receptor and in agonistic as well as antagonistic activity. Passador *et al.* (129) performed a similar series of studies with homologs of the OdDHL signal on recombinant *Escherichia coli* with plasmid-borne *ptac–lasR* (for competitive binding assays) or *lasR, PlasB–lacZ* fusions (for agonistic activity). For interaction with the LasR protein, the authors found that the lower limit for the acyl chain length consisted of eight carbons. The substitution at the C(3′)-position (see Figure 4.1), which is often referred to as 3-oxo-, is important for the agonistic activity of AHLs, but so far no clear prediction on the antagonistic effect of a modification of this position can be made (96, 69, 128).

The C($1'$), on the other hand, is sensitive to substitution with sulfur such as the *n*-pentylsulfonyl-homoserine lactone shown in Figure 4.1. Sulfonamide derivatives of AHLs show QSI activity against the LuxR protein (73) but not the LasR or RhlR proteins (authors' unpublished observations). However, substitution of the C($3'$) atom with sulfur, as in the 2-heptylthioacetyl-homoserine lactone (Figure 4.1) shows strong activity against both LuxR- and LasR-controlled QS systems (84).

With respect to chirality, natural AHL signals are L-isomers whereas D-isomers are generally devoid of biological activity (8, 50, 69). Importantly, D-isomers do not function as antagonists, indicating that they do not bind to the LuxR-type receptor (50).

With respect to (ii), the effects of changes in the composition and size of the homoserine lactone ring with either a 12-carbon (carrying a 3-oxo-substitution at the C($3'$), or a four-carbon acyl side chain were recently investigated by Smith *et al.* (102, 103). From the screening and testing of combinatory libraries it was concluded that ring size, the keto group (at C($1'$)) and the presence of saturated carbons in the ring strongly affected the inhibitory activity of the molecule on the LasR system of *P. aeruginosa*. Only slight variations in these key positions, such as a change from a saturated ring to an aromatic benzene ring, transformed the molecule from an agonist into an antagonist (102). Based on these data, it was suggested that the presence of an aromatic ring interferes with the ability to activate LasR. Accordingly, we recently found that the compound 4-nitropyridine-N-oxide inhibited the *P. aeruginosa* QS systems (91). Substitution of the O(1) with sulfur creates a thiolactone structure, which also antagonizes LuxR activation as demonstrated by Schaefer *et al.* (96); we have recently isolated the garlic compound C-7, carrying two sulfur atoms in the ring, with similar activity (84).

The natural *D. pulchra* furanones were also used as scaffolds for synthetic libraries. Manefield *et al.* (65) found that the longer the aliphatic side chain protruding from the furan ring, the less active are the furanone compounds. On the other hand, removal of the side chain of the natural furanones such as C-30 and C-56 increased their inhibitory activity (Figure 4.1). Concurrent with this ranking was the finding that C-30 promotes a more rapid LuxR turnover than C-8 (65). Furthermore, C-56 and C-30 were the only furanone compounds capable of significant repression of QS-regulated gene expression in *P. aeruginosa*. This was initially based on measurement of the virulence factors and later verified by DNA-microarray-based transcriptomics (42, 43). The inhibitory effect was competitive with that of QdDHL. The bromine atoms also influence inhibitory activity (Figure 4.1). C-8 and C-56 both lack the C(4) Br-atom in comparison with C-2 and C-30, the result of which was found to be a decrease in the QSI activity. However, replacing the single Br atom in C-56 with a methyl group completely abolishes the QSI function.

SCREENS FOR QS INHIBITORS

Throughout the past decade, a number of genetic constructs have been employed to test compounds for their effect on QS systems. The majority of these studies have been performed on the *V. fischeri, Erwinia carotovora, P. aeruginosa, Chromobacterium violaceum* and *Agrobacterium tumefaciens* quorum sensors. The experimental tools and scenarios include binding of ³H-labeled AHL signals and their homologs to recombinant *Escherichia coli* cells overexpressing the respective receptor proteins as well as activation or competitive inhibition of QS target genes and their reporter fusions. A classic reporter system is QS-controlled violacein production in *C. violaceum* (69). Violacein is the purple compound with antibiotic properties produced by certain derepressed strains of *C. violaceum*. The *I*-mutant CVO26 has played an important role as AHL monitor for a number of laboratories and has also been used to describe *D. pulchra* and furanone action as illustrated on the front cover of *Microbiology* by Manefield *et al.* (63). *Escherichia coli* containing the plasmid pSB401 has been used to measure the induction of luminescence by various AHL analogs as well as reduction in luminescence in the presence of both AHL signal and a QSI compound. In pSB401, the *luxR* and the P_{luxI} (*luxI* promoter from *V. fischeri*) have been coupled to the entire *lux* structural operon (*luxCDABE*) from *Photorhabdus luminescens* (122). However, there are limitations to these two systems: the CVO26 only responds to a few AHL signals such as BHL and HHL (*N*-hexanoyl-L-homoserine lactone) and may therefore also be narrow range with respect to various natural QS inhibitors. Expression of bioluminescence is a complex phenotype sensitive to almost anything that affects the physiology of the host bacterium. The *A. tumefaciens* reporter composed by a *traG::lacZ* fusion and *traR* is relatively promiscuous in its choice of AHL molecules and therefore widely used for thin-layer chromatography overlays to identify AHL molecules. With the exception of BHL, this monitor detects AHLs with 3-oxo-, 3-hydroxy-, and 3-unsubstituted side chains of all lengths tested (99).

We have constructed a number of green fluorescent quorum sensors, which consist of *luxR–P*$_{luxI}$ fused to a modified version of the *gfpmut3** gene encoding an unstable green fluorescent protein (GFP) variant of the jellyfish *Aequorea victoria* as a reporter for non-invasive, real-time detection

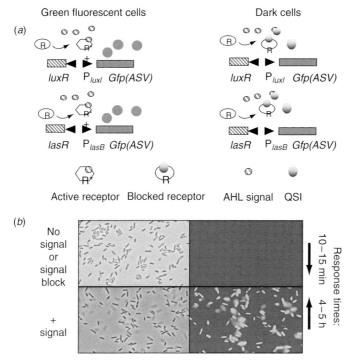

Figure 4.2. (*a*) Molecular details of two green fluorescent QS monitors. The LuxR-based monitor responds to a variety of AHL signals except for BHL, whereas the LasR-based monitor primarily responds to OdDHL. (*b*) Microscopic inspection of *Escherichia coli* cells carrying the LuxR-based monitor (right panels show fluorescent microscopy). The cells respond to the addition of 10 nM OHHL by producing detectable amounts of the unstable GFP protein. If the OHHL is removed by washing, or if a QSI compound is added which blocks GFP synthesis, growing cells lose the GFP signal in 4–5 hours. (See also color plate section)

of gene expression without the addition of chemical substrates (see Figure 4.2) (1). This system opened up unprecedented possibilities for studies *in situ* of QS and QS inhibition in ecologically and clinically relevant scenarios at the macroscopic as well as at the microscopic, single-cell level. Unstable versions of the GFP had previously been constructed to enable *on line* measurements in a non-cumulative fashion (2). Furthermore, it allowed for monitoring of fluctuations in gene expression and even repeated measurements of the same cells to continually assess bacterial activity. Colonies of *E. coli* harboring a *luxR–P$_{luxI}$-gfp*(ASV)-cassette on a high-copy-number pUC-derived plasmid appeared completely dark under the epifluorescence microscope. In striking contrast, the same strain produced bright green fluorescent colonies when cultivated under identical conditions in the presence of 1 nM OHHL. The lesson learned from numerous attempts to construct functional *gfp* gene fusions is that a relatively high level of gene expression followed by efficient translation is required for this to function at the single cell level: 15 min after addition of 10 nM OHHL to a planktonic culture, single cells appeared bright green fluorescent under the epi-fluorescence microscope. This monitor has a relatively broad-spectrum AHL detection limit in the range of 1 nM for OHHL, 10 nM for HHL, 10 nM for OHL, 10 nM for OdDHL and 1000 nM for BHL, with a rapid (15 min) response time at the single-cell level (1).

As with TraR, LuxR overexpression seems to increase agonistic activities of the various AHL analogs (128). The present GFP had a 45 min half-life and consequently the monitor system exhibited a less impressive 4 h lag before the signal had been reduced to below the detection limit (126). QS-regulated gene expression in *Pseudomonas aeruginosa* can be studied in the same way by means of a QS monitor, which consists of a fusion between the *lasB* elastase gene and *gfp* (*PlasB–gfp*(ASV)). By means of a mini-transposon this construct was placed on the chromosome of *P. aeruginosa* to ensure stable segregation and a constant gene dosage of the reporter system (42). Both monitor strains have been further modified to carry a constitutively expressed *dsred* gene encoding the red fluorescent protein RFP (see Figure 4.5). Red fluorescence therefore correlates to bacterial biomass accumulation, green to bacterial communication. These systems have been used to quantify the efficacy of a number of natural and synthetic furanone compounds (42) as well as a number of novel QSI compounds (90, 91).

NOVEL QSI SCREENS

The ability of the above-mentioned reporter fusions to be operated in simple microtiter tray assays offers the researcher a reasonable "high-throughput" facility. As well as being able to detect inducing signals, they can also be used to detect the presence of QSI activity. However, the systems are problematic in that they often require more than one test to verify the result of the screen. The test compound may exert its effect on the reporter itself, or alternatively interfere with precursors, substrates, or environmental conditions for proper function of the reporter, or it might simply affect growth without being bactericidal. Another obstacle is the adjustment of the concentration of the test QSI compound in such a way that it does not lead to any of the above mentioned side effects. This requires that a single test compound must be probed at several different concentrations. A functional QSI is a compound that blocks QS (of a selected QS-controlled monitor system) in the low micromolar range. Furthermore, the inhibitor should be chemically stable and non-toxic. In our view, the last point is essential since new anti-pathogenic drugs against, for example, *P. aeruginosa* should circumvent the problem of development of resistance that is found with conventional antimicrobial compounds.

To meet the above requirements, we recently designed screening systems based on a novel concept, i.e. the bacteria require the presence of a QSI for growth (91). The basic principle of these so-called QSI selectors is that a toxic gene product is placed under QS control (see Figure 4.3). For example, QSI selector 1 (QSIS1) is based on the *luxR* gene and *PluxI* fused to the *phlA* gene from *S. liquefaciens* MG1. Expression of phospholipase leads to lysis of the *E. coli* host cell (35). Therefore, in the presence of externally added OHHL, the *phlA* gene is expressed from *PluxI* and the cells are killed. If, however, a QSI compound is present along with the AHL, the cells survive. QSI selector 2 (QSIS2) is based on the *las* system from *P. aeruginosa*: the *lasB* promoter has been fused to the *sacB* gene. When established on the chromosome in a *rhlI–lasI P. aeruginosa* background, expression causes cell death in the presence of exogenous BHL + OdDHL and sucrose. Presence of a QSI rescues the selector bacteria. The screen is performed as a diffusion assay. Test samples are added to wells made in the agar. The components of the sample diffuse out into the agar, creating a diffusion gradient with the highest concentration closest to the well. If QS is blocked, the cells survive and form a ring of growth around the sample application point (Figure 4.3b). In our experience, the screen works well on everything from pure compounds and extracts with organic solvents to freshly collected samples.

NOVEL NATURAL QSIs

Many plants and fungi have coevolved and established carefully regulated symbiotic associations with bacteria. Exudates from pea (*Pisum sativum*) contain several separable activities able to stimulate, whereas others inhibited, expression of AHL-regulated traits (111). Finding healing powers in plants is an ancient idea. Curiously, since the advent of antibiotics in the 1950s, the use of plant derivatives as antimicrobial agents has

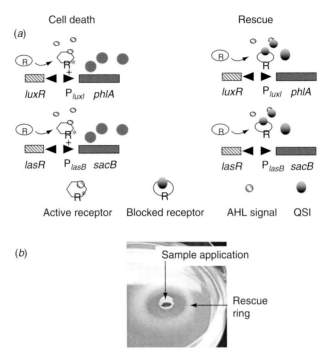

Figure 4.3. (a) Molecular details and principles of the two QSI selectors QSIS1 and QSIS2. QSIS1 is based on the *luxR* gene and the target promoter *PluxI* fused to the *phlA* gene from *Streptococcus liquefaciens* MG1. In *Escherichia coli*, expression of the PhlA phospholipase leads to rapid lysis of the host cell in the presence of exogenous OHHL. QSIS2 is based on the *las* system from *Pseudomonas aeruginosa*; the *lasB* promoter has been fused to the *sacB* gene. When established on the chromosome in a *rhlI – lasI P. aeruginosa* background, expression causes cell death in the presence of exogenous BHL + OdDHL and sucrose. (b) Agar cast with OHHL and *E. coli* cells carrying the QSIS1. A 50 μl volume of C-30 sample was added to the sample application point and incubated overnight. C-30 diffuses out into the agar, creating a diffusion gradient with the highest concentration closest to the well. The clear zone surrounding the application hole indicates that the compound is toxic at high concentrations; this is again surrounded by a rescue ring formed by bacterial growth where the concentration of C-30 is sufficiently high to overcome and block OHHL activation of the phospholipase gene.

been virtually non-existent. The developed world has relied on bacterial and fungal sources for these activities. However, recent studies have shown that a number of foods contain furanone compounds with similarity to the *D. pulchra* compounds described above (101). Inspired by this, we have surveyed various herbs, plants and fungal extracts for AHL-inhibitory activity. It is worth while noting that both plants and fungi are devoid of active immune systems, as seen in mammals, but rather rely on chemical defense systems to deal with bacteria in the environment. Plants and fungi may therefore utilize chemical compounds to inhibit (or in other cases, to stimulate) bacterial AHL-mediated communication.

Plants have an almost limitless ability to synthesize aromatic substances, of which at least 12,000 have been isolated, a number estimated to be less than 10% of the total (97). Bean sprouts, camomile, carrot, chili habañero, the bee wax propolis, yellow pepper, and garlic were recently found by us to hold QSI activity (91). Although garlic (*Allium sativum*) has been used for its medicinal properties for thousands of years, investigations into its mode of action are relatively recent. Garlic has a wide spectrum of actions: not only is it antibacterial, antiviral, antifungal, and antiprotozoal, but it also has beneficial effects on the cardiovascular and immune systems. Garlic extract scored positive with the QSIS1 and QSIS2. This strongly suggests that the extract contains a QSI compound that inhibits QS in *Pseudomonas aeruginosa*. Our result was recently confirmed by Affymetrix GeneChip®-based transcriptomics (91). Some of the garlic QSI compounds were recently identified by us (84) but they did not show activity against *P. aeruginosa* (see Figure 4.1). The structure of the active compound against *P. aeruginosa* is now known by us.

In a recent screening of 55 different *Penicillium* species conducted by us, one third were found to produce QSI activity (90). Two compounds, patulin and penicillic acid (see Figure 4.1) from *Penicillium* species, were isolated and demonstrated to be the biologically active QSI compounds. Their structural similarity to C-30 and C-56 is obvious. Accordingly they were found to block the QS systems of *Pseudomonas aeruginosa* as confirmed by DNA-microarray-based transcriptomics (90).

TRANSCRIPTOMICS: THE ULTIMATE TOOL FOR ANALYZING DRUG SPECIFICITY

DNA microarray technology is a cutting-edge molecular approach to study global gene expression and can be employed to decipher complex regulatory networks in both bacteria and eukaryotes. The technique permits us to take "snapshots" of the involved organisms' global gene expression profile during different stages of growth and treatments. Until recently, comparison of gene expression levels was limited to tracking one or a few genes at a time. The use of reporter fusions is generally accepted as a suitable way to monitor gene expression. It should be kept in mind that such gene fusions are based on hybrid genes and hence produce hybrid mRNAs. Such molecules cannot be expected to have stability similar to that of the native messenger, and therefore information gained from such fusions should not *a priori* be considered more reliable than to the microarray hybridization data that are gained from uninterrupted messengers. Recently, a number of studies using the Affymetrix *P. aeruginosa* microarray have been published (3, 43, 98, 115, 127). In several of these reports, the microarray data have been verified by an independent method such as real-time PCR. It is noteworthy that in all studies the data obtained from microarray experiments and the independent methods were found to correlate well. These reports provide a solid proof of concept for the future application of GeneChips[R].

For example, the drug specificity of the furanone C-30 was evaluated by using the Affymetrix *P. aeruginosa* GeneChip[R] (43). The study was performed as a parallel identification of QS-controlled genes and furanone-target genes and included sample points from $OD_{450} = 0.5$ to approximately $OD_{450} = 3$. In total, 163 genes were found to be controlled by QS. This corresponds to 2.9% of all the predicted genes in *P. aeruginosa* PAO1. Among these genes, most previously reported QS genes were present. Growing *P. aeruginosa* in the presence of the furanone compound caused a change in expression of 93 genes, among which 85 genes (1.5%) were repressed and 8 genes (0.1%) were activated in response to C-30. A comparative analysis of the QS regulon and C-30 target genes showed that 80% of the furanone-repressed genes are in fact controlled by QS. Furanone-repressed genes are not restricted to genes regulated by either the *las* or the *rhl* encoded systems but are found throughout the continuum of QS-induced genes. In essence, the analysis showed a clear overlap between QS-controlled genes and furanone-repressed genes (Figure 4.4.) The study demonstrated that by using furanone compounds it is possible to synthesize QS-inhibitors that specifically target the QS circuit and cause a dramatic reduction in expression of the major *P. aeruginosa* virulence factors, including elastase, hydrogen cyanide, rhamnolipid, chitinase, and phenazine.

Performed under similar growth conditions, Affymetrix GeneChip[R]-based analysis demonstrated that garlic extract affected the expression of 167 (3%) genes in total (91). At the single sample point ($OD_{450} = 2$) 34% of the QS regulon was downregulated by the garlic treatment. Affected genes were found among LasR-, RhlR-, and LasR + RhlR-controlled genes, many of which encode the major *P. aeruginosa* virulence factors. It is encouraging that relatively crude extracts, provided they do not affect the growth rate of the culture, can be applied for GeneChip[R]-based transcriptomics. Along either drug isolation or drug development pipelines, where natural as well

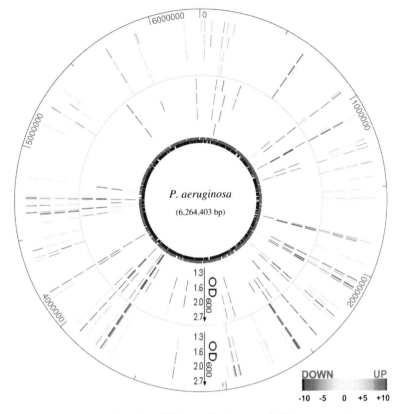

Figure 4.4. Gene expression atlas of QS-controlled genes and QSI target genes in *Pseudomonas aeruginosa* PAO1. Upregulated genes are shown in green; downregulated genes are red. QS-controlled gene expression is shown in the outer half of the atlas; furanone C-30 target genes are indicated in the inner half. QS-controlled genes were identified by growing a *P. aeruginosa* PAO1 Δ*lasIrhlI* mutant with or without exogenous AHL signals (2 μM OdDHL and 5 μM BHL) and retrieving samples for microarray analysis at several culture densities. C-30 target genes were identified by growing *P. aeruginosa* PAO1 with or without 10 μM furanone C-30. *P. aeruginosa* PAO1 ORFs are shown in red (sense strand) and blue (antisense strand) in the innermost circle. The outermost scale gives the gene localization (in base pairs) in the genome. The color scale bar gives the correlation between color coding and fold-change in gene expression. (see also color plate section)

as synthetic compounds are employed as scaffolds for drug design, the GeneChip® delivers the ultimate answer to the target specificity of the compounds in question. This information is highly valuable at an early stage of pipelines. In the former case, it allows a decision to be made as to whether to embark on the time-consuming process of purification of natural compounds.

BIOFILM MODELS: WHAT QSI DRUGS DO TO A BIOFILM

During biofilm formation, cells aggregate and embed themselves in an EPS (exopolymeric substances) matrix, which gives rise to a rigid 3-D structure. Nickel *et al.* (72) demonstrated that the biofilm mode of growth substantially increased *P. aeruginosa*'s tolerance to tobramycin treatment; Costerton *et al.* (12) later expanded that to cover a number of other organisms and antibiotics. Since then numerous studies have generalized this finding; for reviews, see (15, 25, 47). Although several mechanisms (among them diffusion barriers and special physiological conditions) have been postulated to explain reduced susceptibility to antimicrobials in bacterial biofilms, it is becoming evident that biofilm tolerance to antimicrobial treatments is multifactorial.

To study the effect of QSI compounds on biofilm development, the typical experimental scenario used by us is *P. aeruginosa* biofilms that have been allowed to establish in flow chambers in the presence or absence of, for example, furanone compounds. Scanning confocal laser microscopy (SCLM)-based inspection of the GFP-based QS reporter systems revealed that the compounds C-56 and C-30 penetrated the microcolonies, and repressed QS in the majority of the biofilm cells. Importantly, the concentration used was similar to what was found to inhibit QS in planktonic cells without affecting growth or general gene expression (42, 43). By comparing our sets of accumulated data we have found that, similar to the biofilms formed by QS mutants (5), there is (under our experimental settings) no obvious difference in the early stages of biofilm formation in the presence or absence of furanones or any other QSI compound. The effect of QS deficiency (either by mutation or by QSI treatment) becomes more obvious at later stages. For example, at day 7 furanone-treated biofilms were less persistent and more similar to the QS mutant biofilms (39, 42, 43) with the impression of flat, less differentiated biofilms at day 10, which eventually sloughed off (42).

P. aeruginosa biofilms grown in the presence of furanones were also subsequently exposed to 0.1% SDS or tobramycin. Tobramycin is an antibiotic frequently used in the treatment of CF patients. Three days' pretreatment with furanones sensitized the biofilms to dissolution such that both the detergent and tobramycin were able to kill 90%–95% of *P. aeruginosa* cells (43). Similar data have been obtained with regulatory

QS mutants (5). In comparison, the effects of SDS and tobramycin on untreated (or wild-type) biofilms were superficial. We have obtained similar results if the biofilm was treated with both garlic and tobramycin or patulin and tobramycin (90, 91). These results strongly suggest that biofilm sloughing and tolerance to antibiotics and disinfectants can be controlled with QSI drugs. This synergistic effect of QSI drugs and antibiotics is somewhat reminiscent of antibiofilm treatment based on electrical current, which also enhances the efficacy of antibiotics (51).

The biofilm mode of growth also offers protection against the activity of polymorphonuclear leucocytes (PMNs). A PMN–*P. aeruginosa* biofilm model was recently established by us (5). The typical experiment proceeds thus. Biofilms three days old are exposed to freshly isolated PMNs by inoculation into the biofilm flow chambers. The sets of biofilms with PMNs are then inspected by means of SCLM; gradually, differences in the biomass of the types of biofilm can be observed. By using this model, we recently found that QS mutants as well as QSI-treated wild-type biofilms are susceptible to PMN phagocytosis (5, 90).

Activated PMNs liberate oxygen radicals in the form of H_2O_2, which has a bactericidal effect. *P. aeruginosa* biofilm cells tolerate high concentrations (50–100 mM), mainly owing to two protective mechanisms: the presence of *katA*-encoded catalase activity (28) and failure of H_2O_2 to fully penetrate the biofilm owing to a reaction–diffusion interaction as described by Stewart *et al.* (107). These authors concluded that some protective, yet unknown mechanism other than incomplete penetration is operative in *P. aeruginosa* biofilms treated with H_2O_2. We have assessed tolerance of biofilms to H_2O_2 by introducing 100 mM to the influent medium to biofilms 3 days old (5). LIVE/DEAD BacLight staining demonstrated that the wild-type biofilm cells were left unharmed by the treatment in contrast to the QS mutant, which was highly sensitive. Furthermore, we also generated data that suggested this tolerance could result from the amount of oxygen radicals produced by the PMNs, which in turn depends on the magnitude of activation caused by the bacteria. We found that only QS mutants, in contrast to wild-type biofilms, caused detectable activation of the PMNs (5). Furthermore, wild-type biofilms treated with C-30 caused detectable activation of the PMNs similar to that of the QS mutant. Our data support a model where QS signals directly affect the magnitude of the oxidative PMN burst. In other words, PMN activation can be blocked by adding OdDHL and BHL to *lasR–rhlR* double mutants. On the other hand, PMN activation can be promoted by treatment of wild-type biofilm with QSI compounds.

In concurrence with these results, Wu *et al.* (125) found that pulmonary infections caused by a *lasI–rhlI* mutant induced a faster and stronger immune response against the bacterial infection in the early phase, as judged from the severity of lung pathology, higher lung IFN-α production, stronger oxidative burst of blood PMNs, and faster antibody response compared with the wild-type counterpart. Similar data have recently been obtained *in vivo* with the *lasR–rhlR* double mutant (5).

PULMONARY DOSE–RESPONSE MODELS

For drug design and development, it is obvious that a direct study of QSI functionality under complex conditions *in vivo* has tremendous potential. The QS monitors based on GFP reporters offered that possibility. A pseudo-chronic lung infection can be established in mice with *E. coli* carrying a LuxR-controlled *PluxI–gfp*(ASV) system (126) or *P. aeruginosa* equipped with a *lasB-gfp*(ASV) system (43). Healthy mice normally clear introduced bacteria efficiently by means of the mucociliary escalator. The intratracheal introduction of bacteria in the form of alginate beads (see Figure 4.5a) reduces the ability to rapidly clear the bacteria by means of the mucociliary escalator. Instead, PMNs will be recruited by the immune system and surround the infected areas in the lung (see Figure 4.5b). As a result, inflammation and partial tissue destruction occur. For a period of up to a two weeks, this is reminiscent of the situation in the infected human CF lung (45, 71, 83).

Usually the mice are infected one or two days before the dose–response experiment. On the day of the experiment combinations of OHHL and QSI drugs are injected in the tail vein and animals are sacrificed at different time points after the treatments. Sections of frozen lung tissue are prepared for SCLM. Both types of QS monitor express RFP for easy detection and localization in the lung tissue and additionally express GFP in response to AHL-dependent QS (Figure 4.5c, d) If, however, QS is blocked, no or only little GFP signals can be recorded. As shown by Hentzer *et al.* (43) introduction of OHHL into the mouse blood circulation caused activation of the *E. coli* LuxR QS monitor in a concentration-dependent manner. This demonstrates that OHHL is transported by the blood, penetrates the lung tissue, and induces LuxR-controlled *PluxI–gfp*(ASV) expression in the infecting bacteria. This model system has been used by us to evaluate the efficacy of the furanones as well as a number of other potential QSI drugs *in vivo*. For example, furanone C-30 (given as 2 μg/g body mass (BM) or 10 μM if the entire mouse volume is considered to be 20 ml), co-administered

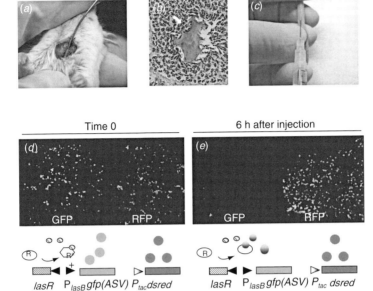

Figure 4.5. The pulmonary dose–response model. (*a*) Mice are challenged intratracheally with alginate beads containing *P. aeruginosa*. (*b*) Photomicrographs of mouse lung tissue infected with *P. aeruginosa*. The arrow points at an alginate bead surrounded by numerous PMNs. Adapted from reference (126). (*c*) QSI drugs can be injected intravenously in the tail vein. (*d, e*) Mouse lung tissue infected with *P. aeruginosa* carrying the LasR-based monitor *PlasB–gfp* for detection of cell-to-cell signaling (green fluorescence) and a tag for simple identification in tissue samples (red fluorescence) examined by SCLM. (*d*) Mice were administered C-30 via intravenous injection at time zero. (*e*) Infected animals were sacrificed in groups of three at the time point indicated and the lung tissue samples were examined by SCLM according to Hentzer *et al.* (43). Parts (*d*) and (*e*) are adapted from (43). (See also color plate section)

with OHHL, caused repression of OHHL-dependent activation of the QS sensor (43). The furanone inhibition could be partly relieved by increasing dosages of OHHL. This experimental setup shows whether a QSI test drug is transported by the blood circulation to the lungs, penetrates the lung tissues, enters the bacteria and in turn represses the OHHL-controlled QS monitor, and it allows for calibration of the effective concentration of the drug.

As shown by Hentzer *et al.* (43), the ability of a test QSI drug to suppress *P. aeruginosa* QS *in vivo* can be tested by infecting mouse lungs with the *P. aeruginosa* QS monitor. The infection was allowed to establish for 2 days before furanone C-30 (2 μg/g-BM) was introduced through the tail vein (Figure 4.5*c*). In this model, no QS signals are introduced because of the endogenous *P. aeruginosa* synthesis (Figure 4.5*d*). Over a time span of 4–6 h after administration of C-30, the GFP signal from the *P. aeruginosa* equipped with a *lasB–gfp*(ASV) monitor was significantly reduced (Figure 4.5*e*). After 8–10 h, the GFP signal reappeared, indicating that the furanone was cleared from the mouse blood circulation and, hence, *de novo* GFP synthesis recommenced. This experimental scenario reveals important information about the mode of action of the test compound: first, the effective concentration required to block wild-type *P. aeruginosa* QS can be determined; and second, the stability of the compounds in the mouse organism can be assessed. For example, this type of experiment told us that a single furanone C-30 injection lasts approximately 6 h (43). This indicates that three doses per day are required for continuous blocking of QS-controlled gene expression of the infecting bacteria. There are many more questions that need to be answered by this model before the treatment model is employed, such as whether the test drug can or should be administered intravenously, intraperitoneally or subcutaneously.

TREATMENT MODEL

The previous dose–response models can be used to establish the conditions required for QS inhibition *in vivo*. To assess the potential of a test QSI compound in the treatment of *P. aeruginosa* infections, we have developed a mouse pulmonary infection model. Groups of mice are infected with *P. aeruginosa* as described above. After this, the mice are typically split in two groups, which receive injections of either the test QSI compound or placebo for the following 3–5 days. The dosages and daily intervals are as determined by the dose–response models above. Groups of animals are sacrificed on a daily basis; the lungs are removed and homogenized and contents spread on agar plates for determination of the number of bacteria. Hentzer *et al.* (43) demonstrated convincing treatment with C-30: the furanone-treated groups of animals displayed a bacterial content on average three orders of magnitude lower than that of the placebo group, and the efficiency of bacterial clearing correlated positively with the concentration of the drug. We have tested several other potential QSI compounds in this model, together with garlic extract and a patulin; the former proved even more efficient than C-30 (6).

ANTIBIOTICS AND QSI DRUGS: ALONE OR TOGETHER?

More than half of the infectious diseases that affect compromised patients involve bacterial species that are commensal or are common in the environment, such as *Staphylococcus epidermidis* and *P. aeruginosa* (13, 14). Biofilm-based infections share clinical characteristics. Biofilms develop commonly on inert surfaces of medical devices or on dead tissue, but they can also form on living tissues, as in endocarditis. Numerous investigations of the surfaces of medical devices (arteriovenous shunts, urinary and other types of catheters, orthopedic devices, and mechanical heart valves) that have been surgically removed owing to device-related infections show the presence of exopolymer-embedded bacteria. Tissues taken from non-device-related chronic infections also show the presence of biofilm bacteria surrounded by an exopolymeric matrix. Such diseases include dental caries, periodontitis, otitis media, biliary tract infections, and bacterial prostatitis (for reviews see (15, 25, 47)).

The best investigated example of a disease in which biofilms play a prominent role is the occurrence of chronic endobronchiolitis caused by mucoid (alginate over-producing) variants of *P. aeruginosa* in patients suffering from CF. CF is caused by mutations in the *CFTR* gene encoding the 1480 amino acid CF transmembrane conductance regulator (CFTR) (16). CFTR functions as a chloride ion channel at the apical membrane of epithelial cells of the sweat ducts, pancreatic ducts, the crypts of the small intestine, the reproductive organs, and the respiratory epithelial cells, which impairs ciliary function (16). Although the endobronchial chronic infection of the biofilm-growing *P. aeruginosa* is impossible to eradicate, repetitive, aggressive antibiotic chemotherapy at the early stages (as practiced at the Danish CF-Centre) has proven to be efficient in postponing the onset of the chronic infection, which is accompanied by deterioration of pulmonary function (31, 112).

The flip side of the coin is the development of antibiotic-resistant bacteria (10). The frequency with which resistant mutants develop in the CF lung may be higher than normal. Oliver *et al.* (78) reported that 36% of a group of CF patients were colonized by hypermutable (mutator) strains in contrast to non-CF patients acutely infected with *P. aeruginosa*. Mutator strains can appear due to inactivating mutations in DNA repair genes, e.g. *mutS*, *mutT* (76, 77, 78). Such mutations are in turn caused by reactive oxygen metabolites released from the activated PMNs in the lungs (123). The high mutation frequency, 3×10^{-6} in CF patients versus 3×10^{-8} in non-CF patients, were associated with a significantly higher frequency of resistance to several groups of antibiotics (78). Such increased antibiotic resistance adds to the protection already offered by the biofilm mode of growth (106).

This highlights the importance of development of non-pathogenic therapies that limit the formation of persistent biofilms in the lungs. In the biofilm mode of growth the colonizing bacteria are protected from the host defense system and the action of antibiotics. It is therefore highly interesting that QSI compounds that were found to efficiently eradicate pulmonary infections also possess PMN-activating properties. The administration of QSI compounds will then be expected to lead to development of less persistent biofilms but also inhibit expression of bacterial exoenzymes that actively degrade components of the immune system. Taken together with the synergistic effect of QS blockage and PMN activity, this might promote clearance, which in turn will reverse the severity of infection and improve the lung function. In addition, the synergistic effect of antibiotics and QSI drugs might turn out to be a useful combination in future chemotherapies. In relation to the insertion of medical implants, short-term prophylactic therapy or alternatively special surface coatings liberating QSI compounds might prove efficient in reducing the risk of developing detrimental biofilms on the devices. The recent finding by us that QSI drugs are effective in treatment of vibriosis in a fish model system illustrates that such compounds might be useful alternatives to antibiotics in fish farming (89).

ACKNOWLEDGEMENT

Our work on QS inhibition has been supported by grants from the Danish Technical Research Council and the Villum Kann Rasmussen Foundation to MG.

REFERENCES

1 Andersen, J. B., A. Heydorn, M. Hentzer *et al.* 2001. *gfp* based *N*-acyl-homoserine-lactone monitors for detection of bacterial communication. *Appl. Environ. Microbiol.* **67**: 575–85.

2 Andersen, J. B., C. Sternberg, L. K. Poulsen *et al.* 1998. New unstable variants of green fluorescent protein for studies of transient gene expression in bacteria. *Appl. Environ. Microbiol.* **64**: 2240–46.

3 Bagge, N., M. Schuster, M. Hentzer *et al.* 2004. *Pseudomonas aeruginosa* biofilms exposed to imipenem exhibit changes in global gene expression and beta-lactamase and alginate production. *Antimicrob. Agents Chemother.* **48**: 1175–87.

4 Belas, M. R. 2003. The swarming phenomenon of *Proteus mirabilis. ASM News* **58**: 15–22.

5 Bjarnsholt, T., P. Ø. Jensen, M. Burmølle *et al.* 2005. *Pseudomonas aeruginosa* tolerance to tobramycin, hydrogen peroxide and polymorphonuclear leucocytes is quorum sensing depended. *Microbiology* **151**: 373–83.

6 Bjarnsholt, T., P. Ø. Jensen, T. B. Rasmussen *et al.* 2005. Garlic jams *Pseudomonas aeruginosa* communication and cures pulmonary infections. *Infect. Immun.* (in press.)

7 Camara, M., P. Williams and A. Hardman 2002. Controlling infection by tuning in and turning down the volume of bacterial small-talk. *Lancet Infect. Dis.* **2**: 667–76.

8 Chhabra, S. R., P. Stead, N. J. Bainton *et al.* 1993. Autoregulation of carbapenem biosynthesis in *Erwinia carotovora* by analogues of *N*-(3-oxohexanoyl)-L-homoserine lactone. *J Antibiot.(Tokyo)* **46**: 441–54.

9 Chun, C. K., E. A. Ozer, M. J. Welsh, J. Zabner and E. P. Greenberg 2004. Inactivation of a *Pseudomonas aeruginosa* quorum-sensing signal by human airway epithelia. *Proc. Natn. Acad. Sci. USA* **101**: 3587–90.

10 Ciofu, O., B. Giwercman, S. S. Pedersen and N. Hoiby 1994. Development of antibiotic resistance in *Pseudomonas aeruginosa* during two decades of antipseudomonal treatment at the Danish CF Center. *Acta Pathol. Microbiol. Immunol. Scand.* **102**: 674–80.

11 Correa, J. A. 1996. Diseases in seaweeds: an introduction. *Hydrobiologia* **326**: 87–8.

12 Costerton, J. W., K. J. Cheng, G. G. Geesey *et al.* 1987. Bacterial biofilms in nature and disease. *A. Rev. Microbiol.* **41**: 435–64.

13 Costerton, J. W., Z. Lewandowski, D. E. Caldwell, D. R. Korber and H. M. Lappin-Scott 1995. Microbial biofilms. *A. Rev. Microbiol.* **49**: 711–45.

14 Costerton, J. W., Z. Lewandowski, D. DeBeer *et al.* 1994. Biofilms, the customized microniche. *J. Bacteriol.* **176**: 2137–42.

15 Costerton, J. W., P. S. Stewart and E. P. Greenberg 1999. Bacterial biofilms: a common cause of persistent infections. *Science* **284**: 1318–22.

16 Davidson, D. J. and D. J. Porteous 1998. Genetics and pulmonary medicine. 1. The genetics of cystic fibrosis lung disease. *Thorax* **53**: 389–97.

17 Davies, D. 2003. Understanding biofilm resistance to antibacterial agents. *Nat. Rev. Drug Discov.* **2**: 114–22.

18 Davies, D. G., M. R. Parsek, J. P. Pearson *et al.* 1998. The involvement of cell-to-cell signals in the development of a bacterial biofilm. *Science* **280**: 295–8.

19 Davis, A. R., Targett, N. M, O. J. McConnell and C. M. Young 1989. Epibiosis of marine algae and benthic invertebrates: natural products chemistry and other mechanisms inhibiting settlement and overgrowth. *Bioorg. Mar. Chem.* **3**: 86–114.

20 De Kievit, T. R., R. Gillis, S. Marx, C. Brown and B. H. Iglewski 2001. Quorum-sensing genes in *Pseudomonas aeruginosa* biofilms: their role and expression patterns. *Appl. Environ. Microbiol.* **67**: 1865–73.

21 de Nys, R., P. Steinberg, C. N. Rogers, T. S. Charlton and M. W. Duncan 1996. Quantitative variation of secondary metabolites in the sea hare *Aplysia parvula* and its host plant, *Delisea pulchra*. *Mar. Ecol. Prog. Ser.* **130**: 135–146.

22 de Nys, R., P. D. Steinberg, P. Willemsen *et al.* 1995. Broad spectrum effects of secondary metabolites from the red alga *Delisea pulchra* in antifouling assays. *Biofouling* **8**: 259–71.

23 de Nys, R., A. D. Wright, G. M. König and O. Sticher 1993. New halogenated furanones from the marine alga *Delisea pulchra*. *Tetrahedron* **49**: 11213–20.

24 Dong, Y. H., L. H. Wang, J. L. Xu *et al.* 2001. Quenching quorum-sensing-dependent bacterial infection by an *N*-acyl homoserine lactonase. *Nature* **411**: 813–17.

25 Drenkard, E. 2003. Antimicrobial resistance of *Pseudomonas aeruginosa* biofilms. *Microbes Infect.* **5**: 1213–19.

26 Dworjanyn, S., R. de Nys, and P. D. Steinberg. 1999. Localization of secondary metabolites in the red alga *Delisea pulchra*. *Mar. Biol.* **133**: 727–36.

27 Eberl, L., M. K. Winson, C. Sternberg *et al.* 1996. Involvement of *N*-acyl-L-homoserine lactone autoinducers in controlling the multicellular behaviour of *Serratia liquefaciens*. *Molec. Microbiol.* **20**: 127–36.

28 Elkins, J. G., D. J. Hassett, P. S. Stewart, H. P. Schweizer and T. R. McDermott 1999. Protective role of catalase in *Pseudomonas aeruginosa* biofilm resistance to hydrogen peroxide. *Appl. Environ. Microbiol.* **65**: 4594–600.

29 Favre-Bonte, S., J. C. Pache, J. Robert *et al.* 2002. Detection of *Pseudomonas aeruginosa* cell-to-cell signals in lung tissue of cystic fibrosis patients. *Microb. Pathogen.* **32**: 143–7.

30 Fenical, W. 1997. New pharmaceuticals from marine organisms. *Trends Biotechnol.* **15**: 339–41.

31 Frederiksen, B., C. Koch and N. Hoiby 1997. Antibiotic treatment of initial colonization with *Pseudomonas aeruginosa* postpones chronic infection and prevents deterioration of pulmonary function in cystic fibrosis. *Pediatr. Pulmonol.* **23**: 330–5.

32 Fuqua, C., S. C. Winans and E. P. Greenberg 1994. Quorum sensing in bacteria: the LuxR-LuxI family of cell density-responsive transcriptional regulators. *J. Bacteriol.* **176**: 269–75.

33 Geesey, G. G., W. T. Richardson, H. G. Yeomans, R. T. Irvin and J. W Costerton 1997. Microscopic examination of natural sessile bacterial populations from an alpine stream. *Can. J. Microbiol.* **23**: 1733–6.

34 Givskov, M., R. de Nys, M. Manefield *et al.* 1996. Eukaryotic interference with homoserine lactone-mediated prokaryotic signaling. *J. Bacteriol.* **178**: 6618–22.

35 Givskov, M. and S. Molin 1993. Secretion of *Serratia liquefaciens* phospholipase from *Escherichia coli*. *Molec. Microbiol.* **8**: 229–42.

36 Hanzelka, B. L. and E. P. Greenberg 1995. Evidence that the N-terminal region of the *Vibrio fischeri* LuxR protein constitutes an autoinducer-binding domain. *J. Bacteriol.* **177**: 815–17.

37 Heilmann, C., C. Gerke, F. Perdreau-Remington and F. Gotz 1996. Characterization of Tn917 insertion mutants of *Staphylococcus epidermidis* affected in biofilm formation. *Infect. Immun.* **64**: 277–82.

38 Henschel, J. R. and P. A. Cook 1990. The development of a marine fouling community in relation to the primary film of microorganisms. *Biofouling* **2**: 1–11.

39 Hentzer, M., L. Eberl and M. Givskov 2004. Quorum sensing in biofilms: Gossip in the slime world? In M. Ghannoum and G. O'Toole (eds.), *Microbial Biofilms*, pp. Washington, DC: ASM Press.

41 Hentzer, M., L. Eberl and M. Givskov 2005. Transcriptome analysis of *Pseudomonas aeruginosa* biofilms: anaerobic respiration and iron limitation. *Biofilms.* (In press.)

42 Hentzer, M., K. Riedel, T. B. Rasmussen *et al.* 2002. Inhibition of quorum sensing in *Pseudomonas aeruginosa* biofilm bacteria by a halogenated furanone compound. *Microbiology* **148**: 87–102.

43 Hentzer, M., H. Wu, J. B. Andersen *et al.* 2003. Attenuation of *Pseudomonas aeruginosa* virulence by quorum sensing inhibitors. *EMBO J.* **22**: 1–13.

44 Heydorn, A., B. Ersboll, J. Kato *et al.* 2002. Statistical analysis of *Pseudomonas aeruginosa* biofilm development: impact of mutations in genes involved in twitching motility, cell-to-cell signaling, and stationary-phase sigma factor expression. *Appl. Environ. Microbiol.* **68**: 2008–17.

45 Hoiby, N. 1993. Antibiotic therapy for chronic infection of *Pseudomonas* in the lung. *A. Rev. Med.* **44**: 1–10.

46 Høiby, N. and B. Frederiksen 2000. Microbiology of cystic fibrosis. In M. E. Hodson and D. M. Geddes (eds.), *Cystic Fibrosis*, pp. 83–107. London: Edward Arnold.

47 Hoiby, N., J. H. Krogh, C. Moser *et al.* 2001. *Pseudomonas aeruginosa* and the in vitro and in vivo biofilm mode of growth. *Microbes Infect.* **3**: 23–35.

48 Huber, B., K. Riedel, M. Hentzer *et al.* 2001. The *cep* quorum-sensing system of *Burkholderia cepacia* H111 controls biofilm formation and swarming motility. *Microbiology* **147**: 2517–28.

49 Huber, B., K. Riedel, M. Kothe *et al.* 2002. Genetic analysis of functions involved in the late stages of biofilm development in *Burkholderia cepacia* H111. *Molec. Microbiol.* **46**: 411–26.

50 Ikeda, T., K. Kajiyama, T. Kita *et al.* 2001. The synthesis of optically pure enantiomers of *N*-acyl-homoserine lactone autoinducers and their analogues. *Chem. Lett.* **30**: 314–15.

51 Jass, J., J. W. Costerton and H. M. Lappin-Scott 1995. The effect of electrical currents and tobramycin in *Pseudomonas aeruginosa* biofilms. *J. Ind. Microbiol.* **15**: 234–42.

52 Kjelleberg, S., P. D. Steinberg, M. Givskov et al. 1997. Do marine natural products interfere with prokaryotic AHL regulatory systems? *Aquat. Microb. Ecol.* **13**: 85–93.

53 Klausen, M., A. Aaes-Jorgensen, S. Molin and T. Tolker-Nielsen 2003. Involvement of bacterial migration in the development of complex multicellular structures in *Pseudomonas aeruginosa* biofilms. *Molec. Microbiol.* **50**: 61–8.

54 Kleerebezem, M., L. E. Quadri, O. P. Kuipers and W. M. de Vos 1997. Quorum sensing by peptide pheromones and two-component signal-transduction systems in Gram-positive bacteria. *Molec. Microbiol.* **24**: 895–904.

55 Kushmaro, A., Y. Loya, E. Fine and E. Rosenberg 1996. Bacterial infection and coral bleaching. *Nature* **380**: 396.

56 Labatte, M., S. Y. Queck, S. A. Rice, M. Givskov and S. Kjelleberg 2004. Quorum sensing controlled biofilm development in *Serratia liquefaciens* MG1. *J. Bacteriol.* **186**: 692–8.

57 Latifi, A., M. Foglino, K. Tanaka, P. Williams and A. Lazdunski 1996. A hierarchical quorum-sensing cascade in *Pseudomonas aeruginosa* links the transcriptional activators LasR and RhIR (VsmR) to expression of the stationary-phase sigma factor RpoS. *Molec. Microbiol.* **21**: 1137–46.

58 Lawrence, J. R., D. R. Korber, B. D. Hoyle, J. W. Costerton and D. E. Caldwell 1991. Optical sectioning of microbial biofilms. *J. Bacteriol.* **173**: 6558–67.

59 Littler, M. M. and D. S. Littler 1995. Impact of CLOD pathogen on pacific coral reefs. *Science* **267**: 1356–60.

60 Lyczak, J. B., C. L. Cannon and G. B. Pier 2002. Lung infections associated with cystic fibrosis. *Clin. Microbiol. Rev.* **15**: 194–222.

61 Lynch, M. J., S. Swift, D. F. Kirke et al. 2002. The regulation of of biofilm development by quorum sensing in *Aeromonas hydrophila. Environ. Microbiol.* **4**: 18–28.

62 Mack, D., M. Nedelmann, A. Krokotsch et al. 1994. Characterization of transposon mutants of biofilm-producing *Staphylococcus epidermidis* impaired in the accumulative phase of biofilm production: genetic identification of a hexosamine-containing polysaccharide intercellular adhesin. *Infect. Immun.* **62**: 3244–53.

63 Manefield, M., R. de Nys, N. Kumar et al. 1999. Evidence that halogenated furanones from *Delisea pulchra* inhibit acylated homoserine lactone (AHL)-mediated gene expression by displacing the AHL signal from its receptor protein. *Microbiology* **145**: 283–91.

64 Manefield, M., L. Harris, S. A. Rice, R. de Nys and S. Kjelleberg 2000. Inhibition of luminescence and virulence in the black tiger prawn (*Penaeus monodon*) pathogen *Vibrio harveyi* by intercellular signal antagonists. *Appl. Environ. Microbiol.* **66**: 2079–84.

65 Manefield, M., T. B. Rasmussen, M. Henzter et al. 2002. Halogenated furanones inhibit quorum sensing through accelerated LuxR turnover. *Microbiology* **148**: 1119–27.

66 Manefield, M., M. Welch, M. Givskov, G. P. Salmond and S. Kjelleberg 2001. Halogenated furanones from the red alga, *Delisea pulchra*, inhibit carbapenem antibiotic synthesis and exoenzyme virulence factor production in the phytopathogen *Erwinia carotovora*. *FEMS Microbiol. Lett.* **205**: 131–8.

67 Maximilien, R., R. de Nys, C. Holmstrom 1998. Chemical mediation of bacterial surface colonisation by secondary metabolites of the red alga *Delisea pulchra*. *Aquat. Microb. Ecol.* **15**: 233–46.

68 Maximilien, R. R., R. de Nys, C. Holmström *et al.* 1998. Chemical mediation of bacterial surface colonisation by secondary metabolites from the red alga *Delisea pulchra*. *Aquat. Microb. Ecol.* **15**: 233–46.

69 McClean, K. H., M. K. Winson, L. Fish *et al.* 1997. Quorum sensing and *Chromobacterium violaceum*: exploitation of violacein production and inhibition for the detection of *N*-acylhomoserine lactones. *Microbiology* **143**: 3703–11.

70 McKnight, S. L., B. H. Iglewski and E. C. Pesci 2000. The *Pseudomonas* quinolone signal regulates *rhl* quorum sensing in *Pseudomonas aeruginosa*. *J. Bacteriol.* **182**: 2702–8.

71 Moser, C., H. K. Johansen, Z. J. Song *et al.* 1997. Chronic *Pseudomonas aeruginosa* lung infection is more severe in Th2 responding BALB/c mice compared to Th1 responding C3H/HeN mice. *Acta Pathol. Microbiol. Immunol. Scand.* **105**: 838–42.

72 Nickel, J. C., I. Ruseska, J. B. Wright and J. W. Costerton 1985. Tobramycin resistance of *Pseudomonas aeruginosa* cells growing as a biofilm on urinary catheter material. *Antimicrob. Agents Chemother.* **27**: 619–24.

73 Nielsen, J. and Givskov, M. 2003. Compounds and methods for controlling bacterial virulence. PCT WO 03/106445. [Patent]

74 O'Toole, G. A., K. A. Gibbs, P. W. Hager, P. V. Phibbs, Jr. and R. Kolter 2000. The global carbon metabolism regulator Crc is a component of a signal transduction pathway required for biofilm development by *Pseudomonas aeruginosa*. *J. Bacteriol.* **182**: 425–31.

75 O'Toole, G. A. and R. Kolter. 1998. Initiation of biofilm formation in *Pseudomonas fluorescens* WCS365 proceeds via multiple, convergent signaling pathways: a genetic analysis. *Molec. Microbiol.* **28**: 449–61.

76 Ochsner, U. A., M. L. Vasil, E. Alsabbagh, K. Parvatiyar and D. J. Hassett 2000. Role of the *Pseudomonas aeruginosa* oxyR-recG operon in oxidative stress defense and DNA repair: OxyR-dependent regulation of katB-ankB, ahpB, and ahpC-ahpF. *J. Bacteriol.* **182**: 4533–44.

77 Oliver, A., F. Baquero and J. Blazquez 2002. The mismatch repair system (mutS, mutL and uvrD genes) in *Pseudomonas aeruginosa*: molecular characterization of naturally occurring mutants. *Molec. Microbiol.* **43**: 1641–50.

78 Oliver, A., R. Canton, P. Campo, F. Baquero and J. Blazquez 2000. High frequency of hypermutable *Pseudomonas aeruginosa* in cystic fibrosis lung infection. *Science* **288**: 1251–4.

79 Palmer, R. J. Jr. and D. C. White 1997. Developmental biology of biofilms: implications for treatment and control. *Trends Microbiol.* **5**: 435–40.

80 Parsek, M. R., D. L. Val, B. L. Hanzelka, J. E. J. Cronan and E. P. Greenberg 1999. Acyl homoserine-lactone quorum-sensing signal generation. *Proc. Natn. Acad. Sci. USA* **96**: 4360–5.

81 Passador, L., J. M. Cook, M. J. Gambello, L. Rust and B. H. Iglewski 1993. Expression of *Pseudomonas aeruginosa* virulence genes requires cell-to-cell communication. *Science* **260**: 1127–30.

82 Pearson, J. P., M. Feldman, B. H. Iglewski and A. Prince 2000. *Pseudomonas aeruginosa* cell-to-cell signaling is required for virulence in a model of acute pulmonary infection. *Infect. Immun.* **68**: 4331–4.

83 Pedersen, S. S., G. H. Shand, B. L. Hansen and G. N. Hansen 1990. Induction of experimental chronic *Pseudomonas aeruginosa* lung infection with *P. aeruginosa* entrapped in alginate microspheres. *Acta Pathol. Microbiol. Immunol. Scand.* **98**: 203–11.

84 Persson, P., H. H. Hansen, T. B. Rasmussen *et al.* 2005. Rational design and synthesis of new quorum sensing inhibitors derived from acylated homoserine lactones and natural products from garlic. *Org. Biomolec. Chem.* **3**: 253–62.

85 Pesci, E. C., J. B. Milbank, J. P. Pearson *et al.* 1999. Quinolone signaling in the cell-to-cell communication system of *Pseudomonas aeruginosa*. *Proc. Natn. Acad. Sci. USA* **96**: 11229–34.

86 Pesci, E. C., J. P. Pearson, P. C. Seed and B. H. Iglewski 1997. Regulation of *las* and *rhl* quorum sensing in *Pseudomonas aeruginosa*. *J. Bacteriol.* **179**: 3127–32.

87 Pratt, L. A. and R. Kolter 1998. Genetic analysis of *Escherichia coli* biofilm formation: roles of flagella, motility, chemotaxis and type I pili. *Molec. Microbiol.* **30**: 285–93.

88 Purevdorj, B., J. W. Costerton and P. Stoodley 2002. Influence of hydrodynamics and cell signaling on the structure and behavior of *Pseudomonas aeruginosa* biofilms. *Appl. Environ. Microbiol.* **68**: 4457–64.

89 Rasch, M., C. Buch, B. Austin *et al.* 2004. An inhibitor of bacterial quorum sensing reduces mortalities caused by Vibriosis in rainbow trout (Oncorhynchus mykiss, Walbaum). *Syst. Appl. Microbiol.* **27**: 350–9.

90 Rasmussen, T., M. E. Skindersø, T. Bjarnsholt *et al.* 2005. Idendity and effects of quorum sensing inhibitors produced by *Penicillum* species. *Microbiology.* (in press.)

91 Rasmussen, T. B., T. Bjarnsholt, M. E. Skindersø *et al.* 2005. Screening for quorum sensing inhibitors using a novel genetic system – the QSI selector. *J. Bacteriol.* **187**: 1799–814.

92 Reichelt, J. L. and M. A. Borowitzka. 1984. Antimicrobial activity from marine algae: results of a large-scale screening programme. *Hydrobiology* **116/117**: 158–68.

93 Rice, S. A., M. Givskov, P. Steinberg and S. Kjelleberg 1999. Bacterial signals and antagonists: the interaction between bacteria and higher organisms. *J. Molec. Microbiol. Biotechnol.* **1**: 23–31.

94 Riedel, K., T. Ohnesorg, K. A. Krogfelt *et al.* 2001. N-acyl-L-homoserine lactone-mediated regulation of the lip secretion system in *Serratia liquefaciens* MG1. *J. Bacteriol.* **183**: 1805–9.

95 Sauer, K., A. K. Camper, G. D. Ehrlich, J. W. Costerton and D. G. Davies 2002. *Pseudomonas aeruginosa* displays multiple phenotypes during development as a biofilm. *J. Bacteriol.* **184**: 1140–54.

96 Schaefer, A. L., B. L. Hanzelka, A. Eberhard and E. P. Greenberg 1996. Quorum sensing in *Vibrio fischeri*: probing autoinducer-LuxR interactions with autoinducer analogs. *J. Bacteriol.* **178**: 2897–901.

97 Schultes, R. E. 1978. The kingdom of plants, p. 208. In W. A. R. Thompson (ed.), *Medicines from the Earth*, p. 208. New York, NY: McGraw-Hill.

98 Schuster, M., C. P. Lostroh, T. Ogi and E. P. Greenberg 2003. Identification, timing, and signal specificity of *Pseudomonas aeruginosa* quorum-controlled genes: a transcriptome analysis. *J. Bacteriol.* **185**: 2066–79.

99 Shaw, P. D., G. Ping, S. L. Daly *et al.* 1997. Detecting and characterizing N-acyl-homoserine lactone signal molecules by thin-layer chromatography. *Proc. Natn. Acad. Sci. USA* **94**: 6036–41.

100 Slattery, M., J. B. McClintoch and J. N. Heine 1995. Chemical defences in Antarctic soft corals: evidence for antifouling compounds. *J. Exp. Mar. Biol. Ecol.* **190**: 61–77.

101 Slaughter, J. C. 1999. The naturally occurring furanones: formation and function from pheromone to food. *Biol. Rev.* **74**: 259–76.

102 Smith, K. M., Y. Bu and H. Suga 2003. Library screening for synthetic agonists and antagonists of a *Pseudomonas aeruginosa* autoinducer. *Chem. Biol.* **10**: 563–71.

103 Smith, K. M., Y. Bu and H. Suga 2003. Induction and inhibition of *Pseudomonas aeruginosa* quorum sensing by synthetic autoinducer analogs. *Chem. Biol.* **10**: 81–9.

104 Smith, R. S., E. R. Fedyk, T. A. Springer *et al.* 2001. IL-8 production in human lung fibroblasts and epithelial cells activated by the *Pseudomonas* autoinducer N-3-oxododecanoyl homoserine lactone is transcriptionally regulated by NF-kappa B and activator protein-2. *J. Immunol.* **167**: 366–74.

105 Steidle, A., K. Sigl, R. Schuhegger 2001. Visualization of *N*-acylhomoserine lactone-mediated cell-cell communication between bacteria colonizing the tomato rhizosphere. *Appl. Environ. Microbiol.* **67**: 5761–70.

106 Stewart, P. S. and J. W. Costerton 2001. Antibiotic resistance of bacteria in biofilms. *Lancet* **358**: 135–8.

107 Stewart, P. S., F. Roe, J. Rayner *et al.* 2000. Effect of catalase on hydrogen peroxide penetration into *Pseudomonas aeruginosa* biofilms. *Appl. Environ. Microbiol.* **66**: 836–8.

108 Stoodley, P., F. Jørgensen, P. Williams and H. M. Lappin-Scott 1999. The role of hydrodynamics and AHL signaling molecules as determinants of the structure of *Pseudomonas aeruginosa* biofilms. In R. Bayston, M. Brading, P. Gilbert,

J. Walker and J. W. T. Wimpenny (eds), *Biofilms: the Good, the Bad, and the Ugly*, pp. 323–30. Cardiff: J. W. T. Bioline.

109 Sutherland, I. W. 2001. Biofilm exopolysaccharides: a strong and sticky framework. *Microbiology* **147**: 3–9.

110 Telford, G., D. Wheeler, P. Williams *et al.* 1998. The *Pseudomonas aeruginosa* quorum-sensing signal molecule N-(3-oxododecanoyl)-L-homoserine lactone has immunomodulatory activity. *Infect. Immun.* **66**: 36–42.

111 Teplitski, M., J. B. Robinson and W. D. Bauer 2000. Plants secrete substances that mimic bacterial N-acyl homoserine lactone signal activities and affect population density-dependent behaviors in associated bacteria. *Molec. Plant Microbe Interact.* **13**: 637–48.

112 Valerius, N. H., C. Koch and N. Hoiby 1991. Prevention of chronic *Pseudomonas aeruginosa* colonisation in cystic fibrosis by early treatment. *Lancet* **338**: 725–6.

113 Vallet, I., J. W. Olson, S. Lory, A. Lazdunski and A. Filloux 2001. The chaperone/usher pathways of *Pseudomonas aeruginosa*: identification of fimbrial gene clusters (cup) and their involvement in biofilm formation. *Proc. Natn. Acad. Sci. USA* **98**: 6911–16.

114 Van Delden, C. and B. H. Iglewski 1998. Cell-to-cell signaling and *Pseudomonas aeruginosa* infections. *Emerg. Infect. Dis.* **4**: 551–60.

115 Wagner, V. E., D. Bushnell, L. Passador, A. I. Brooks and B. H. Iglewski 2003. Microarray analysis of *Pseudomonas aeruginosa* quorum-sensing regulons: effects of growth phase and environment. *J. Bacteriol.* **185**: 2080–95.

116 Wahl, M. 2003. Marine epibiosis. Fouling and antifouling: some basic aspects. *Mar. Ecol. Prog. Ser.* **58**: 175–89.

117 Webb, J. S., L. S. Thompson, S. James *et al.* 2003. Cell death in *Pseudomonas aeruginosa* biofilm development. *J. Bacteriol.* **185**: 4585–92.

118 Whitchurch, C. B., T. Tolker-Nielsen, P. C. Ragas and J. S. Mattick 2002. Extracellular DNA required for bacterial biofilm formation. *Science* **295**: 1487.

119 Whiteley, M., M. G. Bangera, R. E. Bumgarner *et al.* 2001. Gene expression in *Pseudomonas aeruginosa* biofilms. *Nature* **413**: 860–4.

120 Wimpenny, J. W. T. and R. Colasanti 1997. A more unifying hypothesis for biofilm structures – a reply. *FEMS Microbiol. Ecol.* **24**: 185–6.

121 Winson, M. K., M. Camara, A. Latifi *et al.* 1995. Multiple N-acyl-L-homoserine lactone signal molecules regulate production of virulence determinants and secondary metabolites in *Pseudomonas aeruginosa*. *Proc. Natn. Acad. Sci. USA* **92**: 9427–31.

122 Winson, M. K., S. Swift, L. Fish *et al.* 1998. Construction and analysis of luxCDABE-based plasmid sensors for investigating N-acyl homoserine lactone-mediated quorum sensing. *FEMS Microbiol. Lett.* **163**: 185–92.

123 Worlitzsch, D., G. Herberth, M. Ulrich and G. Doring 1998. Catalase, myeloperoxidase and hydrogen peroxide in cystic fibrosis. *Eur. Respir. J.* **11**: 377–83.

124 Wozniak, D. J., T. J. O. Wyckoff, M. Starkey *et al.* 2003. Alginate is not a significant component of the extracellular polysaccharide matrix of PA14 and

PAO1 *Pseudomonas aeruginosa* biofilms. *Proc. Natn. Acad. Sci. USA* **100**: 7907–12.

125 Wu, H., Z. Song, M. Givskov *et al.* 2001. *Pseudomonas aeruginosa* mutations in *lasI* and *rhlI* quorum sensing systems result in milder chronic lung infection. *Microbiology* **147**: 1105–13.

126 Wu, H., Z. Song, M. Hentzer *et al.* 2000. Detection of *N*-acylhomoserine lactones in lung tissues of mice infected with *Pseudomonas aeruginosa*. *Microbiology* **146**: 2481–93.

127 Yoon, S. S., R. F. Hennigan, G. M. Hilliard *et al.* 2002. *Pseudomonas aeruginosa* anaerobic respiration in biofilms: relationships to cystic fibrosis pathogenesis. *Dev. Cell* **3**: 593–603.

128 Zhu, J., J. W. Beaber, M. I. More *et al.* 1998. Analogs of the autoinducer 3-oxo-octanoyl-homoserine lactone strongly inhibit activity of the TraR protein of *Agrobacterium tumefaciens*. *J. Bacteriol.* **180**: 5398–405.

129 Passador, L., K. D. Tucker, K. R. Guertin *et al.* 1996. Functional analysis of the *Pseudomonas aeruginosa* autoinducer PAI. *J. Bacteriol.* **178**: 5995–6000.

CHAPTER 5

Quorum-sensing-mediated regulation of biofilm growth and virulence of *Vibrio cholerae*

Jun Zhu
University of Pennsylvania School of Medicine

John J. Mekalanos
Harvard Medical School, Boston, MA, USA

INTRODUCTION

Many species of bacterium exchange chemical signals to help them monitor their population densities, a phenomenon referred to as quorum sensing. Quorum sensing was first described over two decades ago in two luminescent marine bacterial species, *Vibrio fischeri* and *V. harveyi* (40), which have served as model species for studies of cell-density-dependent gene expression. In both species, the enzymes responsible for light production are encoded by the luciferase structural operon *luxCDABE* (13, 39) and light emission occurs only at high cell density in response to the accumulation of secreted autoinducer signaling molecules (40). In the 1980s, Eberhard *et al.* (11) purified the first homoserine lactone autoinducer from *V. fischeri* and showed that it was indeed a specific inducer of luminescence. In 1983, the basic features of the autoinduction system were revealed at the molecular level when the *lux* genes of *V. fischeri* were successfully cloned and expressed in *Escherichia coli* (12).

Although quorum sensing regulation has been analyzed in great detail in *V. harveyi* and *V. fischeri*, the study of quorum sensing phenotypes in the clinically important *Vibrio* species *V. cholerae* was virtually non-existent until quite recently. This was partly because, unlike *V. fischeri* and *V. harveyi*, *V. cholerae* does not possess luciferase genes and it was therefore unclear whether it possessed any genes that were regulated by quorum sensing. However, when the *V. cholerae* genome sequence was completed (22) it was revealed that *V. cholerae* contains several quorum-sensing genes similar to those of *V. harveyi*. In this chapter, we summarize recent studies on *V. cholerae* quorum sensing from various laboratories and discuss the significance of quorum-sensing-mediated regulation of virulence gene expression and biofilm formation.

Bacterial Cell-to-Cell Communication: Role in Virulence and Pathogenesis, ed. D. R. Demuth and R. J. Lamont. Published by Cambridge University Press. © Cambridge University Press 2005.

VIBRIO CHOLERAE AS A HUMAN PATHOGEN AND ITS INFECTIOUS CYCLE

Cholera is an acute dehydrating diarrheal disease caused principally by a potent enterotoxin (cholera toxin) produced by toxigenic *V. cholerae* during infection of the human host (15). Epidemic cholera caused by toxigenic *V. cholerae* belonging to the O1 or O139 serogroup is a major public health problem in developing countries, causing an estimated 120,000 deaths worldwide and many non-fatal cases each year, the vast majority of which occur in children (41). Cholera is also endemic in many parts of developing countries, where outbreaks occur widely in a seasonal pattern and are particularly associated with poverty and poor sanitation.

The pathogenesis of cholera is a complex process and involves a number of genes encoding virulence factors that aid the pathogen in reaching and colonizing the epithelium of the small intestine and in producing cholera toxin (CT) that disrupts ion transport in intestinal epithelial cells (42). In *V. cholerae*, the major virulence genes exist in clusters (27, 46). These include the CTX genetic element, which is the genome of a lysogenic bacteriophage (CTXΦ) that carries the CT genes, and the TCP pathogenicity island, which carries the genes encoding a pilus colonization factor known as the toxin coregulated pilus (TCP).

Virulence gene expression in *V. cholerae* is controlled by multiple regulatory systems (46). Several critical virulence genes are coordinately regulated by a cascading system of regulatory factors to respond in a concerted fashion to specific environmental conditions (Figure 5.1). The ToxRS and TcpPH membrane complexes are believed to sense environmental cues and to modulate the expression of the AraC-like transcriptional activator ToxT in response to these cues. ToxT in turn controls the transcription of genes involved in the synthesis of virulence factors such as CT and TCP. Recent work has identified AphA and AphB as additional activators of *tcpPH* expression (30, 46).

Analysis of these regulatory cascades has identified pH, osmolarity, and temperature as environmental signals that affect virulence gene expression in vitro. Several other regulatory genes and processes (e.g. motility and iron acquisition) also influence expression of the ToxR regulon (32). These different regulatory systems presumably allow *V. cholerae* bacteria to sense their environment and accordingly vary gene expression to optimize survival under different conditions. Because *V. cholerae* is both a human pathogen and part of the normal aquatic flora in estuarine and brackish waters, the functional repertoire of the *V. cholerae* genome must be

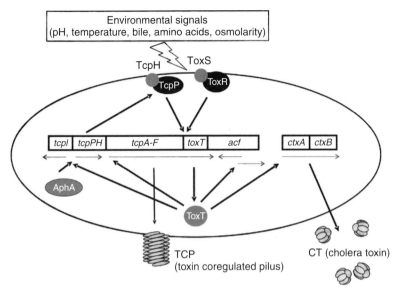

Figure 5.1. Regulatory pathways influencing the expression of virulence genes in *Vibrio cholerae*. Signal transduction cascades coordinately regulate virulence in *V. cholerae*. The ToxR–ToxS and TcpP–TcpH signaling circuits detect and respond to environmental stimuli and activate the expression of *toxT*. ToxT, in turn, activates the transcription of a variety of genes required for virulence, including cholera toxin genes and genes in the TCP pathogenicity island involved in TCP pilus biosynthesis.

unusually broad to accommodate two very different lifestyles: as a pathogen in the human intestine and as a long-term resident of aquatic habitats during interepidemic periods (45). However, little is known about the extracellular signals that affect virulence in vivo or survival in the aquatic environment.

The ToxRS, TcpPH, and ToxT regulators were previously thought to regulate only virulence factor expression. However, a recent study (3) attempted to define the entire ToxR regulon by comparing the transcriptional profiles of *toxRS*, *tcpPH*, and *toxT* mutants with those of their isogenic parent strain under conditions *in vitro* that are optimized for production of virulence factors. This study found that TcpPH and ToxRS control genes involved in metabolism and nutrient uptake, as well as many genes of unknown function, suggesting that these regulators are involved in adaptive responses to the host environment as well as in virulence gene regulation. In contrast, ToxT appears to regulate only the CT and TCP genes, suggesting that ToxT is involved only in virulence processes.

V. cholerae infections begin with the ingestion of contaminated food or water. The bacteria that survive passage through the acid barrier of the stomach then colonize epithelial surfaces in the small intestine. They multiply extensively at this site and produces toxic proteins that alter the normal function of cells lining the intestine, causing these cells to secrete water and solutes into the lumen. The resulting diarrhea is believed to facilitate the spread of *V. cholerae* back into the environment to reinitiate the cycle. *V. cholerae* also maintains a permanent environmental reservoir in aqueous environments in association with various aquatic organisms. More recently, Merrell and colleagues (36) reported that *V. cholerae* present in stools shed by cholera patients were "hyperinfectious" for infant mice compared with a laboratory strain grown in vitro to stationary phase. This work provides evidence that passage of *V. cholerae* through the human host may alter critical virulence phenotypic properties that promote the transmission of the organism to new human hosts.

QUORUM SENSING REGULATION IN *V. CHOLERAE*

Bacteria utilize quorum-sensing regulation systems to control a variety of physiological functions (17, 37). Many Gram-negative bacteria use acyl-homoserine lactones, which are synthesized by LuxI-family proteins, as interbacterial signals and LuxR-type proteins as transcriptional regulators. In these systems, the autoinducer is usually synthesized by a protein related to the LuxI protein of *V. fischeri* and binds to an intracellular receptor protein that resembles the *V. fischeri* LuxR protein. Receptor–autoinducer complexes are thought to bind to target promoters to activate their transcription. These autoinducers diffuse rapidly across the cell envelope and accumulate intracellularly only at high bacterial population densities. In Gram-positive species, modified oligopeptides are secreted and serve as signals when they accumulate at high cell density. The detectors for the oligopeptide signals are two-component response regulator proteins.

V. harveyi possesses a complicated, hybrid quorum-sensing regulation system (1) that has characteristics of both Gram-negative and Gram-positive QS systems. *V. harveyi* uses an acyl-HSL autoinducer (AI-1) similar to that of other Gram-negative quorum sensing bacteria, but the signal detection and relay apparatus consists of two-component proteins similar to the quorum sensing systems of Gram-positive bacteria (2). AI-1 is an acyl-HSL (3-hydroxyl-C4-HSL), but its synthesis is dependent on LuxLM, which does not share any homology with the LuxI family of autoinducer synthases. *V. harveyi* also produces and responds to a second quorum-sensing autoinducer (AI-2) (48), a furanosyl borate diester that is not closely related to the homoserine lactones found in other bacteria (6, 43). Synthesis of AI-2 is dependent on the LuxS protein (55).

Bassler and colleagues used the *V. harveyi* luciferase operon as a heterologous reporter to examine quorum sensing in *V. cholerae* and found that, like *V. harveyi*, *V. cholerae* has at least two quorum sensing systems (38). System 2 (LuxSPQ) is similar to the *V. harveyi* system 2, whereas system 1 (CqsAS) is novel (Figure 5.2). At low cell density, signal concentration is low, and CqsS and LuxQ act as autophosphorylating kinases. These phosphates are then transferred to the cytoplasmic phosphotransferase LuxU, which in turn transfers the phosphates to LuxO, a s^{54}-dependent activator. Upon phosphorylation, LuxO is activated and functions to repress *hapR* expression (see below). HapR belongs to the TetR-family proteins and is a homolog of *V. harveyi* LuxR (although unrelated to LuxR from *V. fischeri*). At high cell density, the CAI-1 and AI-2 signal molecules are produced at a high level and interact with their cognate sensors. This interaction converts CqsS and LuxQ from kinases to

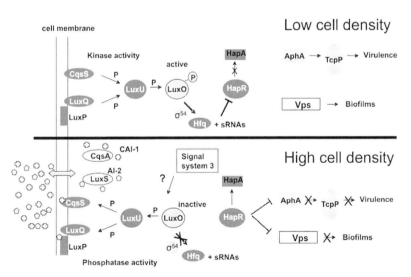

Figure 5.2. Current model for quorum-sensing regulation in *Vibrio cholerae*. At low cell density, LuxO protein is phosphorylated and actively represses *hapR* expression through Hfq and 4 sRNAs. Virulence genes are expressed. At high cell density, accumulation of autoinducers leads to LuxO dephosphorylation; HapR is produced and represses virulence gene expression and biofilm formation, and activates H/A protease production. A putative signal system 3 may also be involved in this process. P: phosphate group.

phosphatases, resulting in the loss of phosphate from LuxU and LuxO. Consequently, LuxO is inactivated and HapR is expressed. A gene called cqsA is responsible for the production of CAI-1 molecules. As predicted by domain conservation, cqsA likely encodes an aminotransferase, but the structure and biosynthesis of the CAI-1 autoinducer are not similar to those of previously identified autoinducer molecules.

The quorum sensing apparatus in V. cholerae is unusually complex. Study of system 1 and system 2 double mutants revealed that a third sensory circuit exists in V. cholerae (38). This system 3, which is hypothesized to respond to an intracellular signal, does not seem to transmit its signal through LuxU but rather directly through LuxO, the central regulator for all three systems.

In a recent study, Lenz et al. (35) found that repression of hapR by phosphorylated LuxO is indirect. At low cell density, phosphorylated LuxO activates the expression of the small RNA binding protein Hfq (47) and four redundant small regulatory RNAs (sRNAs). The Hfq–sRNA repressor complexes then act to destabilize V. cholerae hapR mRNAs. All four sRNAs must be inactivated to eliminate Hfq-mediated quorum-sensing repression, whereas overexpression of only one sRNA is sufficient for repression. Control via sRNAs may permit an ultrasensitive response to the concentration of active LuxO, as base pairing of a sRNA with its target message is known to promote degradation of both the sRNA and the mRNA. The use of sRNAs to accomplish an ultrasensitive response may be particularly apt for processes such as quorum sensing in which an all-or-nothing response is required.

QUORUM SENSING NEGATIVELY REGULATES VIRULENCE GENE EXPRESSION

Unlike in V. harveyi, there is no luciferase operon in V. cholerae. In an effort to determine the targets of quorum sensing in V. cholerae, mutations in quorum-sensing regulators luxO and hapR were tested for their roles in V. cholerae pathogenesis (60). The hapR mutant colonizes as well as the wildtype, but the luxO mutant is profoundly defective in colonization in an infant mouse model. V. cholerae colonization of the infant mouse (4–6 d old) intestine is an animal model commonly used for the human diarrheal disease cholera. This model has been extremely useful in the identification and characterization of proven and putative virulence factors involved in human cholera (28). In addition, the expression of the ToxR regulon was significantly depressed in the luxO mutant, as determined by a whole-genome microarray analysis. Overexpression of TcpP or ToxT, but not of ToxR, restores CT production in a *luxO* mutant, indicating that LuxO controls the ToxR regulon by repressing TcpP. As predicted, because LuxO represses *hapR*, HapR acts downstream of LuxO to repress *tcpP* expression, leading to repression of the ToxR virulence regulon. A follow-up study (31) suggests that, at high cell density, HapR decreases *tcpPH* transcription indirectly by binding directly to the *aphA* promoter and repressing *aphA* transcription. Owing to decreased intracellular levels of AphA, *hapR*$^+$ strains exhibit reduced virulence gene expression at high cell density.

Quorum-sensing-mediated virulence repression is achieved through the autoinducers CAI-1 and AI-2 (38). Addition of cell-free culture supernatant from the wild type, but not from an autoinducer-deficient mutant, represses *tcpP* expression. Autoinducers do not affect *tcpP* expression when added to a recipient lacking HapR, showing that CAI-1 and AI-2 contribute to repression of virulence factor expression specifically through the quorum sensing circuit described above. Each autoinducer is capable of partly inhibiting virulence gene expression, but together the autoinducers have a more than additive repressive effect, suggesting that the *V. cholerae* autoinducers function synergistically.

QUORUM SENSING INHIBITS BIOFILM FORMATION

A biofilm is a surface-associated microbial community embedded in a self-produced, extracellular polymeric matrix (9). These compact microbial 3-D structures are found in numerous environmental sites, such as aquatic reservoirs and tooth surfaces. Biofilm formation is recognized as a bacterial developmental process that requires a series of discrete and well-regulated steps. Biofilms are also clinically significant, as biofilm-associated bacteria are less susceptible to host immune responses and antimicrobial agents than are free-swimming bacteria (10). In addition, biofilms are often associated with chronic infections, most notably in *P. aeruginosa* infections of cystic fibrosis patients and catheter-associated biofilms of *Staphylococcus epidermidis* (18).

Several studies have suggested that biofilm-mediated attachment to abiotic surfaces may be important for *V. cholerae* survival in the environment (51, 58). Biofilm formation in *V. cholerae* is a multistep developmental process that is controlled by several regulatory pathways (52, 53). In the first step of biofilm formation, motility is reduced as the bacterium approaches the surface. The bacterium may then form a transient association with the surface. The next step is the formation of a microcolony, and exopolysaccharide is subsequently produced to form a three-dimensional biofilm. Finally, when environmental conditions change, some of the bacteria may detach and swim away to find a surface in a more favorable environment.

The two-component response regulator VpsR is required for expression of the *Vibrio* polysaccharide synthesis (*vps*) genes, which are essential for *V. cholerae* biofilm formation (56). VpsT, another regulator, also is required for *vps* gene expression (5). VpsT and VpsR positively regulate their own expression and also form a complex regulatory network by positively regulating each other's expression. However, the environmental signals that govern activation by VpsR and VpsT have not been identified. Interestingly, a report showed that the CytR protein negatively regulates biofilm formation by repressing *vps* gene expression (21). CytR also regulates nucleoside catabolism genes in *E. coli* in response to cytidine concentrations. Cyclic diguanylate (c-di-GMP) synthetase and phosphodiesterase domain-containing proteins have also been found to regulate biofilm formation by controlling c-di-GMP concentration (49). *Vibrio* polysaccharide synthesis and biofilm formation are affected by mutation of flagellar structural genes and a sodium-driven motor, suggesting that *V. cholerae* may monitor flagellar torque to sense when a surface is encountered (33, 54). Biofilm formation may therefore be controlled by both extracellular and intracellular cues.

Quorum sensing plays a role in biofilm formation in various microorganisms (26, 44), including *V. cholerae*. Quorum-sensing-deficient *V. cholerae hapR* mutants produce biofilms thicker than those formed by wild-type bacteria (20, 50, 59, 60). Microarray analysis and *lacZ* reporter fusions of biofilm-associated bacteria demonstrated that this effect results from enhanced expression of the *Vibrio* polysaccharide synthesis operons in *hapR* mutants (20, 59). Deletion of the AI-2 synthase gene *luxS* does not affect biofilm formation, whereas deletion of the CAI-1 synthase gene *cqsA* results in the formation of thicker biofilms. These results suggest that AI-2 signals are largely dispensable, whereas CAI-1 signaling is important for regulating biofilm formation. *cqsA* mutants grown in the presence of CAI-1 restore *hapR* expression and form wild-type biofilms; *vps* expression in *cqsA* mutant biofilms is higher than in wild-type biofilms at high cell densities (59). These data suggest that quorum sensing in *V. cholerae* also negatively regulates transcription of the *vps* genes and keeps biofilm synthesis in check. Whole-genome expression profiling revealed that HapR may indirectly repress *vps* expression by negatively regulating expression of the two positive *vps* regulators VpsR and VpsT (57).

There are two points during *V. cholerae* infection of the host that may be particularly affected by biofilm formation. First, the bacteria must survive the acidic environment of the stomach before reaching their site of replication in the upper intestine; inclusion of the bacterial cells in a biofilm may provide protection from this harsh environment. In fact, biofilm-associated *V. cholerae* are strikingly more resistant to low-pH shock than are planktonic cells (59). Second, *V. cholerae* may need to release from the biofilm in order to colonize the intestinal surface individually. Quorum-deficient mutants form thicker biofilms and may detach from biofilms less efficiently than do wild-type bacteria. Because virulence genes are repressed at high cell density, individual cells released from the biofilm may be more capable of colonizing the intestinal epithelium than are cells remaining associated with the biofilm. In fact, the bacteria in *hapR* mutant biofilms show a 10-fold lower colonization defect than wild-type *V. cholerae*, whereas planktonic *hapR* mutants can colonize as well as free-living wild-type cells, suggesting that quorum sensing may affect intestinal colonization by a mechanism that involves a HapR-dependent phenotype expressed in biofilm, such as detachment (59). Altogether, the comparable resistance of mutant and wild-type biofilms to acid shock, and the HapR-dependent ability of *V. cholerae* to detach from biofilms, might contribute to the efficiency of infection of mammalian hosts via the oral–gastric route.

Taken together, we have proposed a working model for the role of quorum sensing in the *V. cholerae* infectious cycle (Figure 5.3) (59). The biofilm structure may be critical during entry into the host in order to protect against acid shock in the stomach. After the bacteria reach the intestine, dispersal of individual cells from the biofilm enables these cells to shut off quorum-sensing-mediated repression of the CT and TCP genes, thus permitting maximal colonization of intestinal sites. The dispersal of cells from biofilms in the intestine may be caused by unknown host signals, or by the natural dynamic equilibrium of bacterial cells entering and leaving biofilm structures, or by the combination of the above two. Failure to disassemble the biofilm reduces the overall level of colonization. Later in the infection, the number of *V. cholerae* in the intestine increases and quorum sensing again represses CT and TCP production and activates protease production. It has been shown that HapR activates hemagglutinin protease (H/A protease) at high cell density (25, 50, 60). This protease was proposed to serve as a detachase during *V. cholerae* intestinal colonization (16). Detachment from the epithelium could permit individual cells to establish new infection foci in the intestine or to exit the host.

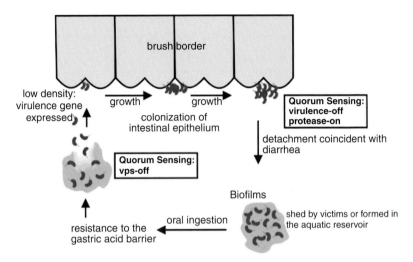

Figure 5.3. Models for the function of quorum sensing regulation and biofilm formation in *Vibrio cholerae* infectious cycles. *V. cholerae* resides in various aquatic environments in the form of biofilms. Upon ingestion, the biofilm structure protects *V. cholerae* cells from acid shock in the gastric environment. After passing through the stomach, quorum sensing represses *vps* production and individual cells dispersed from the biofilm show increased virulence gene expression, which enhances intestinal colonization. As bacteria multiply to high density on the intestinal epithelium, quorum sensing promotes detachment by repressing virulence gene expression and increasing H/A protease expression. Detached vibrios are shed and reinduce biofilm formation at low cell density.

Quorum sensing and biofilm formation may also be important in the environmental phase of the *V. cholerae* life cycle. During the environmental phase, the bacterium resides in diverse aquatic environments, often in association with marine plankton (24). Plankton "blooms" often precede cholera outbreaks, suggesting a role for these blooms in the epidemiology of disease (7). In a recent report (8), a simple, sari-cloth filtration procedure that removes particles larger than 20 μm from water was found to reduce cholera cases in Bangladesh by over 50%. This report suggests that particulate material plays an important role in *V. cholerae* transmission; the particulate material removed by sari-cloth filtration may in fact be *V. cholerae* biofilm material. This biofilm material may be associated with zooplankton (7, 23, 24), chironomid egg masses (4, 19) or other organic material. If so, the removal of acid-resistant *V. cholerae* biofilms might explain the utility of sari-cloth filtration. Finally, *Vibrio* biofilms may also be shed by cholera patients, and this in turn could help explain the interesting finding that *V. cholerae* from human stool appears more infectious than *V. cholerae* from stationary-phase cultures grown in vitro (36).

The formation of *V. cholerae* biofilms might also improve the environmental fitness of *V. cholerae* in other ways. In addition to protecting bacteria from physical–chemical insults such as changes in pH or the presence of antibiotics, biofilms likely make bacteria more resistant to biological threats as well. Recent data support the conclusion that environmental bacteriophages play a role in the epidemiology of cholera in Bangladesh (14). Thus, the sensitivity of *V. cholerae* to environmental vibriophages or predatory eukaryotic protozoa may be modulated by biofilm formation. If so, biofilm formation may play complex roles in the seasonality of cholera, its endemicity, and its transmission.

PERSPECTIVES AND FUTURE STUDIES

Quorum sensing negatively regulates *V. cholerae* virulence gene expression *in vitro* (38, 60). There remain, however, numerous unanswered questions regarding how quorum signals and other environmental cues are integrated to regulate virulence *in vivo*. For example, in the induction medium *in vitro*, HapR must be produced through quorum sensing pathways in early log phase to exert its inhibitory effects on virulence (60). Otherwise, once the ToxR-regulon is activated, the autoregulation of *toxT* overrides the HapR effect. Lee *et al.* employed a recombination-based *in vivo* expression technology (RIVET) to show that the temporal expression patterns of critical *V. cholerae* virulence genes *in vivo* are clearly different from expression patterns *in vitro* (34). Thus, it will be critical to examine the temporal profiles of quorum sensing and virulence gene regulation in the context of true infection.

Interestingly, several toxigenic *V. cholerae* strains (e.g. El Tor strain N16961 and classical strain O395) possess a natural frameshift mutation in the *hapR* gene (60). These strains express low levels of H/A protease. In addition, a mutation in *luxO* in these strains does not result in a CT production defect (29). A high mutation rate at the *hapR* locus was also observed in *V. cholerae* biofilms, thus promoting the formation of more robust biofilms (20). We speculate that elimination of HapR, either by mutation or by LuxO repression, may improve the adaptation of *V. cholerae* to different environments. However, in the infant mouse model, planktonic *hapR* mutant cells colonize normally. This is consistent with our observation that HapR is not involved in long-term colonization (59). These findings suggest that the role of HapR needs to be tested in some other animal models. Also, as we have demonstrated, *hapR* mutants form thicker biofilms, and this may aid their survival in various environments.

Another important and unresolved question is the chemical nature of the *Vibrio* autoinducers. It has been established that *V. cholerae* produces at least two distinct autoinducers, one of which is AI-2, a borate diester produced by the LuxS protein that is conserved in many Gram-positive and Gram-negative bacteria, including *V. cholerae*. Although the other *V. cholerae* autoinducer has yet to be identified, it is more critically involved than AI-2 in regulating virulence and biofilm formation (59). Therefore, it is important to characterize the chemical properties of this autoinducer and the exact roles it plays in regulating virulence. The second autoinducer may also represent a potential therapeutic molecule, since it can decrease virulence gene expression by causing elevated HapR expression and thus decreased *tcpPH* expression (38, 60).

Finally, quorum sensing apparently controls multiple cellular functions in *V. cholerae*. In addition to regulating virulence-factor expression and biofilm formation, LuxO and HapR also control genes involved in chemotaxis and motility, protein secretion, and protease production, as well as a number of genes encoding conserved hypothetical proteins (60). Moreover, comparison of the transcriptional profiles of a *V. cholerae cqsA* mutant grown under different conditions in the presence and absence of CAI-1 (J. Zhu and J. J. Mekalanos, unpublished) revealed that oxygen content apparently affects gene regulation of CAI-1 signal pathways. For example, ABC transporters of lipoproteins are induced by CAI-1 during aerobic growth, but repressed during anaerobic growth. More interestingly, transcriptional profiling of *cqsA* mutants attached to the intestinal epithelial cells shows that CAI-1 signals repress the virulence regulatory genes *tcpPH* and genes involved in biofilm formation, but activate genes involved in twitching motility and fimbrial assembly (authors' unpublished data).

In conclusion, further study of quorum sensing in *V. cholerae* should enable us to understand better the natural role of quorum sensing during *V. cholerae* infection. We believe that characterizing the relationships among biofilm development, quorum sensing, infectivity, and pathogenesis in *V. cholerae* will advance our goal of understanding the host–pathogen interaction and discovering potential novel treatments for cholera disease.

ACKNOWLEDGEMENTS

We thank Drs Su Chiang and Deb Hung for critical reading of the manuscript. Research in the Mekalanos laboratory is supported by NIH grants AI18045 and AI26289. JZ is supported by a NRSA fellowship.

REFERENCES

1 Bassler, B. L. 1999. How bacteria talk to each other: regulation of gene expression by quorum sensing. *Curr. Opin. Microbiol.* **2**: 582–7.

2 Bassler, B. L., M. Wright and M. R. Silverman 1994. Multiple signaling systems controlling expression of luminescence in *Vibrio harveyi*: sequence and function of genes encoding a second sensory pathway. *Molec. Microbiol.* **13**: 273–86.

3 Bina, J., J. Zhu, M. Dziejman *et al.* 2003. ToxR regulon of *Vibrio cholerae* and its expression in vibrios shed by cholera patients. *Proc. Natn. Acad. Sci. USA* **100**: 2801–6.

4 Broza, M. and M. Halpern 2001. Pathogen reservoirs. Chironomid egg masses and *Vibrio cholerae. Nature* **412**: 40.

5 Casper-Lindley, C. and F. H. Yildiz 2004. VpsT is a transcriptional regulator required for expression of *vps* biosynthesis genes and the development of rugose colonial morphology in *Vibrio cholerae* O1 El Tor. *J. Bacteriol.* **186**: 1574–8.

6 Chen, X., S. Schauder, N. Potier 2002. Structural identification of a bacterial quorum sensing signal containing boron. *Nature* **415**: 545–9.

7 Colwell, R. R. 1996. Global climate and infectious disease: the cholera paradigm. *Science* **274**: 2025–31.

8 Colwell, R. R., A. Huq, M. S. Islam *et al.* 2003. Reduction of cholera in Bangladeshi villages by simple filtration. *Proc. Natn. Acad. Sci. USA* **100**: 1051–5.

9 Costerton, J. W., Z. Lewandowski, D. E. Caldwell, D. R. Korber and H. M. Lappin-Scott 1995. Microbial biofilms. *A. Rev. Microbiol.* **49**: 711–45.

10 Costerton, J. W., P. S. Stewart and E. P. Greenberg 1999. Bacterial biofilms: a common cause of persistent infections. *Science* **284**: 1318–22.

11 Eberhard, A., A. L. Burlingame, C. Eberhard *et al.* 1981. Structural identification of autoinducer of *Photobacterium fischeri* luciferase. *Biochemistry* **20**: 2444–9.

12 Engebrecht, J., K. Nealson and M. Silverman 1983. Bacterial bioluminescence: isolation and genetic analysis of functions from *Vibrio fischeri. Cell* **32**: 773–81.

13 Engebrecht, J. and M. Silverman 1984. Identification of genes and gene products necessary for bacterial bioluminescence. *Proc. Natn. Acad. Sci. USA* **81**: 4154–8.

14 Faruque, S., I. B. Nasser, M. J. Islam *et al.* 2005. Seasonal epidemics of cholera are inversely correlated with the prevalence of environmental cholera phages. *Proc. Natn. Acad. Sci. USA* **102**: 1702–7.

15 Faruque, S. M., M. J. Albert and J. J. Mekalanos 1998. Epidemiology, genetics, and ecology of toxigenic *Vibrio cholerae. Microbiol. Molec. Biol. Rev.* **62**: 1301–14.

16 Finkelstein, R. A., M. Boesman-Finkelstein, Y. Chang and C. C. Hase 1992. *Vibrio cholerae* hemagglutinin/protease, colonial variation, virulence, and detachment. *Infect. Immun.* **60**: 472–8.

17 Fuqua, C. and E. P. Greenberg 2002. Listening in on bacteria: acyl-homoserine lactone signaling. *Nat. Rev. Molec. Cell Biol.* **3**: 685–95.

18 Hall-Stoodley, L., J. W. Costerton and P. Stoodley 2004. Bacterial biofilms: from the natural environment to infectious diseases. *Nat. Rev. Microbiol.* **2**: 95–108.

19 Halpern, M., Y. B. Broza, S. Mittler, E. Arakawa, and M. Broza 2004. Chironomid egg masses as a natural reservoir of *Vibrio cholerae* non-O1 and non-O139 in freshwater habitats. *Microb. Ecol.* **47**: 341–9.

20 Hammer, B. K. and B. L. Bassler 2003. Quorum sensing controls biofilm formation in *Vibrio cholerae*. *Molec. Microbiol.* **50**: 101–4.

21 Haugo, A. J. and P. I. Watnick 2002. *Vibrio cholerae* CytR is a repressor of biofilm development. *Molec. Microbiol.* **45**: 471–83.

22 Heidelberg, J. F., J. A. Eisen, W. C. Nelson *et al.* 2000. DNA sequence of both chromosomes of the cholera pathogen *Vibrio cholerae*. *Nature* **406**: 477–83.

23 Huq, A., S. A. Huq, D. J. Grimes *et al.* 1986. Colonization of the gut of the blue crab (*Callinectes sapidus*) by *Vibrio cholerae*. *Appl. Environ. Microbiol.* **52**: 586–8.

24 Huq, A., E. B. Small, P. A. West *et al.* 1983. Ecological relationships between *Vibrio cholerae* and planktonic crustacean copepods. *Appl. Environ. Microbiol.* **45**: 275–83.

25 Jobling, M. G. and R. K. Holmes 1997. Characterization of *hapR*, a positive regulator of the *Vibrio cholerae* HA/protease gene *hap*, and its identification as a functional homologue of the *Vibrio harveyi luxR* gene. *Molec. Microbiol.* **26**: 1023–34.

26 Kjelleberg, S. and S. Molin 2002. Is there a role for quorum sensing signals in bacterial biofilms? *Curr. Opin. Microbiol.* **5**: 254–8.

27 Klose, K. E. 2001. Regulation of virulence in *Vibrio cholerae*. *Int. J. Med. Microbiol.* **291**: 81–8.

28 Klose, K. E. 2000. The suckling mouse model of cholera. *Trends Microbiol.* **8**: 189–91.

29 Klose, K. E., V. Novik and J. J. Mekalanos 1998. Identification of multiple sigma54-dependent transcriptional activators in *Vibrio cholerae*. *J. Bacteriol.* **180**: 5256–9.

30 Kovacikova, G., W. Lin and K. Skorupski 2004. *Vibrio cholerae* AphA uses a novel mechanism for virulence gene activation that involves interaction with the LysR-type regulator AphB at the *tcpPH* promoter. *Molec. Microbiol.* **53**: 129–42.

31 Kovacikova, G. and K. Skorupski 2002. Regulation of virulence gene expression in *Vibrio cholerae* by quorum sensing: HapR functions at the *aphA* promoter. *Molec. Microbiol.* **46**: 1135–47.

32 Krukonis, E. S. and V. J. DiRita. 2003. From motility to virulence: sensing and responding to environmental signals in *Vibrio cholerae*. *Curr. Opin. Microbiol.* **6**: 186–90.

33 Lauriano, C. M., C. Ghosh, N. E. Correa and K. E. Klose 2004. The sodium-driven flagellar motor controls exopolysaccharide expression in *Vibrio cholerae*. *J. Bacteriol.* **186**: 4864–74.

34 Lee, S. H., D. L. Hava, M. K. Waldor and A. Camilli 1999. Regulation and temporal expression patterns of *Vibrio cholerae* virulence genes during infection. *Cell* **99**: 625–34.

35 Lenz, D. H., K. C. Mok, B. N. Lilley *et al.* 2004. The small RNA chaperone Hfq and multiple small RNAs control quorum sensing in *Vibrio harveyi* and *Vibrio cholerae*. *Cell* **118**: 69–82.

36 Merrell, D. S., S. M. Butler, F. Qadri *et al.* 2002. Host-induced epidemic spread of the cholera bacterium. *Nature* **417**: 642–5.

37 Miller, M. B. and B. L. Bassler 2001. Quorum sensing in bacteria. *A. Rev. Microbiol.* **55**: 165–99.

38 Miller, M. B., K. Skorupski, D. H. Lenz, R. K. Taylor and B. L. Bassler 2002. Parallel quorum sensing systems converge to regulate virulence in *Vibrio cholerae. Cell* **110**: 303–14.

39 Miyamoto, C. M., M. Boylan, A. F. Graham and E. A. Meighen 1988. Organization of the lux structural genes of *Vibrio harveyi.* Expression under the T7 bacteriophage promoter, mRNA analysis, and nucleotide sequence of the *luxD* gene. *J. Biol. Chem.* **263**: 13393–9.

40 Nealson, K. H. and J. W. Hastings 1979. Bacterial bioluminescence: its control and ecological significance. *Microbiol. Rev.* **43**: 496–518.

41 World Health Organization 1995. Report meeting: *The Potential Role of New Cholera Vaccine in the Prevention and Control of Cholera During Acute Emergencies.* World Health Organization.

42 Reidl, J. and K. E. Klose 2002. *Vibrio cholerae* and cholera: out of the water and into the host. *FEMS Microbiol. Rev.* **26**: 125–39.

43 Schauder, S., K. Shokat, M. G. Surette and B. L. Bassler 2001. The LuxS family of bacterial autoinducers: biosynthesis of a novel quorum-sensing signal molecule. *Molec. Microbiol.* **41**: 463–76.

44 Schembri, M. A., M. Givskov and P. Klemm 2002. An attractive surface: gram-negative bacterial biofilms. *Sci. Signal Transduct. Knowl. Environ.* **2002**: RE6.

45 Schoolnik, G. K. and F. H. Yildiz 2000. The complete genome sequence of *Vibrio cholerae*: a tale of two chromosomes and of two lifestyles. *Genome Biol* **1**: Reviews, 1016 1–3.

46 Skorupski, K. and R. K. Taylor 1997. Control of the ToxR virulence regulon in *Vibrio cholerae* by environmental stimuli. *Molec. Microbiol.* **25**: 1003–9.

47 Storz, G., J. A. Opdyke and A. Zhang 2004. Controlling mRNA stability and translation with small, noncoding RNAs. *Curr. Opin. Microbiol.* **7**: 140–4.

48 Surette, M. G., M. B. Miller and B. L. Bassler 1999. Quorum sensing in *Escherichia coli, Salmonella typhimurium,* and *Vibrio harveyi*: a new family of genes responsible for autoinducer production. *Proc. Natn. Acad. Sci. USA* **96**: 1639–44.

49 Tischler, A. D. and A. Camilli 2004. Cyclic diguanylate (c-di-GMP) regulates *Vibrio cholerae* biofilm formation. *Molec. Microbiol.* **53**: 857–69.

50 Vance, R. E., J. Zhu and J. J. Mekalanos 2003. A constitutively active variant of the quorum-sensing regulator LuxO affects protease production and biofilm formation in *Vibrio cholerae. Infect. Immun.* **71**: 2571–6.

51 Wai, S. N., Y. Mizunoe, A. Takade, S. I. Kawabata and S. I. Yoshida 1998. *Vibrio cholerae* O1 strain TSI-4 produces the exopolysaccharide materials that determine colony morphology, stress resistance, and biofilm formation. *Appl. Environ. Microbiol.* **64**: 3648–55.

52 Watnick, P. and R. Kolter 2000. Biofilm, city of microbes. *J. Bacteriol.* **182**: 2675–9.

53 Watnick, P. I. and R. Kolter 1999. Steps in the development of a *Vibrio cholerae* El Tor biofilm. *Molec. Microbiol.* **34**: 586–95.

54 Watnick, P. I., C. M. Lauriano, K. E. Klose, L. Croal and R. Kolter 2001. The absence of a flagellum leads to altered colony morphology, biofilm development and virulence in *Vibrio cholerae* O139. *Molec. Microbiol.* **39**: 223–35.

55 Xavier, K. B. and B. L. Bassler 2003. LuxS quorum sensing: more than just a numbers game. *Curr. Opin. Microbiol.* **6**: 191–7.

56 Yildiz, F. H., N. A. Dolganov and G. K. Schoolnik 2001. VpsR, a member of the response regulators of the two-component regulatory systems, is required for expression of vps biosynthesis genes and EPS(ETr)-associated phenotypes in *Vibrio cholerae* O1 El Tor. *J. Bacteriol.* **183**: 1716–26.

57 Yildiz, F. H., X. S. Liu, A. Heydorn and G. K. Schoolnik 2004. Molecular analysis of rugosity in a *Vibrio cholerae* O1 El Tor phase variant. *Molec. Microbiol.* **53**: 497–515.

58 Yildiz, F. H. and G. K. Schoolnik 1999. *Vibrio cholerae* O1 El Tor: identification of a gene cluster required for the rugose colony type, exopolysaccharide production, chlorine resistance, and biofilm formation. *Proc. Natn. Acad. Sci. USA* **96**: 4028–33.

59 Zhu, J. and J. J. Mekalanos 2003. Quorum sensing-dependent biofilms enhance colonization in *Vibrio cholerae. Dev. Cell* **5**: 647–56.

60 Zhu, J., M. B. Miller, R. E. Vance *et al.* 2002. Quorum-sensing regulators control virulence gene expression in *Vibrio cholerae. Proc. Natn. Acad. Sci. USA* **99**: 3129–34.

CHAPTER 6

LuxS in cellular metabolism and cell-to-cell signaling

Kangmin Duan
Molecular Microbiology Laboratory, Northwest University, Xián, China

Michael G. Surette
Faculty of Medicine, University of Calgary, Canada

INTRODUCTION

Many bacteria regulate gene expression in a cell-density-dependent manner; this behaviour has been collectively referred to as quorum sensing or cell-to-cell communication. In its simplest form this process results from the production and accumulation of signaling molecules in the surrounding environment. At some threshold concentration, the signaling molecules (also referred to as autoinducers) bind to receptors on or in the bacterial cell that regulate gene expression. In recent years the significance of cell-to-cell signaling in bacteria has become widely appreciated and has been extensively reviewed (2, 21, 25, 64, 84, 106).

The concept that bacterial cells can communicate through small-molecule chemical signals first arose in the pioneering studies in the Hastings laboratory in the early 1970s. *Vibrio fischeri* produces light; the light production is induced as a culture reaches high cell density, i.e. in a cell-density-dependent manner. In the seminal paper by Nealson *et al.* (69) it was demonstrated that cell-free cultures of *V. fischeri* contained a substance (termed an autoinducer) which, when added back to cell cultures, induced light production in the cells at a much earlier stage in the culture. Since that time the field of cell-to-cell signaling (quorum sensing) has grown dramatically; in recent years there has been an explosion of research in this area. In general there remain two well-established paradigms of cell-to-cell signaling: the *N*-acyl homoserine lactones used by some Gram-negative bacteria (see Chapters 1 and 3) and the oligopeptide signals used by some Gram-positive bacteria (see Chapters 9 and 10). These systems are described in detail elsewhere in this book and briefly outlined below.

Bacterial Cell-to-Cell Communication: Role in Virulence and Pathogenesis, ed. D. R. Demuth and R. J. Lamont. Published by Cambridge University Press. © Cambridge University Press 2005.

HOMOSERINE LACTONE QUORUM SENSING SYSTEMS

The classical example of quorum sensing is the autoinduction of luminescence in the marine bacterium *V. fischeri* (32, 69). In this system an *N*-acyl homoserine lactone (AHL) interacts directly with a regulatory protein, LuxR, to regulate expression of luminescence genes. The LuxI protein carries out synthesis of the signaling molecule. Over 40 members of the *luxI/luxR* gene family have been identified in Gram-negative bacteria and are involved in a wide range of responses (30, 53, 64, 74, 79, 102). A second family of AHL biosynthetic enzymes is known; these probably employ the same biosynthesis mechanism as LuxI but share no sequence homology with it. The best-studied examples of this family are the LuxLM proteins from *V. harveyi* (5). Since bacteria produce species-specific autoinducers and each species usually responds to its own autoinducers, the AHL-mediated quorum sensing is believed to be an intraspecies signaling system.

OLIGOPEPTIDE QUORUM SENSING PATHWAYS

Many Gram-positive bacteria communicate in a cell-density-dependent manner by using oligopeptides (see recent reviews (52, 94)). These systems involve synthesis of a precursor polypeptide that is cleaved during or after export; in some cases the mature signaling peptide may be further modified. Receptors for the peptides are typically found in the inner membrane and belong to the histidine kinase family of two-component signal transduction systems. In *Bacillus subtilis* (58) and *Streptococcus pneumoniae* (35), oligopeptide signaling is involved in regulation of competence. In *Enterococcus faecalis* and related species, oligopeptide signaling regulates aggregation and conjugational transfer between strains (68). In *Enterococcus* both stimulatory and inhibitory peptides are made and the active signaling occurs between related but not identical cells. In *Staphylococcus aureus*, a modified peptide cyclized into a thiolactone ring is involved in cell-to-cell signaling; many virulence factors are under control of this global regulatory system (44). Different groups of *S. aureus* produce different signaling peptides. Each group is distinguished by the observation that the peptides produced by isolates within the group are recognized as signaling molecules but the peptides from other groups are competitive inhibitors (43). Similar competition has been demonstrated between *S. aureus* and *S. epidermidis* strains (71, 72).

Although the AHL and oligopeptide cell-to-cell signaling systems remain the most recognized and studied signaling systems, there are other very well established signaling pathways such as the butyryl lactone signals of *Streptomyces* (26, 38) and the two signaling systems (A-signal and C-signal) of *Myxococcus xanthus* (41, 47, 48). There seems also to be an increasing list of cell-to-cell signaling molecules of distinct chemical properties.

The focus of this chapter will be on another pathway that involves the production of a signal, called autoinducer-2 (AI-2), by the enzyme LuxS. The identification of the *luxS* gene and a readily available bioassay for AI-2 has resulted in the identification of AI-2 production by a wide variety of bacteria. The elucidation of the biochemical pathway involved in AI-2 production and the lack of a *luxS* phenotype in many bacteria has also led to the question as to whether AI-2 is indeed involved in cell-to-cell signaling. The chapter will review the biology of *luxS* and AI-2 production and examine the recent evidence for and against a role in cell-to-cell signaling.

IDENTIFICATION OF LuxS

The identification of the *luxS*/AI-2 pathway originated in the marine bacterium *V. harveyi*. In this bacterium it was recognized that two independent signals were produced and detected by the organism. Although the species is Gram-negative, *V. harveyi* has a cell-to-cell signaling system with elements of both Gram-negative and Gram-positive paradigms, as described above. Two signaling molecules, autoinducer-1 (AI-1) and autoinducer-2 (AI-2), are involved. The first signal is β-hydroxy-butyryl homoserine lactone (10), typical of the Gram-negative bacteria. However, the synthesis of this signal is not by a LuxI protein homologue but by a distinct class of enzymes, designated LuxLM (5). The detection of the signals was found to be through two-component histidine kinase receptors; the mechanism is similar to that of oligopeptide sensing in the Gram-positive bacteria. The homoserine lactone AI-1 is sensed at the cell surface by the histidine kinase receptor LuxN (5). The detection of AI-2 is through a second histidine kinase (LuxQ); in this instance the ligand is a complex of a periplasmic binding protein, LuxP, and AI-2 (6). The signal transduction pathways for the two histidine kinases converge on a common phosphorelay cascade, outlined in Figure 6.1 (67).

The identification of the signal pathway led to the generation of *V. harveyi* reporter strains that could detect either AI-1 or AI-2 (*luxN* and *luxQ* mutants, respectively) (6). These reporter strains or variants thereof remain the primary method of detection of AI-2 today. Using these reporter

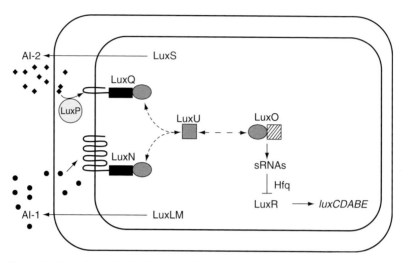

Figure 6.1. The schematic phosphorelay of the *V. harveyi* quorum sensing system. The cascade of phosphate transfer is influenced by cell density through signal-molecule concentration. AI-1 and AI-2 are produced by LuxLM and LuxS, respectively. At low cell density LuxQ and LuxN act as kinase and transfer ∼P to LuxU, and in turn to the response regulator LuxO, which regulates gene transcription through small regulatory RNA molecules (sRNA) and Hfq. The sRNA–Hfq complex is a negative regulator of *luxR*, the activator of the luciferase gene expression. At high cell density LuxQ and LuxN bind with cognate autoinducers, increasing their phosphatase activity. Phosphate flow is reverted from LuxO to LuxU to LuxQ and LuxN. Dephosphorylation of LuxO relieves the inhibition of *luxR* expression, leading to light production. The histidine kinase and response regulator domains are shown as black rectangles and grey circles, respectively. The response regulator LuxO contains a DNA binding domain (hatched). AI-1 (HSL) and AI-2 are depicted as filled circles and squares, respectively. Adapted from Lenz *et al.* (54a).

strains, Bassler and collaborators demonstrated that, among a wide variety of bacteria samples, none produced a detectable AI-1 signal but many produced measurable AI-2 activity (4). This led to the hypothesis that in *V. harveyi* System 1 is a species-specific signaling system and System 2 is a more general signaling system. This has been to some extent borne out and is reflected in the nature of species-specific acyl homoserine lactone signals, although some crosstalk between organisms can be observed together with the apparent universal detection of AI-2 activity by *V. harveyi*. This will be elaborated on below.

Despite many years of screening, a gene responsible for AI-2 production in *V. harveyi* had not been identified. That changed when it was discovered that both *Salmonella typhimurium* and many *Escherichia coli* strains (but not the common laboratory strain DH5α) produce AI-2 activity in a growth-condition and phase-dependent manner (95, 97). Significantly no AI-2 activity was detected in cell-free culture supernatants of these bacteria in the original strain survey by Bassler *et al.* (4). By using the *V. harveyi* AI-2 specific reporter strains, a transposon mutant of *S. typhimurium* was identified that did not produce AI-2 (95). The insertion was in a gene of unknown function (*ygaG*); it was shown that the corresponding gene in DH5α contained a 1 bp deletion inactivating the gene. AI-2 production was restored in the *Salmonella* transposon mutant and in *E. coli* DH5α was restored by adding back the *Salmonella* gene *in trans*. This complementation assay was used to screen for the *V. harveyi* gene responsible for AI-2 production, which was, not surprisingly, in a *ygaG* homolog. Given the role of this enzyme in regulating light production in *V. harveyi*, it was designated *luxS* (96, 97). The *luxS* gene was found to be widespread in many bacterial species, including both Gram-negative and Gram-positive species (97) (Figure 6.2); to our knowledge, AI-2 activity can be detected from culture-free supernatants of all species with a functional *luxS*.

METABOLIC PATHWAY OF AI-2 PRODUCTION

Despite the identification of the gene responsible for AI-2 production, the chemical identity of AI-2 and the nature of its production remained elusive. In most bacteria, *luxS* is in a single-gene operon, giving no clues to its function. The first hint at the pathway came from a comparative genomics analysis. In the spirochete *Borrelia burgdorferi* it was observed that the *luxS* gene was in an operon with *metK* and *pfs* (80). MetK is responsible for the production of the universal metabolite S-adenosylmethionine (SAM); Pfs is a bifunctional enzyme with methylthioadenosine/S-adenosylhomocysteine nucleotidase activity (19). SAM is utilized by a number of methyltransferases in the cell; the by-product of methyltransferase reaction is S-adenosylhomocysteine (SAH). SAH is a potent inhibitor of methyltransferases and is actively degraded in all cells. In eukaryotes and some bacteria, the removal of SAH is mediated by a SAH hydrolase generating homocysteine and adenosine. In other bacteria SAH is converted to S-ribosylhomocysteine (SRH) and adenine by Pfs (33, 103). Based on the comparative genomics, Schauder *et al.* reasoned that LuxS catalyzed the cleavage of SRH to 4,5-dihydroxy-2,3-pentanedione (DPD) and homocysteine. This enzymatic activity was known to be present in *E. coli* (24, 33, 62), but the gene responsible had not been identified. Similar reasoning by Winzer *et al.* (103) came to the same

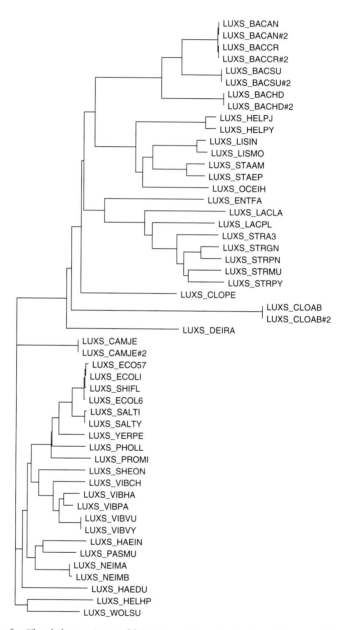

Figure 6.2. The phylogenetic tree of the LuxS protein, indicating the relatedness of the LuxS protein from 46 bacteria. The corresponding organisms are: BACAN, *Bacillus anthracis*; BACCR, *B. cereus* strain ATCC 14579 / DSM 31; BACHD, *B. halodurans*; BACSU, *B. subtilis*; BORBU, *Borrelia burgdorferi*; CAMJE, *Campylobacter jejuni*; CLOAB, *Clostridium acetobutylicum*; CLOPE, *C. perfringens*; DEIRA, *Deinococcus radiodurans*;

conclusion. The reconstitution of the production *in vitro* of AI-2 from SAH by using Pfs and LuxS or from SRH by using only LuxS confirmed the predictions (80, 103). Homocysteine was shown not to be the AI-2 signal, clearly implicating DPD or a derivative. DPD is chemically unstable and spontaneously cyclizes to a mixture of furanone rings that are more stable than the linear DPD. Knowing the biochemical pathway still left the chemical identity unknown.

There has been significant progress on the characterization of LuxS and its enzymatic mechanism in the past few years (73, 75). There are at least four LuxS orthologs for which X-ray crystal structures have been reported (37, 55, 78). The enzyme is a homodimer with two active sites. The crystal structures contain a zinc ion near the active site, suggesting that the molecule is a zinc metalloenzyme. Recent enzymatic studies, however, suggest that Fe may be preferred over zinc, with the metal playing an active role in catalysis (108). A more detailed understanding of the reaction mechanism (108, 109) has led to the synthesis of LuxS inhibitors (1), which will prove useful in further characterization of LuxS.

THE STRUCTURES OF AI-2

The chemical identity of AI-2 had remained elusive for many years. The chemical properties associated with the activity, a small polar molecule, contributed to the difficulty in purifying the activity from complex

---

Figure 6.2. (cont.)

ECO57, *Escherichia coli O157:H7*; ECOL6, *E. coli O6*; ECOLI, *E. coli*; ENTFA, *Enterococcus faecalis* (*Streptococcus faecalis*); HAEDU, *Haemophilus ducreyi*; HAEIN, *H. influenzae*; HELHP, *Helicobacter hepaticus*; HELPJ, *H. pylori J99* (*Campylobacter pylori J99*); HELPY, *H. pylori (C. pylori)*; LACLA, *Lactococcus lactis* (subsp. *lactis*) (*Streptococcus lactis*); LACPL, *L. plantarum*; LISIN, *Listeria innocua*; LISMO, *L. monocytogenes*; NEIMA, *Neisseria meningitidis* (serogroup A); NEIMB, *N. meningitidis* (serogroup B); OCEIH, *Oceanobacillus iheyensis*; PASMU, *Pasteurella multocida*; PHOLL, *Photorhabdus luminescens* (subsp. *laumondii*); PROMI, *Proteus mirabilis*; SALTI, *Salmonella typhi*; SALTY, *S. typhimurium*; SHEON, *Shewanella oneidensis*; SHIFL, *Shigella flexneri*; STAAM, *Staphylococcus aureus* (strain Mu50/ATCC 700699, N315, and MW2); STAEP, *S. epidermidis*; STRA3, *Streptococcus agalactiae* (serotype III and serotype V); STRGN, *S. gordonii*; STRMU, *S. mutans*; STRPN, *S. pneumoniae* (strain ATCC BAA-255/R6); STRPY, *S. pyogenes* (serotype M3 and serotype M18); VIBCH, *Vibrio cholerae*; VIBHA, *V. harveyi*; VIBPA, *V. parahaemolyticus*; VIBVU, *V. vulnificus*; VIBVY, *V. vulnificus* (strain YJ016); WOLSU, *Wolinella succinogenes*; YERPE, *Yersinia pestis*.

culture media. With hindsight, this was further complicated by the heterogeneity in molecular species. Surprisingly, the structure(s) of the active signal molecule were resolved not by small molecule purification and identification but by crystallization and structure determination of the receptor proteins in two seminal papers by the groups of Fred Hughson and Bonnie Bassler at Princeton University (11, 66). Each of these studies yielded quite unexpected results.

Using LuxP, which is the ligand-binding receptor of AI-2, Chen *et al.* (11) trapped AI-2 out of the mixture of furanone molecules and successfully determined the structure of AI-2 by crystallizing LuxP in a complex with the autoinducer. As shown in Figure 6.3, the AI-2 in *V. harveyi* is a borate containing (2S,4S)-2-methyl-2,3,3,4- tetrahydroxytetrahydrofuran (*S*-THMF-borate). The identification of borate in the LuxP structure was surprising. The use of borate in biological systems is very uncommon; however, given that the concentration of borate in seawater is in millimolar ranges, it would be an abundant anion for *V. harveyi* to utilize.

By using the same strategy, the crystal structure of *S. typhimurium* AI-2 binding protein LsrB was determined together with the bound AI-2. Unlike that in *V. harveyi* the active form of AI-2 is the unborated form of (2R,4S)-2-methyl-2,3,3,4-tetrahydroxytetrahydrofuran (*R*-THMF) (66) (Figure 6.3). Importantly, the *R*-THMF is in equilibrium with DPD, which is in turn in equilibrium with *S*-THMF-borate, and the equilibrium may change toward one of the active AI-2 molecules depending on the environmental conditions. The chemical equilibrium between DPD and its derivatives also indicated that these AI-2 molecules, whether borated or not, are interconvertible. This at least in part explains why AI-2 produced in many other bacteria, including *S. typhimurium*, can be detected by the *V. harveyi* reporter system even though it only responds to the borated *S*-THMF form of AI-2.

It is important to note two clearly distinctive characteristics of LuxP and LsrB. The former appears to be involved solely in sensing the extracellular concentration of signal whereas the latter is involved in transport. This is likely reflected in the nature of the ligand–receptor interactions, with many more coordinating interactions in the case of LuxP. This tighter interaction coupled with the complexing of *S*-THMF with borate in the *V. harveyi* sensory pathway may be incompatible with the transport function in the *Salmonella* system. In *Salmonella* the response to AI-2 in the *lsr* system is mediated by the cytoplasmic protein LsrR (98); the nature of AI-2 in its interaction with LsrR is not known but may be a phosphorylated derivative (98).

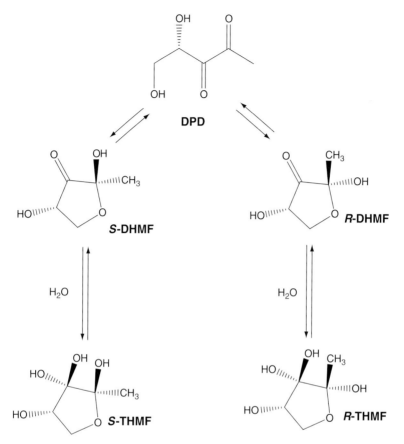

Figure 6.3. Two chemically distinctive AI-2 structures. Proposed solution structures of the AI-2 signaling molecules recognized by *Vibrio harveyi* (*S*-THMF) and *Salmonella typhimurium* (*R*-THMF). In *V. harveyi*, the *S*-THMF complexes with borate in the binding pocket of the receptor LuxP. *R*-THMF interacts with the receptor LsrB in *S. typhimurium*. DPD, 4,5-dihydroxy-2,3-pentanedione; *S*-DHMF and *R*-DHMF, stereoisomers of 2,4-dihydroxy-2-methyldihydrofuran-3-one; *S*-THMF and *R*-THMF, stereoisomers of 2-methyl-2,3,3,4-tetrahydroxyfuran. Adapted from Miller *et al.* (66).

It has become clear that at least two derivatives of the LuxS product DPD are sensed in different bacteria. The identification of chemically distinct forms of AI-2 in *S. typhimurium* raises the question about the chemical entity of the AI-2 in other bacteria. It is plausible that the *S*-THMF-borate AI-2 signal molecule is common in marine bacteria whose growth environment contains borate. However, whether the AI-2 in other bacteria is the same as that used by *S. typhimurium* remains to be determined. Nevertheless, it is evident that what has generically been termed "AI-2" is actually a mixture of molecules, more than one of which can be used for interspecies signaling. In this chapter we use *luxS*/AI-2-mediated quorum sensing to include the AI-2 in *V. harveyi*, that in *S. typhimurium*, and other active DPD derivatives yet to be identified.

THE CONTROVERSY OF AI-2 AND CELL-TO-CELL SIGNALING

The identification of the biochemical pathway leading to AI-2 production has highlighted controversy regarding the role of LuxS in bacterial physiology. Based on the metabolic pathway (Figure 6.4) its role in the removal of SAH from the cell places it clearly in primary metabolism, whereas more traditional cell-to-cell signaling pathways such as AHLs are more typical of secondary metabolites. The biochemical pathway has two implications regarding possible physiological roles for LuxS. The first is in the removal of SAH. The strong inhibitory effect of SAH on methyl transferase makes this an important step; *pfs* mutants show significantly reduced growth rates in many bacteria, and second-site revertants that restore growth rates are readily isolated. In contrast in many bacteria *luxS* has no observable growth phenotypes. The second possible role is in the production of homocysteine for methionine recycling. This is further hinted at by the organization of operons with which *luxS* is associated in some bacteria (Figure 6.5).

As the number of sequenced genomes increases, information from comparative genomic studies can add to our hypothetic roles for genes. Just as the organization with *metK* and *pfs* implicated *luxS* in the SAH degradation pathway, its association with *metB* and *cysK* in some bacteria further supports a role for LuxS in primary metabolism (see Figure 6.4). However, the lack of a *luxS* growth phenotype in *Salmonella* even in methionine-free media (A. L. Beeston and M. G. Surette, unpublished data) makes the significance of this pathway unclear. The importance of *luxS* in methionine recycling through production of homocysteine from S-ribosylhomocysteine remains to be determined in most if not all bacteria. It is likely that under some growth conditions this pathway will be important and that pleiotropic effects on metabolism in a *luxS* mutant may account for many of the observed phenotypes in some bacteria. This is likely to be particularly true in those examples where complementation by exogenous AI-2 does not restore the wild-type phenotypes. It is not clear how addition of exogenous AI-2 can complement the *luxS* mutant with respect to methionine recycling

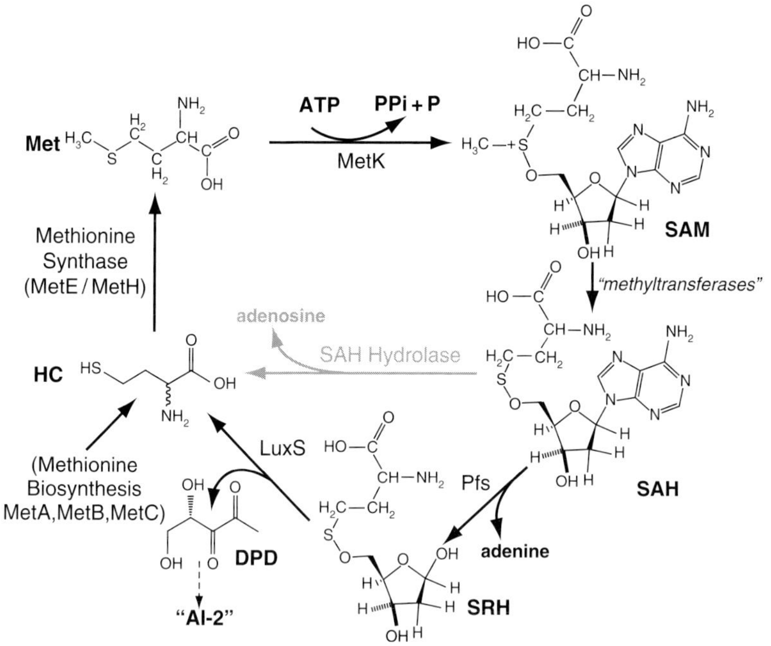

Figure 6.4. The metabolic pathways involved in AI-2 production and methionine metabolism. The pathway is based on *Escherichia coli* and *Salmonella*. Not all genes are present in all bacteria. Methionine (Met) is converted to S-adenosylmethionine (SAM) by the enzyme MetK. Numerous methyltransferases in the cell use SAM, generating S-adenosylhomocysteine (SAH) as a by-product. Removal of SAH is important, because it is a potent inhibitor of methyltransferases. The pathway of SAH degradation in eukaryotes and some bacteria by SAH hydrolase to homocysteine and adenosine is shown in gray. In other bacteria, SAH is removed by the action of Pfs to generate S-ribosylhomocysteine (SRH) and adenine. SRH is converted to homocysteine (HC) and 4,5-dihydroxy-2,3-pentanedione (DPD) by LuxS. DPD spontaneously cyclizes into "AI-2" (see Figure 6.3). Recycling of homocysteine into methionine is carried out by methionine synthase (MetE or MetH). In the biosynthesis of methionine, homocysteine is generated from oxaloacetate by MetA, MetB, and MetC. Adapted from Greene (33) and the Ecocyc database and Metacyc database (49, 50).

since the DPD would not be involved in the salvage pathway for homocysteine to methionine. One possible explanation may be that homocysteine is released into the media with AI-2 in conditioned media that is used for complementation, but to our knowledge this has not been tested. However, this seems unlikely, as it would be significantly diluted. It is not surprising that many bacteria can utilize AI-2 as a sole carbon source

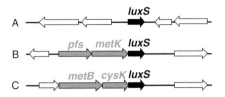

Figure 6.5. Gene location and operon context of *luxS*. (*a*) In most bacterial strains, luxS appears to be in a single-gene operon. (*b*) In at least two bacterial strains *luxS* is found in an operon with *pfs* and *metK*. (*c*) In at least three bacterial strains *luxS* is found in an operon with *metB* and *cysK*.

(as has been observed for AHLs as well) as many bacteria are efficient at utilizing a large variety of compounds for growth. However, most experiments where exogenous AI-2 restores phenotypes are carried out in media in which the exogenous AI-2 would contribute a very minor amount to the available nutrients and it seems unlikely that the observed effects are mediated through altered metabolism. When examining *luxS*/AI-2-mediated quorum sensing in a bacterium, one of the key results to look for is complementation of the *luxS* mutant by exogenous AI-2 or culture supernatant.

AI-2 AS A CLASSICAL QUORUM SENSING SIGNAL

The role of AI-2 as a *bona fide* cell-to-cell signaling molecule is well established among the *Vibrio* spp.; this is not surprising, because it was first characterized for its role in cell-density-dependent light production in *V. harveyi* as described above. The *V. harveyi* bioassay remains the primary method of detecting AI-2 activity in culture supernatants. The role of AI-2 in signaling is now well established in several *Vibrio* species in addition to *V. harveyi*. The general architecture of the signaling network seems to be conserved, with two or three signaling systems apparently acting in each species.

The sequencing of the *V. cholerae* genome (36) revealed the presence of an intact AI-2 signaling pathway homologous to that in *V. harveyi* and that the regulators in this pathway play an important role in regulating virulence gene expression (110). The realization that the *V. harveyi* and *V. cholerae* systems were very similar suggested that the *V. harveyi* *luxCDABE* operon could be used as a heterologous reporter of quorum sensing in *V. cholerae*; this was indeed the case (65). By using this strategy it was revealed that, in addition to a *luxS*/AI-2 pathway and convergent AHL pathway, named CqsA/CqsS as in *V. harveyi*, a third cell-to-cell signaling pathway was also present (65). The genes involved in this third pathway have yet to be identified. In addition to the regulation of virulence genes, cell-to-cell signaling has been shown to control biofilm formation in *V. cholerae* (34).

V. fischeri, the strain in which the LuxI/R homoserine lactone system and cell-to-cell signaling was initially described, has two AHL systems and *luxS*. These signals regulate light production and symbiotic colonization of the bobtail squid. The predominant signal is produced by LuxI; however, both AinS (the second AHL pathway) and LuxS contribute to the timing and magnitude of the response. Colonization of the host and light production is dependent on the LuxIR pathway and a modest reduction is observed in an *ainS* mutant. A *luxS* mutant had minimal effect on colonization of the host; however, an *ainS–luxS* double mutant had a synergistic decrease in colonization and light production (57).

In *V. vulnificus* the expression of extracellular virulence determinants is under control of the two systems SmcR (an AHL system) and LuxS. Mutations in either system reduce the production of extracellular cytotoxic factors (51, 81). Spent media could complement the *luxS* defects. The *luxS* mutation also resulted in attenuation of lethality in a mouse infection model (51).

AI-2 AS A QUORUM SENSING SIGNAL IN BACTERIA OTHER THAN *VIBRIO* SPP.

In *E. coli*, AI-2 signaling was initially reported together with that in *V. harveyi* and *S. typhimurium* (97). A role for LuxS/AI-2 in pathogenic *E. coli* has been well demonstrated. Both enterohemorrhagic *E. coli* (EHEC) and enteropathogenic *E. coli* (EPEC) produce a lesion on epithelial cells called the attaching and effacing (AE) lesion; the genes responsible for the lesion are located in the locus of enterocyte effacement (LEE). Kaper and colleagues have reported the involvement of AI-2 signals in the expression of type III secretion genes and virulence genes in EPEC and EHEC (31, 82, 86, 87; see also Chapter 7). Conditioned media were shown to increase the transcription of the LEE1 and LEE2 operons from EPEC and of the LEE1 operon from EHEC 2–3-fold. In EPEC, AI-2-mediated quorum sensing regulates the expression of a number of virulence-associated genes (82, 90) through *luxS*-dependent regulation of per, an early transcriptional regulator of the LEE virulence genes.

The spectrum of *luxS*-mediated gene regulation in *E. coli* was further characterized by using genome-scale examinations (18, 87). From DNA microarrays, Sperandio *et al.* (87) reported that approximately 10% of the array genes of *E. coli* O157:H7 were regulated by *luxS* at least 5-fold. Widespread effects on the regulation of flagella and motility genes by AI-2 signaling were observed. A total of 32 genes including the master regulator FlhDC, genes encoding flagella biosynthetic genes, motility proteins, and chemotaxis proteins were all downregulated significantly in the *luxS* mutant compared with the wild-type strain (87). These results suggest that AI-2 signaling may have diverse roles in cell division, growth, SOS response, shiga toxin production, and motility and chemotaxis (87).

However, a further report from Kaper and colleagues has demonstrated that the signal-dependent responses on LEE and flagellar gene expression are not, as previously reported, due directly to AI-2 (89). Another uncharacterized signal, termed AI-3, regulates the expression of the type III secretion genes and the two-component system *qseBCA*, which in turn regulates the flagella regulon (90) through *flhDC*, the master regulator for the flagella and motility genes. Interestingly, AI-3 signal production is also dependent on *luxS* (89; see also Chapter 7). It remains to be seen what chemical entity AI-3 is and how it is synthesized via LuxS.

Another microarray study by DeLisa *et al.* (18) using *E. coli* strain W3110 indicated that 242 genes, or approximately 5.6% of the *E. coli* genome, are regulated by AI-2. This report also shows AI-2 regulation of genes whose products are involved in diverse cellular processes including virulence, biofilm formation, motility, surface and membrane-associated functions, and small molecule metabolism, as well as a number of genes of unknown function. Although these findings also show AI-2 regulation of genes involved in cell division, only one gene (*ftsE*) was reported by both Sperandio *et al.* (87) and DeLisa *et al.* (18); the remaining genes on both lists are mutually exclusive. Changes in transcription ranged from 2.5-fold to 33-fold; the majority of activated or repressed genes showed less than a 5-fold difference.

In *Serratia marcescens, luxS* mutants are defective in antibiotic and virulence factor production (81); importantly, the defects can be complemented by exogenous AI-2 by using conditioned culture media. The mutant shows reduced virulence in a *Caenorhabditis elegans* infection model (81). Similarly, Coulthurst *et al.* (14) reported the detection of AI-2 in *S. marcescens* ATCC 274 and *Serratia* ATCC 39006 and demonstrated a role for AI-2 in the regulation of carbapenem antibiotic production in *Serratia* ATCC 39006 and of prodigiosin and secreted haemolysin production in *S. marcescens* ATCC 274. The phenotypic defect of *luxS* mutants in the strains was complemented by the addition of the culture supernatant of a wild-type strain. The complementation is effective even if the supernatant is substantially diluted. It is clear that *luxS*-mediated signaling operates in these bacteria as a classic quorum sensing system. A low amount of signal molecule regulates gene expression, and the regulation defects in signal synthesis mutants can be complemented by addition of exogenous signal molecule.

In *Shigella flexneri* the *luxS* mutation decreases in the expression of *virB*, a transcriptional regulator of invasion gene loci; however, invasiveness and dissemination of *Shigella* to adjacent cells remained unaffected by the lack of AI-2 (15). The addition of AI-2 upregulated the expression of *virB* in this bacterium.

AI-2-mediated quorum sensing has also been reported in many Gram-positive bacteria. In *Clostridium perfringens*, extracellular toxin production is reduced in a *luxS* mutant; this defect can be complemented by addition of culture media from a wild-type strain or from DH5α expressing *C. perfringens luxS*. The levels of all three toxins are affected; however, only one of these (*pfoA*) appears to be affected at the level of transcription (70). Interestingly, *luxS* appears to be in an operon with *metB* and *cysK* in *C. perfringens*. These enzymes are involved in the synthesis of methionine and cysteine, respectively. Importantly, the expression of the *pfoA* is not affected by mutations in either *metB* or *cysK*, indicating that *luxS* and presumably AI-2 are responsible for increased expression of toxin (70).

The *luxS* gene in *Actinobacillus actinomycetemcomitans* appears to play a role in aerobic growth in iron limiting conditions (see Chapter 8). Reduced expression of iron scavenging, transport and storage proteins was observed in a *luxS* mutant; a number of phenotypes in a *luxS* mutant are restored by exogenous AI-2 (27). In a search for homolog of the *V. harveyi* AI-2 sensory pathway, ArcB was identified as the closest LuxQ homolog (27). The *arcB* mutant was phenotypically similar to the *luxS* mutant.

Within the genus *Streptococcus*, several species are important human pathogens, and many have been shown to have a functional *luxS*/AI-2-mediated quorum sensing system. In *Streptococcus gordonii* DL1, a common human oral bacterium, inactivation of *luxS* lowered the expression of a number of genes (59) although growth of the bacterium was unaffected under several tested conditions (8). These genes include *gtfG*, encoding glucosyltransferase; *fruA*, encoding extracellular exo-β-D-fructosidase; and *lacD*, encoding tagatose 1,6-diphosphate aldolase (59). The *luxS* mutant also forms altered microcolony architecture within *S. gordonii* biofilms (8) and is unable to form normal mixed-species biofilms with *S. gingivalis* (59) (see below). *S. mutans* is another important component in the human oral biofilm community and a primary pathogen of human dental caries. Similar to *S. gordonii*, inactivation of *luxS* gene in this bacterium does not result in any apparent growth defect in liquid culture (60). No obvious difference in growth rate, nutrient requirements, acid production, or colony morphology could be observed between the wild type and the mutant. Although earlier observations by Wen and Burne indicated that in *S. mutans* strain UA159 the *luxS* mutation had no effect on its ability to form biofilms (100), marked alterations in the biofilm structure and its resistance to adverse agents were seen in the *luxS* mutant of *S. mutans* 25175 strain as a result of loss of AI-2 production (60). This apparent discrepancy was found to be due to the methods and growth medium of biofilm growth and evaluation. In a later study by Wen and Burne (101), inactivation of *luxS* resulted in a downregulation of fructanase, a demonstrated virulence determinant, by more than 50%, and the expression of other genes including *recA*, *smnA* encoding AP endonuclease, and *nth*-encoding endonuclease were downregulated in the *luxS* mutant, especially in early-exponential-phase cells. The *brpA* encoding a biofilm regulatory protein was also downregulated in the mutant. On a different supporting matrix, i.e. hydroxyapatite disks, the ability of the bacterium to form biofilms was shown to be greatly decreased, especially when biofilm was grown in medium with sucrose; the structure of such biofilms was also different from that of the wild type. Importantly, the *luxS* mutant phenotypes were able to be complemented by supernatants from wild-type cultures (101).

BACTERIA WITH AMBIGUOUS ROLES FOR AI-2

In each of the species and experiments described above, *luxS*/AI-2 appears to play an important role in the organism. LuxS defects can be complemented by exogenous AI-2 (or at least by conditioned media from wild-type but not *luxS* mutant strains), strongly suggesting a role for AI-2 in cell-to-cell communication. However, there are many systems in which no such role is apparent or the existing experimental evidence is ambiguous.

The gastric pathogen *Helicobacter pylori* possesses a *luxS* ortholog and produces AI-2 (28, 45). An earlier search for AI-2-regulated genes in this bacterium failed to detect any, either by two-dimensional protein profiling comparing the wild type and *luxS* mutant, or by a test of the known virulence genes (45). However, it has been demonstrated that the growth-phase-dependent expression of *flaA*, the major flagellin gene, is also *luxS*-dependent (56). Inactivation of *luxS* abolished the induction of *flaA* expression in higher cell density, but conditioned media from a wild-type culture resulted in only a slight (though statistically significant) increase in *flaA* expression (56). Biofilm formation in a *luxS* mutant is about two times greater than in a wild-type strain (13), but the effect of conditioned medium was not tested in this study. These results are consistent with AI-2 as a signal as well as with *luxS* playing an important role in metabolism through homocysteine pathways.

Interestingly, *H. pylori* is one of the two bacteria where *luxS* is located downstream of *metB* or *cysK* genes and likely forms an operon with these genes, as discussed previously. The *H. pylori* strain J99 has the gene order of *cysK–metB–luxS* and the strain 26695 has *metB–cysK–luxS* orientation (GenBank). The operon structure of *luxS*, with genes involved in amino acid synthesis pathways, seems also to be in agreement with the dual role of *luxS* in this bacterium.

Other ambiguous examples of AI-2 signaling are found in *Photorhabdus luminescens* and *Campylobacter jejuni*. In the first case, the production of the antibiotic carbapenem has been reported as being repressed by *luxS*/AI-2, but no complementation by exogenous AI-2 was done (20). In *C. jejuni*, the expression of *flaA* was reduced by about half in a *luxS* mutant, but no difference at protein level was observed (42). The motility and autoagglutination also experienced small reductions but, considering the lack of exogenous complementation experiments and the magnitude of the alteration, the signaling role of *luxS*/AI-2 is yet to be established in this bacterium.

Borrelia burgdorferi is the causative agent of Lyme disease. The spirochete contains a *luxS* gene that could complement an *E. coli* mutant (91). The *luxS* system appears to control the expression of a number of bacterial proteins of which some are involved in the infection process; the addition of exogenous AI-2 altered the expression of these genes (63, 91, 92). However, it has recently been reported that *luxS* is not required for tick colonization, transmission to a mammalian host, or the induction of disease (9, 40), strongly suggesting that AI-2 signaling plays no role in disease processes or bacterial virulence in this species. When infections of a susceptible host were initiated by intraperitoneal injection rather than by transmission by a tick, similar results were obtained (i.e. no effect on virulence of the *luxS* mutant) (40). The *luxS* mutants can colonize and cause disease efficiently, but whether they are at a competitive disadvantage compared with wild-type strains was not addressed. Although competition experiments might have been useful to clarify the infectiveness of *luxS* mutants, it should be noted that even if the wild-type strain were more competitive than the mutant this would not necessarily indicate that the mechanism was through cell-to-cell signaling.

Similar to *H. pylori*, the *luxS* gene in *B. burgdorferi* forms an operon with other genes in the genome (40), but in this case with *metK*, encoding the SAM synthase and *pfs*. As the genes in this *pfs–metK–luxS* operon encoding the enzymes catalyze continuous reactions in one metabolic pathway, it is likely that they all have metabolic functions in this bacterium. Consistent with this possibility is the fact that no AI-2 activity could be detected in the culture supernatant or the concentrated cell lysate by using the *V. harveyi* reporter system (9, 40) even though the *luxS* gene has been shown to be active and functional (91). It is important to note, however, that neither *metE* or *metH* homolog are predicted in the *B. burgdorferi* genome. Considering this organism's lack of biosynthetic pathways (29) and of apparent methionine salvage enzymes, it seems unlikely that the role of LuxS would be in methionine recycling.

In *Neisseria menigitidis*, an important Gram-negative bacterium causing septicemia and meningitis, it has been shown that *luxS*/AI-2 plays a role in virulence (104). A *luxS* mutant of Serogroup B of *N. meningitidis* had attenuated virulence in an infant rat model of bacteremia (104); however, it was later demonstrated that AI-2 appears to have no effect on concerted regulation of gene expression, by using microarray analysis (22).

Many of these studies are complicated by the lack of extracellular complementation by AI-2; this clearly is at odds with the conventional view of cell-to-cell signaling. These observations argue against a role for AI-2 in intraspecies signaling, at least in these bacteria. There always remains the caveat that the appropriate conditions have not been examined but the accumulating evidence would suggest that for many bacteria AI-2 may not play a role as a typical quorum-sensing signal.

AI-2 AS AN INTERSPECIES QUORUM SENSING SIGNAL

An important paper indicating that AI-2 activity might be more general than the better characterized species/strain-specific AHL signaling molecules came from Bassler *et al.* in 1997 (4). Using the *V. harveyi* reporter strains specific for *V. harveyi* AI-1 (butyryl homoserine lactone) and AI-2, they observed that, whereas system 1 of *V. harveyi* responded only to culture supernatants from *V. harveyi*, system 2 responded to culture supernatants from a number of different bacterial species, primarily *Vibrio* species but also *Yersinia enterocolytica*. Notable with hindsight was the absence of signal from a number of bacterial species, including *E. coli* and *S. typhimurium*.

This was because the conditions for AI-2 accumulation in culture media were not used in the original study; because AI-2 is actively removed by the bacteria as the culture approaches stationary phase, no or very little detectable activity remains in overnight culture media. The observation that the AI-2 signal was more widely produced than AI-1 led to the hypothesis that perhaps AI-2 was a general signaling mechanism indicating bacterial cell density among different species whereas AI-1 was indicative of the species that produces it itself (3, 4, 25, 79).

Although AHL-mediated quorum sensing is generally species-specific, crosstalk among closely related species through AHL pathways has been demonstrated for a number of bacteria (61, 76, 83); the use of reporter strains for identification of signaling molecules (AHL and AI-2) implies the ability of signals to cross species boundaries. In the case of AI-2 signaling the *V. harveyi* reporter strains remain the primary assay used to detect signal production. Most strains containing a *luxS* gene have been shown to produce a signal under some growth condition that activates the AI-2-specific *V. harveyi* reporters. As described above, however, the nature of the AI-2 signal is not straightforward; in *V. harveyi* the active molecule is a furanyl borate diester (11) whereas in *S. typhimurium* the active signal is *R*-THMF (66). The chemical equilibrium of products will be affected by the chemical environment; different forms of the molecule may predominate under particular assay conditions. Nonetheless, the use of different derivatives of the LuxS product as AI-2 molecules in different species does not disagree with the suggestion that AI-2 may be a generic bacterial signaling molecule shared by a wide variety of species.

The implication of AI-2 as an interspecies signal is that AI-2 may contribute to the establishment and behavior of mixed microbial communities. The hypothesized role of AI-2 in interspecies signaling has been borne out in a number of studies, most significantly in the mixed microbial communities forming dental plaque and also as part of the interspecies interactions between *Pseudomonas aeruginosa* and oropharyngeal flora in cystic fibrosis. There is strong evidence in support of this from studies of these two polymicrobial communities.

Cell-to-cell signaling and AI-2 in dental plaque

Dental plaque is a complex biofilm community that comprises more than 500 bacterial species. The bacteria normally exist in harmony with the host when the dynamic yet stable microbial community is maintained. However, overgrowth by some microorganisms, especially a group of

Gram-negative anaerobic species, e.g. *Porphyromonas gingivalis*, causes the initiation and progression of periodontitis, an oral disease that is among the most prevalent human diseases (85). The biofilm community has a well-differentiated, organized structure formed in a spatiotemporal order by cooperative actions of multispecies bacteria. The biofilm formation on the surface of teeth is initiated by the early colonizer bacteria, among which 60% are *Streptococcus* spp., and develops through coaggregation and coadherence by late-colonizing bacteria via interactions with the early colonizers. It is suggested that the coaggregation and coadherence are essential to communication between species and that the communication between bacterial species are likewise integral to this process and help to establish patterns of spatiotemporal development. An AHL quorum-sensing signal, however, has not been detected and is therefore probably not produced, at least not in a significant quantity, in this community. In contrast, some of the key components in this community have been shown to produce AI-2, including *S. mutans*, *S. gordonii*, *Actinobacillus actinomycetemcomitans*, *Porphyromonas gingivalis*, *Fusobacterium nucleatum*, and *Prevotella intermedia*. It is clear that LuxS-mediated interspecies quorum sensing plays important roles in this complex community.

In *S. mutans*, which is among the early colonizers of dental plaque biofilm, a *luxS* mutant forms a biofilm remarkably different from that formed by the wild type. Instead of being smooth and confluent, the biofilm formed by the *luxS* mutant is rather rough and granular. Furthermore, the biofilm formed by the *luxS* mutant has increased resistance to detergent and antibiotics (60). Inactivation of the *luxS* gene in *Porphyromonas gingivalis* and *A. actinomycetemcomitans* resulted in changes in the expression of genes involved in hemin acquisition and LtxA leukotoxin production, respectively. In addition, the virulence of *P. gingivalis* was attenuated in *luxS* mutants. Importantly, conditioned medium from *E. coli* DH5a expressing the *luxS* gene of *A. actinomycetemcomitans* was able to complement the defects in gene expression in the *P. gingivalis luxS* mutant, indicating that LuxS-dependent signaling potentially mediates interspecies communication in mixed-species biofilms. Experiments with *S. gordonii* and *P. gingivalis*, mixed-species biofilms showed that LuxS-dependent intercellular communication between *P. gingivalis* and *S. gordonii* is essential for biofilm formation (59). *S. gordonii* is one other member of the early colonizers. A substratum of *S. gordonii* attached to solid surfaces supports the adhesion of *P. gingivalis*, which is initiated via the interactions between *S. gordonii* cell-surface proteins and the fimbriae of *P. gingivalis*. The adhered *P. gingivalis* then progresses to form biofilm microcolonies distinctive from the simple adhered microaggregates of *P. gingivalis* itself.

Inactivation of the *luxS* gene in *S. gordonii* altered the expression of a number of genes, among which is a group of genes involved in carbohydrate metabolism in this organism. More importantly, this *luxS* mutant was unable to form normal mixed-species biofilms with LuxS-deficient *P. gingivalis*. However, if one of the partner strains possessed a functional *luxS* gene the mixed-species biofilms microcolonies were formed normally, indicating that the bacteria could respond to a heterologous AI-2 signal. These observations strongly support a role for AI-2-mediated interspecies quorum sensing in mixed-species biofilm formation in dental plaque.

The CF community

The other example of AI-2 playing a role in microbial communities is the mixed-species signaling between *Pseudomonas* and the oropharyngeal flora (OF) in cystic fibrosis (CF) lungs. In patients with CF, altered ion transport gives rise to a more viscous pulmonary mucous layer, thereby impairing ciliary clearance and permitting the existence of a biofilm-forming microbial community in the lungs (99). Among them, the leading pathogenic agent is *P. aeruginosa*, whose infection ultimately causes pulmonary failure, resulting in premature mortality (93). In addition to *P. aeruginosa*, these patients' respiratory tracts commonly experience infections by other pathogens such as *S. aureus, Haemophilus influenzae* and *Burkholderia cepacia*, as well as a variety of other avirulent microorganisms (12, 23) including OF bacteria such as viridans-group streptococci and coagulase-negative staphylococci, which only colonize the upper respiratory tracts in healthy adults. Using an *in vitro* system it has been shown that in this microbial community multifactorial interactions occur among the microorganisms, which significantly influence the pathogenesis of *P. aeruginosa* (23). One of the factors was determined to be AI-2 produced by non-pseudomonad strains. Although *P. aeruginosa* does not possess a *luxS* gene in its genome and does not produce AI-2, a large amount of AI-2 was readily detectable in all the sputum samples of the CF patients tested and in the culture supernatants of most of the OF isolates using the *V. harveyi* assay. Transcriptional profiling of a set of defined *P. aeruginosa* virulence factor promoters indicated that OF and exogenous AI-2 could upregulate overlapping subsets of these genes. Six out of the nine genes that were affected by the presence of OF strains were also regulated by AI-2 added to the media. It seems that *P. aeruginosa* is modulating its behavior by monitoring the environmental conditions and by eavesdropping on the other bacteria via AI-2 and probably other signals. *P. aeruginosa* may have evolved to use AI-2 as a signal indicative of a polymicrobial environment in an animal host. The phenomenon that AI-2 from one microorganism affects the gene expression of the other indicates that the *luxS*-mediated quorum sensing system does indeed play a role in the dynamics of this microbial community.

It has been observed that *P. aeruginosa* could remove AI-2 in the medium (103). This is not surprising: degradation of AI-2 has been observed in a number of bacteria that respond to the signal, including *P. aeruginosa* (39, 77). It is not known, however, whether this represents a strategy that *P. aeruginosa* employs to counter other bacteria or whether it is just one aspect of how *P. aeruginosa* responds to exogenous cell-to-cell signals. Because *P. aeruginosa* could be modulated by AI-2, it is possible that it has a mechanism to control the signal. Indeed, in the classical example of *V. harveyi* the molecule decreases rapidly after it reaches its peak amount. It is normal for a regulatory system to have a component to degrade and turn over its signal (107). In a microbial community it is conceivable that a molecule being used by one species as a signal molecule could be attacked or degraded by another species. Some of the bacteria may have evolved to sense the signals produced by others; some may have evolved to produce the signals just to affect the behavior of others. Because of AI-2's important role in gene regulation and its relatively small amount, this is probably more an aspect of cell-to-cell interaction than of syntrophic interactions.

AI-2 AS A REPORTER OF METABOLIC STATUS

Because the LuxS protein is not only responsible for the synthesis of AI-2 precursor but may also function to recycle S-adenosylhomocysteine, a question has been raised about its role as a signal molecule (103). However, for the same reasons, it has been suggested as being a signal in some bacteria which measures the metabolic status of the cell instead of just cell density (7, 105). Although the role of AI-2 as a cell-to-cell communication signal is clear in many bacteria, as discussed above, there are increasing data that suggest its role as a signal reflecting the metabolic status of the cell instead of simply cell number.

In the classical definition of quorum sensing, autoinducer production increases with the increase in cell numbers, and the signals accumulate in the culture environment. AI-2 accumulation in the extracellular environment is often very growth-phase- and media-dependent. In the original *V. harveyi* studies, AI-2 accumulates throughout growth, reaching maximum accumulation near the end of growth, presumably because there is little or no turnover of the signal. In the original study demonstrating AI-2 activity in a wide variety of *Vibrio* spp. (4), no activity was detected in *E. coli* or *Salmonella*. It was subsequently shown that both these bacteria produce AI-2 but in a growth-phase-dependent manner (96).

AI-2 production peaks before stationary phase and decreases after the peak in a number of bacteria. We have observed significant effects of growth conditions on signal production and degradation in the *luxS* system for numerous bacteria (95, 96). Expression data also show that neither *luxS* nor *pfs* expression is regulated by AI-2, suggesting that AI-2 production is not regulated at the level of *luxS* expression. This result means the AI-2 production differs from classical autoinducers in the sense that it is not autoinduced. It should be noted that autoinduction is a common but not universal feature of cell-to-cell signaling systems. The amount of AI-2 produced is instead regulated at the level of LuxS substrate availability, which reflects the metabolic status of the cells but not the cell number. These results indicate that AI-2-dependent signaling is a reflection of the metabolic state of the cell and not of cell density (7). The concentration of AI-2 accumulating in the extracellular environment is also a function of its degradation or active removal by bacteria; in *Salmonella*, production and removal via the Lsr system seem to be oppositely regulated (A. L. Beeston and M. G. Surette, unpublished results).

As discussed above, quorum sensing is prevalent and important in pathogenic *E. coli*. In non-pathogenic, laboratory-domesticated *E. coli*, it has been demonstrated that the AI-2 signaling pathway communicates the stress or burden of overexpressing heterologous genes in the bacterium (16). The activity of AI-2 decreases significantly following induction of several plasmid-encoded genes in the bacterium at both low- and high-cell-density culture conditions. The AI-2 signaling level was linearly related to the accumulation level of each protein product. In this case the metabolic status that AI-2 signals is reflected as the stress or burden of gene overexpression.

The other example of a metabolic status reporting role for AI-2 comes from the findings of the linkage between *luxS* and *relA* in *S. mutans* (54). In a study investigating the role of the *relA* gene, which codes for a guanosine tetraphosphate and guanosine pentaphosphate [(p)ppGpp] synthetase/hydrolase, in biofilm formation and acid tolerance, Lemos *et al.* found that a *relA* mutant showed significant reductions in biofilm formation after the induction of a stringent response and altered acid resistance in biofilms. Interestingly, the expression of the *luxS* gene was increased as much as 5-fold in the *relA* mutants, suggesting a link between AI-2 quorum sensing and the stringent response. Because of the role of the stringent response in sensing cellular amino acid shortage, etc., this connection strongly suggests a role for *luxS*-mediated QS in monitoring cell metabolic status in this bacterium.

In *S. pneumoniae*, the *luxS* mutant shows a significantly decreased ability to persist in a murine model of nasopharyngeal carriage (46). The mutation also affects at least five operons that are involved in fatty-acid biosynthesis and virulence-factor production. However, *luxS*-mediated regulation does not follow the typical quorum-sensing paradigm: it does not occur at high cell density (46). LuxS activity in this bacterium instead modulates the fitness of the organism in a discrete host niche and LuxS functions in a mechanism independent of cell density. However, as noted above, it is important to distinguish between cell-to-cell signaling and cell density.

Although quorum sensing and cell-to-cell signaling in bacteria are normally interchangeable terms and cell density dependence is one of the hallmarks of this process, it has become clear that cell-to-cell signaling is not simply cell-density-dependent. A more general definition of quorum sensing as a gene regulation coordinated or mediated by secreted small signal molecules which function at low concentrations (i.e. at concentrations lower than that which would mediate physiological changes through the metabolism of the substrate) seems appropriate. The responding bacteria do so when the autoinducer molecule reaches a threshold concentration. In many experimental systems this occurs late in the growth of liquid cultures as the cells reach high cell density. However, this concentration of signal can also be reached at lower cell densities. It is now clear that in many cell-to-cell signaling systems (including AHLs and oligopeptides) both the production and the degradation of the signaling molecule can be tightly regulated and can be influenced by growth conditions. Moreover, in most laboratory studies it is routine practice to vigorously shake liquid cultures, because this results in a homogeneous environment for all the cells. This does remove any local concentration gradients and in effect dilutes signaling molecules. The dynamics of response as well as the growth phase can be significantly altered by simply not shaking the cultures! The concept of cell density should be used cautiously in the context of cell-to-cell signaling. In homogenous growth conditions such as in shaking liquid culture, all cells respond to essentially the same conditions. When the environment is not "homogenized," different cells within the population are exposed to very different environments.

Cell-to-cell signaling is often interpreted under the paradigm established for *V. fischeri*, where cell density in a culture correlates with signal accumulation and response. The culture tube offers a reasonable approximation of *V. fischeri* life as a symbiont in the light organ of the squid with respect to quorum sensing. However, this should not be extrapolated to all circumstances. The importance of growth conditions for production and response is often under-estimated. We have observed significant effects of growth conditions on signal production and degradation in the *luxS* system in numerous bacteria (95, 96) and in the production of acyl homoserine lactone signals in *P. aeruginosa* (K. Duan and M. G. Surette, unpublished results). Localized signaling and coordinated behaviour in small groups of cells in micro-colonies, aggregates, microniches within colonies, biofilms, and consortia of multiple species are likely to be more natural situations for cell-to-cell signaling for most bacteria. Coordinated behaviour can also arise from bacterial manipulation of the environment through the utilization of nutri-ents and the release of toxic metabolites, enzymes (e.g. proteases), and extracellular matrices (e.g. capsular material). This can give the appearance of coordinated behaviour. Bacterial chemotaxis in soft agar provides an excellent example of this. There is "coordination" of cells within the popula-tion at the macroscopic level (well-defined chemotactic rings) arising from the manipulation of the environment and a common response (i.e. chemo-taxis up a gradient of nutrients). However, each cell within the population behaves independently. Similar "microenvironments" will arise under any condition where the cells are not in homogeneous environments (i.e. almost any condition other than well-mixed liquid cultures). This is one mechanism that can give rise to spatial and temporal patterns of gene expression in microniches within bacterial colonies, swarming bacteria, and biofilms. In the natural world, where bacteria exist primarily in mixed populations and often in well-defined consortia of species, exchange of metabolites and scavenging of other cells' waste products may be a central feature of their natural history. Cell-to-cell signaling is distinct from that behavior described above in that it is mediated by signaling molecules that act at low concentrations and not directly as metabolites. This is an important distinc-tion and the two are not always readily distinguished. This is particularly an issue when using "conditioned media" (even more so when using complex media) and an obvious caveat to such experiments that can complicate interpretations.

CONCLUSIONS

For many organisms, in particular the *Vibrio* spp., the role of AI-2 in typical cell-to-cell signaling seems indisputable. For others, there seem to be equally compelling data to suggest that *luxS* functions as part of primary metabolic pathways and that AI-2 production is simply a by-product of metabolism. We should be careful not to be constrained by current paradigms trying to fit AI-2 into one domain or another: biology is full of exceptions to general rules. LuxS/AI-2 is perhaps a little more blatant than others in its disregard for established paradigms. Although we can consider AI-2 as a typical intraspecies signaling molecule for many bacteria, there is evidence (or more often the lack of evidence in support thereof) that argues against that role in many other bacteria. Observations of AI-2 playing a role in interspecies communication in polymicrobial communities are increasing. The current data suggest that AI-2 will be found to be one of many contributing interactions that play a role in community structure and dynamics. Understanding polymicrobial communities will be a rich source of research in cell–cell communication; the evidence to date suggests that interspecies signaling via AI-2 will be a player in many of these systems.

REFERENCES

1   Alfaro, J. F., T. Zhang, D. P. Wynn, E. L. Karschner and Z. S. Zhou 2004. Synthesis of LuxS inhibitors targeting bacterial cell-cell communication. *Org. Lett.* **6**: 3043–6.

2   Bassler, B. L. 1999. How bacteria talk to each other: regulation of gene expression by quorum sensing. *Curr. Opin. Microbiol.* **2**: 582–7.

3   Bassler, B. L. 2002. Small talk. Cell-to-cell communication in bacteria. *Cell* **109**: 421–4.

4   Bassler, B. L., E. P. Greenberg and A. M. Stevens 1997. Cross-species induction of luminescence in the quorum-sensing bacterium *Vibrio harveyi*. *J. Bacteriol.* **179**: 4043–5.

5   Bassler, B. L., M. Wright, R. E. Showalter and M. R. Silverman 1993. Intercellular signaling in *Vibrio harveyi*: sequence and function of genes regulating expression of luminescence. *Molec. Microbiol.* **9**: 773–86.

6   Bassler, B. L., M. Wright and M. R. Silverman 1994. Multiple signaling systems controlling expression of luminescence in *Vibrio harveyi*: sequence and function of genes encoding a second sensory pathway. *Molec. Microbiol.* **13**: 273–86.

7   Beeston, A. L. and M. G. Surette 2002. pfs-Dependent regulation of autoinducer 2 production in *Salmonella enterica* serovar Typhimurium. *J. Bacteriol.* **184**: 3450–6.

8   Blehert, D. S., R. J. Palmer, Jr., J. B. Xavier, J. S. Almeida and P. E. Kolenbrander 2003. Autoinducer 2 production by *Streptococcus gordonii* DL1 and the biofilm phenotype of a luxS mutant are influenced by nutritional conditions. *J. Bacteriol.* **185**: 4851–60.

9   Blevins, J. S., A. T. Revel, M. J. Caimano *et al.* 2004. The *luxS* gene is not required for *Borrelia burgdorferi* tick colonization, transmission to a mammalian host, or induction of disease. *Infect. Immun.* **72**: 4864–7.

10 Cao, J. G. and E. A. Meighen 1989. Purification and structural identification of an autoinducer for the luminescence system of *Vibrio harveyi*. *J. Biol. Chem.* **264**: 21670–6.

11 Chen, X., S. Schauder, N. Potier *et al.* 2002. Structural identification of a bacterial quorum-sensing signal containing boron. *Nature* **415**: 545–9.

12 Coenye, T., J. Goris, T. Spilker, P. Vandamme and J. J. LiPuma 2002. Characterization of unusual bacteria isolated from respiratory secretions of cystic fibrosis patients and description of *Inquilinus limosus* gen. nov., sp. nov. *J. Clin. Microbiol.* **40**: 2062–9.

13 Cole, S. P., J. Harwood, R. Lee, R. She and D. G. Guiney 2004. Characterization of monospecies biofilm formation by *Heliobacter pylori*. *J. Bacteriol.* **186**: 3124–32.

14 Coulthurst, S. J., C. L. Kurz and G. P. Salmond 2004. *luxS* mutants of *Serratia* defective in autoinducer-2-dependent 'quorum sensing' show strain-dependent impacts on virulence and production of carbapenem and prodigiosin. *Microbiology* **150**: 1901–10.

15 Day, W. A., Jr. and A. T. Maurelli 2001. *Shigella flexneri* LuxS quorum-sensing system modulates *virB* expression but is not essential for virulence. *Infect. Immun.* **69**: 15–23.

16 DeLisa, M. P., J. J. Valdes and W. E. Bentley 2001. Quorum signaling via AI-2 communicates the 'Metabolic Burden' associated with heterologous protein production in *Escherichia coli*. *Biotechnol. Bioeng.* **75**: 439–50.

18 DeLisa, M. P., C. F. Wu, L. Wang, J. J. Valdes and W. E. Bentley 2001. DNA microarray-based identification of genes controlled by autoinducer 2-stimulated quorum sensing in *Escherichia coli* . *J. Bacteriol.* **183**: 5239–47.

19 Della Ragione, F., M. Porcelli, M. Carteni-Farina, V. Zappia and A. E. Pegg 1985. *Escherichia coli* S-adenosylhomocysteine/5'-methylthioadenosine nucleosidase. Purification, substrate specificity and mechanism of action. *Biochem. J.* **232**: 335–41.

20 Derzelle, S., E. Duchaud, F. Kunst, A. Danchin and P. Bertin 2002. Identification, characterization, and regulation of a cluster of genes involved in carbapenem biosynthesis in *Photorhabdus luminescens*. *Appl. Environ. Microbiol.* **68**: 3780–9.

21 Donabedian, H. 2003. Quorum sensing and its relevance to infectious diseases. *J. Infect.* **46**: 207–14.

22 Dove, J. E., K. Yasukawa, C. R. Tinsley and X. Nassif 2003. Production of the signaling molecule, autoinducer-2, by *Neisseria meningitidis*: lack of evidence for a concerted transcriptional response. *Microbiology* **149**: 1859–69.

23 Duan, K., C. Dammel, J. Stein, H. Rabin and M. G. Surette 2003. Modulation of *Pseudomonas aeruginosa* gene expression by host microflora through interspecies communication. *Molec. Microbiol.* **50**: 1477–91.

24 Duerre, J. A. and C. H. Miller 1966. Cleavage of S-ribosyl-L-homocysteine by extracts from *Escherichia coli*. *J. Bacteriol.* **91**: 1210–17.

25 Federle, M. J. and B. L. Bassler 2003. Interspecies communication in bacteria. *J. Clin. Invest.* **112**: 1291–9.

26 Folcher, M., H. Gaillard, L. T. Nguyen *et al.* 2001. Pleiotropic functions of a *Streptomyces pristinaespiralis* autoregulator receptor in development, antibiotic biosynthesis, and expression of a superoxide dismutase. *J. Biol. Chem.* **276**: 44297–306.

27 Fong, K. P., L. Gao and D. R. Demuth 2003. *luxS* and *arcB* control aerobic growth of *Actinobacillus actinomycetemcomitans* under iron limitation. *Infect. Immun.* **71**: 298–308.

28 Forsyth, M. H. and T. L. Cover 2000. Intercellular communication in *Helicobacter pylori: luxS* is essential for the production of an extracellular signaling molecule. *Infect. Immun.* **68**: 3193–9.

29 Fraser, C. M., S. Casjens, W. M. Huang *et al.* 1997. Genomic sequence of a Lyme disease spirochaete, *Borrelia burgdorferi. Nature* **390**: 580–6.

30 Fuqua, C., M. R. Parsek and E. P. Greenberg 2001. Regulation of gene expression by cell-to-cell communication: acyl-homoserine lactone quorum sensing. *A. Rev. Genet.* **35**: 439–68.

31 Giron, J. A., A. G. Torres, E. Freer and J. B. Kaper 2002. The flagella of enteropathogenic *Escherichia coli* mediate adherence to epithelial cells. *Molec. Microbiol.* **44**: 361–79.

32 Gray, K. M. and E. P. Greenberg 1992. Physical and functional maps of the luminescence gene cluster in an autoinducer-deficient *Vibrio fischeri* strain isolated from a squid light organ. *J. Bacteriol.* **174**: 4384–90.

33 Greene, R. 1996. Biosynthesis of methionine. In F. C. Neidhardt (ed.), *Escherichia coli* and *Salmonella*: Cellular and Molecular Biology, vol. 1, pp. 542–60. Washington, DC: ASM Press.

34 Hammer, B. K. and B. L. Bassler 2003. Quorum sensing controls biofilm formation in *Vibrio cholerae. Molec. Microbiol.* **50**: 101–4.

35 Havarstein, L. S., G. Coomaraswamy and D. A. Morrison 1995. An unmodified heptadecapeptide pheromone induces competence for genetic transformation in *Streptococcus pneumoniae. Proc. Natn. Acad. Sci. USA* **92**: 11140–4.

36 Heidelberg, J. F., J. A. Eisen, W. C. Nelson *et al.* 2000. DNA sequence of both chromosomes of the cholera pathogen *Vibrio cholerae. Nature* **406**: 477–83.

37 Hilgers, M. T. and M. L. Ludwig 2001. Crystal structure of the quorum-sensing protein LuxS reveals a catalytic metal site. *Proc. Natn. Acad. Sci. USA* **98**: 11169–74.

38 Horinouchi, S. and T. Beppu 1994. A-factor as a microbial hormone that controls cellular differentiation and secondary metabolism in *Streptomyces griseus. Molec. Microbiol.* **12**: 859–64.

39 Huang, J. J., J. I. Han, L. H. Zhang and J. R. Leadbetter 2003. Utilization of acyl-homoserine lactone quorum signals for growth by a soil pseudomonad and *Pseudomonas aeruginosa* PAO1. *Appl. Environ. Microbiol.* **69**: 5941–9.

40 Hubner, A., A. T. Revel, D. M. Nolen, K. E. Hagman and M. V. Norgard 2003. Expression of a *luxS* gene is not required for *Borrelia burgdorferi* infection of mice via needle inoculation. *Infect. Immun.* **71**: 2892–6.

41 Jelsbak, L. and L. Sogaard-Andersen 2003. Cell behavior and cell-cell communication during fruiting body morphogenesis in *Myxococcus xanthus*. *J. Microbiol. Methods* **55**: 829–39.

42 Jeon, B., K. Itoh, N. Misawa and S. Ryu 2003. Effects of quorum sensing on flaA transcription and autoagglutination in *Campylobacter jejuni*. *Microbiol. Immunol.* **47**: 833–9.

43 Ji, G., R. Beavis and R. P. Novick 1997. Bacterial interference caused by auto-inducing peptide variants. *Science* **276**: 2027–30.

44 Ji, G., R. C. Beavis and R. P. Novick 1995. Cell density control of staphylococcal virulence mediated by an octapeptide pheromone. *Proc. Natn. Acad. Sci. USA* **92**: 12055–9.

45 Joyce, E. A., B. L. Bassler and A. Wright 2000. Evidence for a signaling system in *Helicobacter pylori*: detection of a *luxS*-encoded autoinducer. *J. Bacteriol.* **182**: 3638–43.

46 Joyce, E. A., A. Kawale, S. Censini *et al.* 2004. LuxS is required for persistent pneumococcal carriage and expression of virulence and biosynthesis genes. *Infect. Immun.* **72**: 2964–75.

47 Kaiser, D. 2003. Coupling cell movement to multicellular development in myxobacteria. *Nat. Rev. Microbiol.* **1**: 45–54.

48 Kaplan, H. B. 2003. Multicellular development and gliding motility in *Myxococcus xanthus*. *Curr. Opin. Microbiol.* **6**: 572–7.

49 Karp, P. D., M. Arnaud, J. Collado-Vides *et al.* 2004. The *E. coli* EcoCyc database: no longer just a metabolic pathway database. *ASM News* **70**: 25–30.

50 Karp, P. D., M. Riley, S. M. Paley and A. Pellegrini-Toole 2002. The MetaCyc Database. *Nucleic Acids Res.* **30**: 59–61.

51 Kim, S. Y., S. E. Lee, Y. R. Kim *et al.* 2003. Regulation of *Vibrio vulnificus* virulence by the LuxS quorum-sensing system. *Molec. Microbiol.* **48**: 1647–64.

52 Kleerebezem, M., L. E. Quadri, O. P. Kuipers and W. M. de Vos 1997. Quorum sensing by peptide pheromones and two-component signal-transduction systems in Gram-positive bacteria. *Molec. Microbiol.* **24**: 895–904.

53 Lazdunski, A. M., I. Ventre and J. N. Sturgis 2004. Regulatory circuits and communication in Gram-negative bacteria. *Nat. Rev. Microbiol.* **2**: 581–92.

54 Lemos, J. A., T. A. Brown, Jr. and R. A. Burne 2004. Effects of RelA on key virulence properties of planktonic and biofilm populations of *Streptococcus mutans*. *Infect. Immun.* **72**: 1431–40.

54a Lenz, D. H., K. C. Mok, B. N. Lilley, R. V. Kulkarni, N. S. Wingreen and B. L. Bassler (2004). The small RNA chaperone Hfq and multiple small RNAs control quorum sensing in *Vibrio harveyi* and *Vibrio cholerae*. *Cell* **118**: 69–82.

55 Lewis, H. A., E. B. Furlong, B. Laubert *et al.* 2001. A structural genomics approach to the study of quorum sensing: crystal structures of three LuxS orthologs. *Structure (Camb.)* **9**: 527–37.

56 Loh, J. T., M. H. Forsyth and T. L. Cover 2004. Growth phase regulation of *flaA* expression in *Helicobacter pylori* is *luxS* dependent. *Infect. Immun.* **72**: 5506–10.

57 Lupp, C. and E. G. Ruby 2004. *Vibrio fischeri* LuxS and AinS: comparative study of two signal synthases. *J. Bacteriol.* **186**: 3873–81.

58 Magnuson, R., J. Solomon and A. D. Grossman 1994. Biochemical and genetic characterization of a competence pheromone from *B. subtilis*. *Cell* **77**: 207–16.

59 McNab, R., S. K. Ford, A. El-Sabaeny *et al.* 2003. LuxS-based signaling in *Streptococcus gordonii*: autoinducer 2 controls carbohydrate metabolism and biofilm formation with *Porphyromonas gingivalis*. *J. Bacteriol.* **185**: 274–84.

60 Merritt, J., F. Qi, S. D. Goodman, M. H. Anderson and W. Shi 2003. Mutation of *luxS* affects biofilm formation in *Streptococcus mutans*. *Infect. Immun.* **71**: 1972–9.

61 Michael, B., J. N. Smith, S. Swift, F. Heffron and B. M. Ahmer 2001. SdiA of *Salmonella enterica* is a LuxR homolog that detects mixed microbial communities. *J. Bacteriol.* **183**: 5733–42.

62 Miller, C. H. and J. A. Duerre 1968. S-ribosylhomocysteine cleavage enzyme from *Escherichia coli. J. Biol. Chem.* **243**: 92–7.

63 Miller, J. C. and B. Stevenson 2004. Increased expression of *Borrelia burgdorferi* factor H-binding surface proteins during transmission from ticks to mice. *Int. J. Med. Microbiol.* **293** (suppl. 37): 120–5.

64 Miller, M. B. and B. L. Bassler 2001. Quorum sensing in bacteria. *A. Rev. Microbiol.* **55**: 165–99.

65 Miller, M. B., K. Skorupski, D. H. Lenz, R. K. Taylor and B. L. Bassler 2002. Parallel quorum sensing systems converge to regulate virulence in *Vibrio cholerae. Cell* **110**: 303–14.

66 Miller, S. T., K. B. Xavier, S. R. Campagna *et al.* 2004. *Salmonella typhimurium* recognizes a chemically distinct form of the bacterial quorum-sensing signal AI-2. *Molec. Cell* **15**: 677–87.

67 Mok, K. C., N. S. Wingreen and B. L. Bassler 2003. *Vibrio harveyi* quorum sensing: a coincidence detector for two autoinducers controls gene expression. *EMBO J.* **22**: 870–81.

68 Mori, M., Y. Sakagami, Y. Ishii *et al.* 1988. Structure of cCF10, a peptide sex pheromone which induces conjugative transfer of the *Streptococcus faecalis* tetracycline resistance plasmid, pCF10. *J. Biol. Chem.* **263**: 14574–8.

69 Nealson, K. H., T. Platt and J. W. Hastings 1970. Cellular control of the synthesis and activity of the bacterial luminescent system. *J. Bacteriol.* **104**: 313–22.

70 Ohtani, K., H. Hayashi and T. Shimizu 2002. The *luxS* gene is involved in cell-cell signaling for toxin production in *Clostridium perfringens. Molec. Microbiol.* **44**: 171–9.

71 Otto, M. 2001. *Staphylococcus aureus* and *Staphylococcus epidermidis* peptide pheromones produced by the accessory gene regulator agr system. *Peptides* **22**: 1603–8.

72 Otto, M., R. Sussmuth, C. Vuong, G. Jung and F. Gotz 1999. Inhibition of virulence factor expression in *Staphylococcus aureus* by the *Staphylococcus epidermidis* agr pheromone and derivatives. *FEBS Lett.* **450**: 257–62.

73 Pappas, K. M., C. L. Weingart and S. C. Winans 2004. Chemical communication in proteobacteria: biochemical and structural studies of signal synthases and receptors required for intercellular signaling. *Molec. Microbiol.* **53**: 755–69.

74 Parsek, M. R. and E. P. Greenberg 2000. Acyl-homoserine lactone quorum sensing in gram-negative bacteria: a signaling mechanism involved in associations with higher organisms. *Proc. Natn. Acad. Sci. USA* **97**: 8789–93.

75 Pei D. and J. Zhu 2004. Mechanism of action of S-ribosylhomocysteinase (LuxS). *Curr. Opin. Chem. Biol.* **8**: 492–7.

76 Riedel, K., M. Hentzer, O. Geisenberger *et al.* 2001. N-acylhomoserine-lactone-mediated communication between *Pseudomonas aeruginosa* and *Burkholderia cepacia* in mixed biofilms. *Microbiology* **147**: 3249–62.

77 Roche, D. M., J. T. Byers, D. S. Smith *et al.* 2004. Communications blackout? Do N-acylhomoserine-lactone-degrading enzymes have any role in quorum sensing? *Microbiology* **150**: 2023–8.

78 Ruzheinikov, S. N., S. K. Das, S. E. Sedelnikova *et al.* 2001. The 1.2 A structure of a novel quorum-sensing protein, *Bacillus subtilis* LuxS. *J. Molec. Biol.* **313**: 111–22.

79 Schauder, S. and B. L. Bassler 2001. The languages of bacteria. *Genes Dev.* **15**: 1468–80.

80 Schauder, S., K. Shokat, M. G. Surette and B. L. Bassler 2001. The LuxS family of bacterial autoinducers: biosynthesis of a novel quorum-sensing signal molecule. *Molec. Microbiol.* **41**: 463–76.

81 Shin, N. R., D. Y. Lee, S. J. Shin, K. S. Kim and H. S. Yoo 2004. Regulation of proinflammatory mediator production in RAW264.7 macrophage by *Vibrio vulnificus luxS* and *smcR. FEMS Immunol. Med. Microbiol.* **41**: 169–76.

82 Sircili, M. P., M. Walters, L. R. Trabulsi and V. Sperandio. 2004. Modulation of enteropathogenic *Escherichia coli* virulence by quorum sensing. *Infect. Immun.* **72**: 2329–37.

83 Smith, J. N. and B. M. Ahmer 2003. Detection of other microbial species by *Salmonella*: expression of the SdiA regulon. *J. Bacteriol.* **185**: 1357–66.

84 Smith, R. S. and B. H. Iglewski 2003. *P. aeruginosa* quorum-sensing systems and virulence. *Curr. Opin. Microbiol.* **6**: 56–60.

85 Socransky, S. S. and A. D. Haffajee 1992. The bacterial etiology of destructive periodontal disease: current concepts. *J. Periodontol.* **63**: 322–31.

86 Sperandio, V., J. L. Mellies, W. Nguyen, S. Shin and J. B. Kaper 1999. Quorum sensing controls expression of the type III secretion gene transcription and protein secretion in enterohemorrhagic and enteropathogenic *Escherichia coli*. *Proc. Natn. Acad. Sci. USA* **96**: 15196–201.

87 Sperandio, V., A. G. Torres, J. A. Giron and J. B. Kaper 2001. Quorum sensing is a global regulatory mechanism in enterohemorrhagic *Escherichia coli* O157:H7. *J. Bacteriol.* **183**: 5187–97.

89 Sperandio, V., A. G. Torres, B. Jarvis, J. P. Nataro and J. B. Kaper 2003. Bacteria-host communication: the language of hormones. *Proc. Natn. Acad. Sci. USA* **100**: 8951–6.

90 Sperandio, V., A. G. Torres and J. B. Kaper 2002. Quorum sensing *Escherichia coli* regulators B and C (QseBC): a novel two-component regulatory system involved in the regulation of flagella and motility by quorum sensing in *E.coli*. *Molec. Microbiol.* **43**: 809–21.

91 Stevenson, B. and K. Babb 2002. LuxS-mediated quorum sensing in *Borrelia burgdorferi*, the lyme disease spirochete. *Infect. Immun.* **70**: 4099–105.

92 Stevenson, B., K. von Lackum, R. L. Wattier *et al.* 2003. Quorum sensing by the Lyme disease spirochete. *Microbes Infect.* **5**: 991–7.

93 Stover, C. K., X. Q. Pham, A. L. Erwin *et al.* 2000. Complete genome sequence of *Pseudomonas aeruginosa* PAO1, an opportunistic pathogen. *Nature* **406**: 959–64.

94 Sturme, M. H., M. Kleerebezem, J. Nakayama *et al.* 2002. Cell to cell communication by autoinducing peptides in gram-positive bacteria. *Antonie Van Leeuwenhoek* **81**: 233–43.

95 Surette, M. G. and B. L. Bassler 1998. Quorum sensing in *Escherichia coli* and *Salmonella typhimurium*. *Proc. Natn. Acad. Sci. USA* **95**: 7046–50.

96 Surette, M. G. and B. L. Bassler 1999. Regulation of autoinducer production in *Salmonella typhimurium*. *Molec. Microbiol.* **31**: 585–95.

97 Surette, M. G., M. B. Miller and B. L. Bassler 1999. Quorum sensing in *Escherichia coli*, *Salmonella typhimurium* and *Vibrio harveyi*: a new family of genes responsible for autoinducer production. *Proc. Natn. Acad. Sci. USA* **96**: 1639–44.

98 Taga, M. E., S. T. Miller and B. L. Bassler 2003. Lsr-mediated transport and processing of AI-2 in *Salmonella typhimurium*. *Molec. Microbiol.* **50**: 1411–27.

99 Tummler, B. and C. Kiewitz 1999. Cystic fibrosis: an inherited susceptibility to bacterial respiratory infections. *Molec. Med. Today* **5**: 351–8.

100 Wen, Z. T. and R. A. Burne 2002. Functional genomics approach to identifying genes required for biofilm development by *Streptococcus mutans*. *Appl. Environ. Microbiol.* **68**: 1196–203.

101 Wen, Z. T. and R. A. Burne 2004. LuxS-mediated signaling in *Streptococcus mutans* is involved in regulation of acid and oxidative stress tolerance and biofilm formation. *J. Bacteriol.* **186**: 2682–91.

102 Whitehead, N. A., A. M. Barnard, H. Slater, N. J. Simpson and G. P. Salmond 2001. Quorum-sensing in Gram-negative bacteria. *FEMS Microbiol. Rev.* **25**: 365–404.

103 Winzer, K., K. R. Hardie, N. Burgess *et al.* 2002. LuxS: its role in central metabolism and the in vitro synthesis of 4-hydroxy-5-methyl-3(2H)-furanone. *Microbiology* **148**: 909–22.

104 Winzer, K., Y. H. Sun, A. Green *et al.* 2002. Role of *Neisseria meningitidis luxS* in cell-to-cell signaling and bacteremic infection. *Infect. Immun.* **70**: 2245–8.

105 Xavier, K. B. and B. L. Bassler 2003. LuxS quorum sensing: more than just a numbers game. *Curr. Opin. Microbiol.* **6**: 191–7.

106 Yarwood, J. M. and P. M. Schlievert 2003. Quorum sensing in *Staphylococcus* infections. *J. Clin. Invest.* **112**: 1620–5.

107 Zhang, H. B., L. H. Wang and L. H. Zhang 2002. Genetic control of quorum-sensing signal turnover in *Agrobacterium tumefaciens*. *Proc. Natn. Acad. Sci. USA* **99**: 4638–43.

108 Zhu, J., E. Dizin, X. Hu *et al.* 2003. S-Ribosylhomocysteinase (LuxS) is a mono-nuclear iron protein. *Biochemistry* **42**: 4717–26.

109 Zhu, J., X. Hu, E. Dizin and D. Pei 2003. Catalytic mechanism of S-ribosylhomocysteinase (LuxS): direct observation of ketone intermediates by 13C NMR spectroscopy. *J. Am. Chem. Soc.* **125**: 13379–81.

110 Zhu, J., M. B. Miller, R. E. Vance *et al.* 2002. Quorum-sensing regulators control virulence gene expression in *Vibrio cholerae*. *Proc. Natn. Acad. Sci. USA* **99**: 3129–34.

CHAPTER 7

LuxS-dependent regulation of *Escherichia coli* virulence

Marcie B. Clarke and Vanessa Sperandio

University of Texas Southwestern Medical Center, Dallas, TX, USA

INTRODUCTION

Escherichia coli is the most abundant facultative anaerobe found in the human intestinal microbial flora. This organism resides in the mucus layer of the mammalian colon, and typically colonizes the gastrointestinal tract of humans a few hours after birth. However, there are several clones of *E. coli* that have acquired virulence traits that allow them to cause a broad spectrum of disease. These virulence traits are usually encoded within mobile genetic elements, such as plasmids and pathogenicity islands, that have evolved to be stable within these clones. Three general clinical syndromes result from the infection with these pathotypes: diarrheal disease, urinary tract infections, and meningitis/sepsis. Among the intestinal pathogens there are six well-described categories: enterohemorrhagic *E. coli* (EHEC), enteropathogenic *E. coli* (EPEC), enterotoxigenic *E. coli* (ETEC), enteroaggregative *E. coli* (EAEC), enteroinvasive *E. coli* (EIEC), and diffusely adherent *E. coli* (DAEC) (59). This chapter will focus primarily on EHEC and EPEC, given that quorum sensing has been mostly described within these pathotypes.

ENTEROHEMORRHAGIC *E. COLI* (EHEC)

Enterohemorrhagic *E. coli* (EHEC) O157:H7 is responsible for major outbreaks of bloody diarrhea and hemolytic uremic syndrome (HUS) throughout the world. EHEC causes an estimated 73,000 illnesses, 2,000 hospitalizations, and 60 deaths in the United States alone each year. EHEC has a very low infectious dose (as few as 50 cfu); this is one of the major contributing factors to EHEC outbreaks. Treatment and intervention

Bacterial Cell-to-Cell Communication: Role in Virulence and Pathogenesis, ed. D. R. Demuth and R. J. Lamont. Published by Cambridge University Press. © Cambridge University Press 2005.

strategies for EHEC infections are still very controversial, with conventional antibiotics usually having little clinical effect and possibly even being harmful (by increasing the chances of patients developing hemolytic uremic syndrome (HUS))(41, 42).

EHEC colonizes the large intestine, where it causes attaching and effacing (AE) lesions. The AE lesion is characterized by the destruction of the microvilli and the rearrangement of the cytoskeleton to form a pedestal-like structure, which cups the bacteria individually. The genes involved in the formation of the AE lesion are encoded within a chromosomal pathogenicity island named the Locus of Enterocyte Effacement (LEE) (37). The LEE region contains five major operons: *LEE1, LEE2, LEE3, tir (LEE5)*, and *LEE4* (4), which encode a type III secretion system (TTSS), an adhesin (intimin), and this adhesin's receptor (Tir), which is translocated to the epithelial cell through the bacterial TTSS (16, 55). The LEE genes are directly activated by the LEE-encoded regulator (Ler), which is the first gene in the *LEE1* operon (5, 15, 55, 70, 84). Transcription of the LEE genes is further positively and negatively modulated by GrlA and GlrR, respectively, which are encoded in a small operon downstream of *LEE1* (13). EHEC also produces a potent Shiga toxin (Stx) that is responsible for the major symptoms of hemorrhagic colitis and HUS. There are two types of Stx, Stx1 and Stx2, which are most frequently associated with human disease. Both of the genes encoding Stx1 and Stx2 are located within the late genes of a λ-like bacteriophage, and are transcribed when the phage enters its lytic cycle (60). Disturbances in the bacterial membrane, DNA replication, or protein synthesis (which are the targets of conventional antibiotics) may trigger a SOS response in the bacterial cells that signals the bacteriophage to enter the lytic cycle (41, 42). The phage replicates, Shiga toxin is produced, and the phage lyses the bacteria, thereby releasing Shiga toxin into the host.

LuxS AND CELL-TO-CELL SIGNALING IN BACTERIA

The phenomenon of cell-to-cell signaling in bacteria has historically been referred to as quorum sensing. The most widespread quorum sensing system is the *luxS* system, first described as being involved in bioluminescence in *Vibrio harveyi* (90). Among the diverse bacterial species that contain the *luxS* quorum sensing system is *E. coli*, including EHEC serotype O157:H7 (85, 89, 90). LuxS is an enzyme involved in the metabolism of S-adenosylmethionine (SAM); it converts ribose-homocysteine into homocysteine and 4,5-dihydrody-2,3-pentanedione (DPD). DPD is a very

unstable compound that reacts with water and cyclizes into several furanones (73, 87, 98), one of which is thought to be the precursor of auto-
inducer-2 (AI-2) (73). The AI-2 structure has been solved by co-crystallizing
this ligand with its receptor LuxP (a periplasmic protein that resembles
the ribose binding protein RbsB) in *Vibrio harveyi*, and reported to be a
furanosyl-borate-diester (6). However, LuxP homologs, as well as homologs
from this signaling cascade, have only been found in *Vibrio* spp. Several
other bacterial species harbor the *luxS* gene and have AI-2 activity as
measured by a *Vibrio harveyi* bioluminescence assay (72, 100). However,
the only genes shown to be regulated by AI-2 in other species encode for
an ABC transporter in *Salmonella typhimurium* named Lsr (LuxS-regulated),
responsible for the AI-2 uptake (93). This ABC transporter is also present in
E. coli and shares homology with sugar transporters. Once inside the cell,
AI-2 is modified by phosphorylation and proposed to interact with LsrR,
which is a SorC-like transcription factor involved in repressing expression of
the *lsr* operon (92, 93) (Fig. 7.1). Several groups have been unable to detect
the furanosyl-borate-diester, proposed to be AI-2, in purified fractions containing AI-2 activity from *Salmonella* spp. and *E. coli* (as measured by using
the *V. harveyi* bioluminescence assay) (73, 87, 98). These fractions only
yielded the identification of several furanosyl compounds that did not
contain boron. These results can be explained now that AI-2 has been
co-crystallized with its receptor (the periplasmic protein LsrB) in *Salmonella*.
In these studies the LsrB ligand was not a furanosyl-borate-diester, but a
furanone (2R, 4S-2-methyl-2,3,3,4-tetrahydrofuran (R-THMF)), consistent
with what has been observed in AI-2 fractions of *Salmonella* and *E. coli*
(58, 87, 98). This is fundamentally different from AI-2 detection in *Vibrio
harveyi* and raises the question whether all bacteria may actually use AI-2 as a
signaling compound, or whether it is released as a waste product or used as a
metabolite by some bacteria, rather than as a signal.

Diverse roles in signaling have been attributed to AI-2 in other organisms by comparing *luxS* mutants with wild-type strains, and complementing these mutants either genetically or with spent supernatants. Among
these roles are the LEE-encoded type III secretion system and flagellar
expression in EHEC (85, 86), expression of VirB in *Shigella flexneri* (10),
secretion of SpeB cysteine protease in *Streptococcus pyogenes* (49), type III
secretion in *V. harveyi* and *V. parahaemolyticus* (31), etc. The most comprehensive studies concerning the role for a LuxS-dependent autoinducer in
virulence have been performed in enterohemorrhagic *E. coli* (EHEC)
(83, 85–88). However, by using purified and AI-2 synthesized "in vitro", it has
been demonstrated that the signaling molecule activating type III secretion

M. B. CLARKE AND V. SPERANDIO

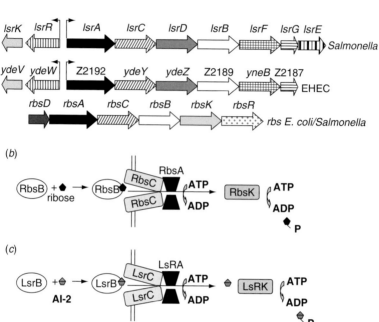

Figure 7.1. The Lsr ABC transporter system (LuxS regulated genes) from *Salmonella typhimurium* and enterohemorrhagic *Escherichia coli* (EHEC) serotype O157:H7. All of the *lsr* genes from *Salmonella* are present in EHEC, with the exception of *lsrE*. The *lsrACDBFGE* genes are transcribed in an operon; in the opposite direction to this operon is *lsrR* (encoding the SorC-like transcription factor that represses expression of the LsrABC transporter) and *lsrK*, which phosphorylates AI-2 upon its entry in to the bacterial cell (92, 93). LsrB shares homology with the ribose-binding periplasmic protein RbsB and is the receptor for AI-2 in Salmonella and *E. coli* (58). Upon binding to LsrB, AI-2 is transported through the LsR ABC transporter, which closely resembles the ribose ABC transporter. Once inside the cell AI-2 is phosphorylated by LsrK. Phospho-AI-2 is thought to interact with LsrR to relieve the repression of the *lsr* operon.

and the flagellar regulon in EHEC is not the AI-2 autoinducer (87). The autoinducer responsible for this signaling is dependent on the presence of the *luxS* gene for its synthesis, but is different from AI-2. AI-2 is a very polar furanone that does not bind to C-18 columns. The signaling compound activating the EHEC virulence genes, which was renamed autoinducer-3 (AI-3), binds to C-18 columns and can only be eluted with methanol (87). Electrospray Mass Spectrometry analysis of the AI-3 fraction showed a major peak with a molecular mass of 213.1 Da and minor peaks at 109.1, 164.9,

176.1, 196.1, 211.1, 214.1 and 222.9 Da (87). All of these are different from those of AI-2 (6), suggesting that AI-3 is a novel compound.

These results suggest that some of the phenotypes attributed to AI-2 signaling need to be revised in light of the fact that LuxS is not devoted to AI-2 production; it is in fact an enzyme involved in the biochemical pathway for metabolism of SAM. Consequently, altered gene expression due to a *luxS* mutation will involve both genes affected by quorum sensing *per se* and genes differentially expressed because of the interruption of this metabolic pathway. Furthermore, one also has to take into consideration that a knockout of *luxS* seems to affect the synthesis of at least two autoinducers, AI-2 and AI-3 (87). The activity of both signals can be differentiated by utilizing biological tests specific to each signal. For example, AI-3 shows no activity for the AI-2 bioassay (87), which is predicated on the production of bioluminescence in *Vibrio harveyi* (89) and is the gold standard for AI-2 production. On the other hand, AI-3 activates the transcription of the EHEC LEE virulence genes, whereas AI-2 has no effect in this assay (87). The only two phenotypes shown to be AI-2-dependent, using either purified or *in vitro* synthesized AI-2, are bioluminescence in *V. harveyi* (73) and expression of the *lsr* operon in *S. typhimurium* (93).

The *luxS* gene is present in an array of bacterial species, including several members of the human commensal microflora (90). It has recently been shown, by using anaerobically cultured stools from healthy human volunteers, that the microbial intestinal flora produce both AI-2 (detected with the *V. harveyi* bioluminescence assay) and AI-3 (detected with the *LEE1* transcription AI-3-dependent bioassay) (87). To obtain further information regarding which intestinal commensals and pathogens are able to produce AI-2 and AI-3, freshly isolated strains from patients were tested (M. P. Sircili and V. Sperandio, unpublished results). Using the bioassays described above, AI-2 and AI-3 activity was observed in spent supernatants from enteropathogenic *E. coli* strains from serogroups O26:H11 and O111ac:H9, *Shigella* sp., and *Salmonella* sp. Activity from both autoinducers was also detected in normal flora bacteria such as a commensal *E. coli*, *Klebsiella pneumoniae*, and *Enterobacter clocae* (M. P. Sircili and V. Sperandio, unpublished results). These results suggest that interspecies signaling may be involved in the pathogenesis of disease caused by these other bacteria, and also in signaling by the intestinal microflora.

CELL-TO-CELL SIGNALING IN EHEC

Sperandio *et al.* (85) reported that transcription of all of the LEE operons is activated by the presence of autoinducers in supernatants from wild-type EHEC, commensal *E. coli*, and MG1655 (K-12) strains, but not from an isogenic EHEC *luxS* mutant or from K-12 strain DH5α (which has a frameshift mutation in the *luxS* gene) (85). Analysis of *luxS* mutants in EPEC and EHEC strains demonstrated an effect on type III secretion in both of them. Type III secretion was diminished in the EPEC *luxS* mutant (85) and could not be detected in the EHEC *luxS* mutant (87). In both cases, this phenotype could be restored by genetic complementation with the *luxS* gene cloned on a plasmid, or by providing AI-3 exogenously (85, 87).

In addition to activation of type III secretion, Sperandio *et al.* (86) reported that about 10% of the common genome between EHEC and *E. coli* K-12 is differentially expressed between a wild-type EHEC and its isogenic *luxS* mutant (EHEC has 1.3 Mb of DNA absent in K-12, and K-12 has 0.53 Mb of DNA that is absent in EHEC (62)) (86). DeLisa and colleagues (12) also reported, by using gene arrays, that about 5.6% of the K-12 genome was differentially regulated between a wild-type K-12 strain and its isogenic *luxS* mutant. The difference in the numbers of genes regulated in both reports may be due to differences in methodology (growth temperature: Sperandio *et al.* (86) used 37 °C whereas DeLisa and colleagues (12) used 30 °C; nutrient availability: Sperandio *et al.* (86) grew their strains in DMEM, which they have previously shown to give better expression of *luxS*-controlled genes (85), whereas DeLisa and colleagues (12) used LB broth) and in strains utilized (Sperandio *et al.* (86) used an EHEC strain, whereas DeLisa and colleagues (12) used a K-12 strain).

The observation that about 10% of the array genes are differentially regulated between an EHEC wild type and its isogenic *luxS* mutant is not surprising if one considers the pleiotropic nature of a *luxS* mutation. LuxS is a metabolic enzyme involved primarily in the conversion of ribosyl-homocysteine into homocysteine and 4,5-dihydroxy-2,3-pentanedione, which is the precursor of AI-2 (73). A *luxS* mutation will interrupt this metabolic pathway, changing the whole metabolism of the bacterium. A *luxS* mutant will accumulate S-ribosyl-homocysteine because it is unable to catalyze its conversion to homocysteine. This could cause the concentration of homocysteine to diminish within the cell. Inasmuch as homocysteine is used for the *de novo* synthesis of methionine, the cell will use a salvage pathway: it will use oxaloacetate to produce homocysteine to synthesize methionine. Given that oxaloacetate is necessary, together with L-glutamate, to synthesize aspartate, by using this salvage pathway for the *de novo* synthesis of methionine, other amino-acid synthetic and catabolic pathways will be changed within the cell (www.ecosal.org/ecosal/index.jsp).

Among the quorum-sensing-regulated genes and phenotypes noted in these studies were the genes encoding flagella, and motility (which may also be involved in pathogenesis) (86). Specifically, it has been shown that transcription of *flhDC* (the master regulator of the flagellar regulon) and the *mot* operon (encoding motility genes) is decreased in a *luxS* mutant compared with wild-type and complemented strains. It has also been shown that transcription of these genes, as well as motility, could be restored by addition of signals exogenously, further confirming that regulation of flagellar expression and motility is being controlled by a quorum sensing signaling mechanism (86, 87). Quorum-sensing regulation of *flhDC* expression has far-reaching implications beyond flagellar expression, given that FlhDC has been shown to also regulate bacterial cell division (65, 66) and several metabolic processes (64). Quorum-sensing regulation of the LEE-encoded type III secretion and the flagellar regulon in EHEC is dependent on the AI-3 signal; the role of AI-2 signaling in EHEC remains to be established (87). Given the widespread nature of the *luxS*/AI-3 system in bacteria, an interesting extrapolation is that the AI-3/*luxS* quorum sensing system might have evolved to mediate microflora–host interactions, but evolved to be exploited by EHEC to activate its virulence genes. In this manner, the AI-3/*luxS* system alerts EHEC as to when it has reached the large intestine, where large numbers of commensal *E. coli*, *Enterococcus*, *Clostridium*, and *Bacteroides*, all of which contain the AI-3/*luxS* quorum sensing system, are resident.

BACTERIAL–HOST CELL-TO-CELL SIGNALING

It is estimated that humans have about 10^{13} eukaryotic cells and 10^{14} prokaryotic cells (comprising our endogenous bacterial flora). The gastrointestinal (GI) tract is the site of the largest and most complex environment in the mammalian host. The density of bacteria along the GI tract can vary greatly, with the majority of the flora residing in the colon (10^{11}–10^{12} bacterial cells ml^{-1}). Given the enormous number and diversity of bacteria in the GI environment, it should not be surprising that the members of this community somehow communicate among themselves and with the host itself to coordinate various processes. The observation that the human bacterial flora is extremely important to development, as well as in shaping the innate immune system, further reinforces this suggestion (34). However, some interactions among eukaryotes and microbes are detrimental and culminate in disease. Given these polar relationships, it may be asked, "at what levels do prokaryotes and eukaryotes communicate"?

The bacterial signal AI-3 is not only involved in bacterial–bacterial communication, but also in bacterial–host communication, crosstalking with the human hormones epinephrine and norepinephrine (87). An EHEC *luxS* mutant is unable to signal to itself in an *"in vitro"* culture and therefore cannot express the LEE genes, which are essential for EHEC virulence. These results suggest that type III secretion may be abrogated in the *luxS* mutant *in vitro*. Based on these data, it was expected that the *luxS* mutant would be unable to produce attaching and effacing (AE) lesions on cultured epithelial cells. However, the *luxS* mutant was still able to produce AE lesions on epithelial cells, indistinguishable from those produced by the wild type. Since quorum sensing in bacteria is a cell-to-cell signaling system, it was hypothesized that a eukaryotic signaling compound could complement the bacterial mutation. Eukaryotic cell-to-cell signaling occurs through hormones. There are three major groups of endocrine hormones: polypeptide hormones, steroid hormones, and hormones derived from the amino acid tyrosine, which include the catecholamines norepinephrine and epinephrine (30). Two of the Gram-negative bacterial autoinducers (acyl-homoserine lactones and the AI-2) are also derived from amino-acid metabolism (72). Norepinephrine has been demonstrated to induce bacterial growth (52) and to be taken into bacteria (43). Using purified epinephrine and norepinephrine, Sperandio *et al.* (87) showed that the *luxS* mutant still responds to these eukaryotic signals.

It has been shown that there is a considerable amount of epinephrine and norepinephrine in the human GI tract (14) and that these hormones induce chloride and potassium secretion in the colon (35). The neuronally mediated response to epinephrine in the distal colon can be suppressed by the non-selective β-adrenergic receptor antagonist propranolol and by the non-selective α-adrenergic receptor antagonist phentolamine in the proximal colon (35). Finally, it has been demonstrated that the EHEC response to epinephrine and norepinephrine signaling is specific, given that it can be blocked by β-adrenergic antagonists (such as propranolol). Epinephrine and norepinephrine can substitute for AI-3 to activate transcription of the LEE genes, type III secretion and AE lesions on epithelial cells. Taken together, these results suggest that AI-3 and epinephrine/norepinephrine crosstalk and that these compounds may use the same signaling pathway. As further evidence, regulation of the flagellar regulon is also under AI-3 and epinephrine/norepinephrine control and one can block EHEC response to both AI-3 and epinephrine/norepinephrine by using the β-adrenergic antagonist propranolol. Specifically, propranolol can prevent formation of the AE lesions by the wild-type EHEC, and by the *luxS* mutant in epithelial cells (87).

Norepinephrine has been reported to induce bacterial growth (19, 52); there are reports in the literature, albeit conflicting, that imply that norepinephrine might function as a siderophore (20, 43). Recently, norepinephrine has been implicated as inducing expression of enterobactin and iron uptake in *E. coli*, suggesting that this is the mechanism involved in growth induction (4). However, the role of norepinephrine in bacterial pathogenesis seems to be more complex, as several reports suggested that this signal also activates virulence gene expression in *E. coli*, such as production of fimbriae and Shiga toxin (50, 51), by an unknown mechanism of induction. Sperandio *et al.* (87) show that both epinephrine and norepinephrine seem to crosstalk with a bacterial quorum sensing system to regulate virulence gene expression in EHEC. This signaling is not due to enterobactin and is TonB-independent, suggesting that it is not dependent on the FepA outer membrane receptor for this siderophore. Finally, this signaling is dependent on a novel autoinducer, AI-3, which is produced by intestinal flora (87). The line dividing quorum-sensing signaling and iron uptake is becoming increasingly blurred, especially with the discovery that the siderophore pyoverdine from *P. aeruginosa* also acts as a signaling molecule (47).

In conclusion, EHEC could respond to both a bacterial quorum sensing signaling system and a mammalian signaling system to "fine tune" transcription of virulence genes at different stages of infection and/or different sites of the gastrointestinal tract (Figure 7.2). Given that eukaryotic cell-to-cell signaling occurs through hormones, and bacterial cell-to-cell signaling occurs through quorum sensing, it is tempting to propose that quorum sensing might be a language by which bacterial and host cells communicate. Inasmuch as the host hormones epinephrine and norepinephrine signal to EHEC, it remains to be determined whether AI-3 exerts any functional effects on eukaryotic cell signaling.

THE EHEC QUORUM-SENSING SIGNALING CASCADE

Quorum-sensing regulatory cascades have been extensively studied in organisms such as *Pseudomonas aeruginosa* and *Vibrio harveyi*, and have proven to be very complex (11, 72). Concerning the EHEC AI-3–epinephrine–norepinephrine signaling cascade, a transcriptional regulator from the LysR family, designated as QS *E. coli* regulator A (QseA) (83) has been recently identified. QseA is transcriptionally activated through quorum sensing and, in turn, binds to and directly activates transcription of the

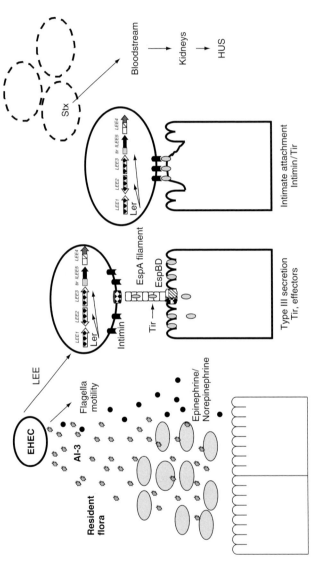

Figure 7.2. Model of cell-to-cell signaling regulation of EHEC virulence during infection. EHEC reaches the large intestine where the microbial flora is in high density and producing AI-3. It combines the AI-3 signal derived from the flora with the epinephrine/norepinephrine signals derived from the host to activate the AI-3 quorum-sensing signaling cascade. It is hypothesized that the flagellar regulon is activated first, allowing EHEC to swim proficiently through the intestinal mucus layer and attain close contact with the enteric epithelia. Then EHEC activates the LEE-encoded type III secretion system, forming the attaching and effacing lesions responsible for the beginning of the diarrheal process. Later in the course of disease, some EHEC cells will activate the Stx phage, undergo lysis and release Shiga toxin. Shiga toxin is known to be translocated through the intestinal barrier and travel to the urinary tract and other organs through the bloodstream. If enough Shiga toxin is released, the patient may develop hemolytic uremic syndrome (HUS).

LEE-encoded regulator (Ler) (encoded within the *LEE1* operon) (F. Sharp and V. Sperandio, unpublished results) (83). Ler is the activator for all of the other genes within the LEE island (55). In addition, QseA also activates transcription of the *grlRA* operon (R. Russell and V. Sperandio, unpublished results). GrlA has been reported to activate transcription of *LEE1* (*ler*), whereas GrlR seems to repress it (13). These results suggest that QseA regulates transcription on *LEE1* at more than one level. Consequently, an EHEC *qseA* mutant has a striking reduction in type III secretion but has no defect in flagellation or motility, suggesting that QseA regulates only the LEE genes and plays no role in the flagellar regulon (83). QseA belongs to the LysR family of transcription factors; it is therefore not surprising that it autorepresses its own transcription (79).

In addition, a two-component system renamed QseBC has been identified. QseBC is responsible for the transcriptional activation of the flagellar regulon in response to quorum sensing (88). The QseBC system is present in UPEC, EPEC, EHEC, *E. coli* K12, *Salmonella typhimurium*, *S. typhi*, *Pasteurella multocida*, and *Haemophilus influenzae*. QseB has three amino-acid changes between EHEC and *E. coli* K12, whereas QseC has eight amino acid changes. In addition, QseB shares a high level of homology with *S. typhimurium* PmrA (46% identity and 62% similarity over 222 amino-acids) and QseC shares homology with *S. typhimurium* PmrB (28% identity and 45% similarity over 269 amino acids). PmrAB is involved in gene regulation in response to extracytoplasmic ferric iron (99) and genes that confer resistance to antimicrobial peptides such as polymyxin (28, 81).

Traditional two-component systems consist of the sensor protein, which acts as a histidine kinase to transfer a phosphoryl group upon sensing of an environmental signal to an aspartate residue of its cognate response regulator, which goes on to act as a transcription factor. Both QseB and QseC contain conserved domains characteristic of a two-component system. QseC, the putative sensor kinase, has two conserved transmembrane domains and a conserved histidine kinase domain, indicating that its membrane localization may allow autophosphorylation upon the recognition of its specific environmental cue. QseC also contains an ATPase domain, which may allow it to exhibit phosphatase activity toward QseB. In addition, QseC has a conserved EAL domain, commonly found in signaling proteins, which consists of several acidic residues that could be important for metal binding and may make up an active site for a phospho-diesterase of cyclic diguanylate (c-di-GMP), a cyclic nucleotide (56, 94). The EAL domain of the VieA response regulator in *V. cholerae* has recently been implicated in the formation of biofilms by controlling c-di-GMP concentration in *V. cholerae* (94). QseB contains typical response regulator and DNA binding domains, which may allow it to receive a phosphate from QseC and undergo a conformational change allowing it to bind efficiently to DNA and regulate gene transcription.

The translational stop codon of *qseB* overlaps with the translational start codon of *qseC*; reverse transcriptase PCR experiments further confirmed that the *qseBC* genes comprise an operon (8). It is well known that many two-component systems act to positively regulate their own transcription (3). QseBC is no exception to this rule and has also been shown to autoactivate its own transcription (8). Transcriptional autoregulation could serve several purposes, including the amplification of signal or providing an additional threshold for gene activation. Signal amplification could allow bacteria to respond extremely quickly to an environmental signal. This scenario has been observed in other two-component systems, such as PmrB/PmrA of *S. typhimurium* (28), CpxAR of *E. coli* (67), BvgAS of *B. pertussis* (69), and PhoQ/PhoP of *S. typhimurium* (82). In addition, Hoffer *et al.* (2001) has suggested that autoregulation of a two-component system may be responsible for a "learning" system in which bacteria can respond quickly and more effectively to a signal that has been seen in the recent past. The PhoB/PhoR two-component system of *Salmonella*, in which previous exposure to a signal appears to boost reaction during the second exposure, appears to exhibit this type of "learning" (33). As a final point, autoregulation of a two-component system could provide an additional threshold for gene activation, as is observed with CpxAR in *E. coli*. In this two-component system, signal persistence is essential for the autoamplification and accumulation of the CpxR response regulator to a threshold concentration before transcription can be activated (67). This additional level of control could allow the bacterial cell to activate the energetically expensive production of flagella through QseBC only under appropriate conditions.

In addition to auto-activating its own transcription, QseBC is involved in activating transcription of the flagellar regulon (88). The expression and synthesis of the flagella and motility genes is a highly complex process, requiring the products of more than 50 genes that are organized in 17 operons (54). These operons are organized into a hierarchy of transcriptional classes: class 1, class 2, and class 3 (46). Class 1 consists solely of the master regulator of the flagellar regulon, FlhDC, which is a required transcriptional activator of the class 2 genes (48). The second class of genes includes proteins that form the hook and basal body of the flagellar apparatus, σ^{28}, and FlgM. During early time points of flagellar activation, FlgM acts as an anti-sigma factor to form a complex with σ^{28} in order to sequester it from RNA polymerase (RNAP). The σ^{28} protein is an alternative sigma factor, which has been shown to associate with RNAP and is necessary for the transcriptional activation of the class 3 genes, which include the motility proteins (*mot* operon) and the flagellin subunit (*fliC*) (7). Upon completion of the hook and basal body, FlgM is exported through the immature flagellar apparatus and depleted from the cytoplasm, freeing σ^{28} to activate class 3 gene expression (22, 36).

The expression of the flagellar regulon has been shown to be regulated through several environmental factors, including cAMP-CRP (78, 102), temperature (1), osmolarity (77), cell-cycle control (61), and bacterial cell density (quorum sensing) (7, 86). Currently, six transcriptional start sites have been mapped for *flhDC* (101). It has recently been reported that the expression of flagella and motility in EHEC and *E. coli* K12 is regulated by quorum sensing through QseBC (88). An isogenic mutant in *qseC* in both EHEC and K12 produced fewer flagella and was less motile than wild-type and complemented strains. In addition, transcriptional fusions of flagella class 1 (*flhDC*), class 2 (*fliA*), and class 3 (*fliC* and *motA*) were reduced compared with the wild type. These data may suggest that QseBC acts to regulate flagellar expression through the master regulator *flhDC*. Additional studies (M. B. Clarke and V. Sperandio, unpublished results) indicate that QseB directly binds to the *flhDC* and *qseBC* promoters in order to regulate flagellar and its own expression.

The QseBC two-component system is activated by quorum sensing through AI-3 (86–88). Early studies indicated that an isogenic mutant in the *qseC* sensor kinase was unable to respond to bacterial autoinducers or epinephrine given exogenously (86–88). Interestingly, the motility of a *luxS* mutant can be restored by the addition of either autoinducers contained in preconditioned supernatants or epinephrine (86, 87). In addition, the transcription of *flhDC* is activated by both epinephrine and AI-3 in the *luxS* mutant. Motility and *flhDC* transcription in a *qseC* mutant, however, are unable to respond to the presence of either AI-3 or epinephrine, indicating that QseC may possibly be sensing the presence of these cross-signaling compounds (87).

Although QseBC regulates both its own transcription and that of *flhDC*, it plays no role in the regulation of other quorum-sensing phenotypes, such as the LEE genes (88). Because flagella and motility are not the only phenotypes controlled by quorum sensing in EHEC, it is hypothesized that there are several other regulators involved in this quorum-sensing system.

Finally, three other genes in this signaling cascade have also been identified recently: *qseD* (encoding another regulator of the LysR family), and *qseE* and *qseF* (encoding a second two-component system), which are involved in regulating expression of the LEE genes (F. Sharp *et al.* and N. Reading *et al.*, unpublished results). These data suggest that both AI-3 and epinephrine/norepinephrine are recognized by the same receptor, which is probably in the outer membrane of the bacterium (owing to the non-polar nature of both AI-3 and epinephrine) (87). These signaling molecules might be imported into the periplasmic space, where they then would interact either with one major sensor kinase (that directs the transcription of other sensor kinases) or with more than one sensor kinase. The latter hypothesis is favored, given the results that a *qseC* mutant, which does not respond to either AI-3 or epinephrine, only affects the quorum-sensing regulation of the flagellar regulon and not of the LEE genes (88), and that a *qseEF* mutation only affects transcription of the LEE genes and not the flagellar regulon (Reading, unpublished results). The interaction of AI-3 and epinephrine with more than one sensor kinase would also impart a "timing" mechanism to this system, which is a desirable feature given that it would be inefficient for EHEC to produce both the LEE type III secretion system and flagella simultaneously. Therefore, it is hypothesized that EHEC activates expression of the flagellar regulon first through QseBC, and then the LEE genes at a later time through QseEF. A model of the EHEC AI-3 quorum sensing signaling cascade is depicted in Figure 7.3. The AI-3-dependent quorum-sensing signaling cascade is present in all Enterobacteriaceae (*E. coli*, *Salmonella* spp., *Shigella* spp., and *Yersinia* spp.). The most striking feature is that the genes encoding the transcriptional factors of this cascade are always in exactly the same context in the chromosome of all these strains and share high levels of identity among these different species, suggesting that this signaling cascade is functionally conserved in Enterobacteriaeceae.

E. coli and *Salmonella* also have a LuxR homolog, SdiA (96), but do not have a *luxI* gene and do not produce acyl-homoserine lactones (57, 91). The *E. coli sdiA* was initially isolated as a regulator of the cell-division genes *ftsQAZ* (96). However, the precise role of SdiA in quorum sensing was elusive for several years until Michael *et al.* (57) recently reported that SdiA is not sensing an autoinducer produced by *Salmonella* itself, but rather acyl-homoserine lactones produced by other bacterial species. Kanamaru *et al.* (39) reported that overexpression of SdiA from a high-copy-number plasmid in EHEC caused abnormal cell division, reduced adherence to cultured epithelial cells, and reduced expression of the intimin adhesion protein and

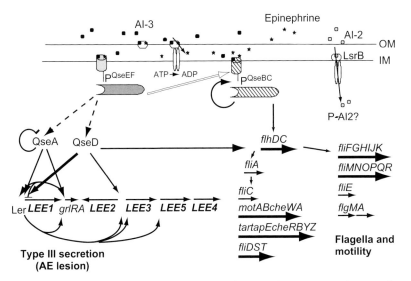

Figure 7.3. Model of the AI-3 quorum-sensing signaling cascade in EHEC. Both AI-3 and epinephrine seem to be recognized by the same receptor, which is probably in the outer membrane of the bacterium, owing to the non-polar nature of both signals. These signals might be imported to the periplasmic space where they interact with two major sensor kinases. QseC might be the sensor kinase transducing these signals towards activation of the flagellar regulon, whereas QseE might transduce these signals to activate transcription of the LEE genes. QseC phosphorylates the QseB response regulator, which binds to the promoter of *flhDC* (encoding the flagellar master regulators FlhDC) to activate expression of the flagellar regulon. QseB also binds to its own promoter to positively autoregulate its own transcription. QseE is the sensor kinase and its predicted response regulator is QseF. At what levels QseF regulates transcription of the LEE genes remains to be established. QseA is one of the transcriptional factors involved in the regulation of *ler* (*LEE1*) transcription at two levels, by binding and activating transcription of LEE1 and by activating transcription of the *grlRA* operon, where GrlA and GrlR positively and negatively regulate expression of *ler*, respectively. Then, in a cascade fashion, Ler activates transcription of the other LEE genes. QseD is a second LysR-like regulator, involved in modulating expression of the LEE and flagellar genes. EHEC also possess an *lsr* operon involved in recognition and uptake of AI-2; however, the role of AI-2 signaling in EHEC remains to be addressed.

the EspD protein, both of which are encoded within the LEE. However, no *sdiA* EHEC mutant was constructed and tested; consequently, the phenotypes observed could be artefacts due to the abnormally high expression of SdiA. Because no *E. coli* genes from either EHEC or K-12 have yet been demonstrated to be regulated by a single chromosomal copy of *sdiA*, Ahmer (2) recently concluded that there are no confirmed members of an SdiA regulon in this species.

QUORUM SENSING IN EPEC

EPEC colonizes the proximal small intestine and causes profuse and persistent watery diarrhea lasting up to 120 days in infants (18, 68). EPEC pathogenesis has several steps. First the bacteria adhere to the intestinal epithelial cells, probably through the EspA filament secreted by the LEE-encoded TTSS (32, 45). Then, Tir is translocated through the LEE-encoded type III secretion system and inserts itself into the mammalian cell membrane, where it serves as the intimin receptor, allowing the intimate attachment characteristic of AE lesion formation (40). Other EPEC cells then interact with each other, forming large microcolonies (32). The successful formation of these microcolonies requires the bundle-forming pili (BFP) and flagella (23, 24). The present knowledge about EPEC pathogenesis suggests that expression of EPEC virulence genes is dependent upon the concerted action of several regulatory factors.

One of the hallmarks of EPEC is its characteristic adherence to epithelial cells, forming microcolonies, usually referred to as localized adherence (LA) (71). EPEC produces a type IV pilus called a bundle-forming pilus (BFP), which is responsible for interbacterial interactions, leading to microcolony formation (23). Recent work from Giron and collaborators (24) suggests that EPEC flagella are also involved in adhesion and are essential for microcolony formation.

EPEC contains a large plasmid, referred to as the EPEC adherence factor (EAF) plasmid. The EAF plasmid encodes a regulator of virulence genes called Per (Plasmid-Encoded Regulator) consisting of three ORFs: *perA*, *perB* and *perC*. PerA is an AraC homolog (26) and activates the expression of the *bfp* operon encoding the bundle-forming pilus (95). The *per* loci also activate the expression of *ler* (LEE-encoded regulator, the first gene in the *LEE1* operon), which activates expression of the *LEE2*, *LEE3*, *LEE5*, and *LEE4* operons in EPEC in a regulatory cascade (29, 55, 84). Transcription of *ler* is also regulated by IHF (21), Fis (25), and BipA (27). Transcription of *per* is auto-activated by Per (53) and modulated by GadX (76).

It must be noted that quorum-sensing regulation differs between EPEC and EHEC. EPEC colonizes the proximal small intestine, which is thought to have very few or no resident flora. Therefore, whereas quorum sensing is primarily an interspecies signaling system during EHEC infection, it seems to be used for intraspecies signaling during EPEC infection. In contrast to EHEC, type III secretion in EPEC is diminished, but never abrogated, in a *luxS* mutant (85, 87). This differential regulation can be explained by the additional control of the LEE genes through Per, which is absent in EHEC (26, 55). In EPEC, GadX also represses transcription of the LEE genes through Per in acid pH (possibly when EPEC is crossing the stomach) and activates their transcription in alkaline pH (possibly when EPEC reaches the small intestine) (76). EPEC has to coordinate transcription of the LEE genes with microcolony formation. This is when quorum-sensing regulation may play an active role. Furthermore, flagellation and motility are also altered in both *luxS* and *qseA* mutants. Disruption of quorum-sensing signaling affects expression of the LEE genes, BFP, and the flagellar regulon, thereby interfering with microcolony formation and adherence to epithelial cells (79). As a result, *luxS* and *qseA* mutants form smaller microcolonies and adhere two and four orders of magnitude, respectively, less than the wild-type strain to cultured epithelial cells. In EPEC, quorum sensing is probably involved in the spatial–temporal regulation of virulence genes, allowing successful colonization of the host.

Microcolony formation is one of the first steps towards biofilm development. Inasmuch as EPEC pathogenesis is modulated by a quorum-sensing regulatory mechanism, and EPEC adheres to epithelial cells to form microcolonies, it can be hypothesized that EPEC may be forming a biofilm during infection. The observation that the *luxS* quorum-sensing system in EPEC regulates expression of antigen 43, type 1 fimbriae, and flagella (C. G. Moreira and V. Sperandio, unpublished results), structures that have been extensively associated with biofilm formation (9, 38, 44, 63, 74, 75, 97), further supports this hypothesis. Localized adhesion could be a step towards biofilm maturation in EPEC pathogenesis, especially given the observation that *espA, bfp*, and *fliC* mutants are altered for biofilm formation compared with the wild-type strain (C. G. Moreira and V. Sperandio, unpublished results). In conclusion, given that EPEC pathogenesis is controlled by a quorum-sensing regulatory mechanism, and EPEC adheres in the small intestine to form microcolonies, it could be hypothesized that EPEC forms a biofilm in the small intestine. This would be an explanation for the persistent diarrhea associated with EPEC infections (17).

CONCLUDING REMARKS

Treatment of EHEC infections with conventional antimicrobials is highly controversial, because it is well documented that antimicrobials activate the Stx phage to enter the lytic cycle, thereby producing and releasing Shiga toxin (41, 42). There are now preliminary data (87) indicating that β-adrenergic antagonists, such as propranolol, can inhibit the entire signaling cascade in EHEC, rendering it unable to induce flagellation, motility, and AE lesion formation in response to either AI-3 and/or epinephrine/norepinephrine (87). These results thus suggest an exciting possible alternative for the treatment of EHEC infections by using β-adrenergic antagonists. In addition, once the AI-3 structure is solved, it will allow the design of antagonists to AI-3. These studies may help generate a whole new class of antimicrobials that can block both AI-3 and epinephrine signaling to bacterial pathogens. Finally, these antimicrobials will be useful not only against EHEC but possibly also against other pathogens such as enteropathogenic *E. coli* (EPEC), *Salmonella*, *Shigella*, and *Yersinia pestis*, all of which harbor this signaling cascade.

REFERENCES

1 Adler, J. and B. Templeton 1967. The effect of environmental conditions on the motility of *Escherichia coli*. *J. Gen. Microbiol.* **46**: 175–84.

2 Ahmer, B. M. 2004. Cell-to-cell signaling in *Escherichia coli* and *Salmonella enterica*. *Molec. Microbiol.* **52**: 933–45.

3 Bijlsma, J. J. and E. A. Groisman 2003. Making informed decisions: regulatory interactions between two-component systems. *Trends Microbiol.* **11**: 359–66.

4 Burton, C. L., S. R. Chhabra, S. Swift *et al.* 2002. The growth response of *Escherichia coli* to neurotransmitters and related catecholamine drugs requires a functional enterobactin biosynthesis and uptake system. *Infect. Immun.* **70**: 5913–23.

5 Bustamante, V. H., F. J. Santana, E. Calva and J. L. Puente 2001. Transcriptional regulation of type III secretion genes in enteropathogenic *Escherichia coli*: Ler antagonizes H-NS-dependent repression. *Molec. Microbiol.* **39**: 664–78.

6 Chen, X., S. Schauder, N. Potier *et al.* 2002. Structural identification of a bacterial quorum-sensing signal containing boron. *Nature* **415**: 545–9.

7 Chilcott, G. S. and K. T. Hughes 2000. Coupling of flagellar gene expression to flagellar assembly in *Salmonella enterica* serovar *typhimurium* and *Escherichia coli*. *Microbiol. Molec. Biol. Rev.* **64**: 694–708.

8 Clarke, M. B. and V. Sperandio 2005. Transcriptional autoregulation by quorum sensing *E. coli* regulators B and C (QseBC) in enterohemorrhagic *E. coli* (EHEC). (in press.)

9 Danese, P. N., L. A. Pratt, S. L. Dove and R. Kolter 2000. The outer membrane protein, antigen 43, mediates cell-to-cell interactions within *Escherichia coli* biofilms. *Molec. Microbiol.* **37**: 424–32.

10 Day, W. A., Jr. and A. T. Maurelli 2001. *Shigella flexneri* LuxS quorum-sensing system modulates *virB* expression but is not essential for virulence. *Infect. Immun.* **69**: 15–23.

11 de Kievit, T. R. and B. H. Iglewski 2000. Bacterial quorum sensing in pathogenic relationships. *Infect. Immun.* **68**: 4839–49.

12 DeLisa, M. P., C. F. Wu, L. Wang, J. J. Valdes and W. E. Bentley 2001. DNA microarray-based identification of genes controlled by autoinducer 2-stimulated quorum sensing in *Escherichia coli. J. Bacteriol.* **183**: 5239–47.

13 Deng, W., J. L. Puente, S. Gruenheid *et al.* 2004. Dissecting virulence: systematic and functional analyses of a pathogenicity island. *Proc. Natn. Acad. Sci.USA* **101**: 3597–602.

14 Eisenhofer, G., A. Aneman, P. Friberg *et al.* 1997. Substantial production of dopamine in the human gastrointestinal tract. *J. Clin. Endocrinol. Metab.* **82**: 3864–71.

15 Elliott, S. J., V. Sperandio, J. A. Giron *et al.* 2000. The locus of enterocyte effacement (LEE)-encoded regulator controls expression of both LEE- and non-LEE-encoded virulence factors in enteropathogenic and enterohemorrhagic *Escherichia coli. Infect. Immun.* **68**: 6115–26.

16 Elliott, S. J., L. A. Wainwright, T. K. McDaniel *et al.* 1998. The complete sequence of the locus of enterocyte effacement (LEE) from enteropathogenic *Escherichia coli* E2348/69. *Molec. Microbiol.* **28**: 1–4.

17 Fagundes Neto U., L. G. Schmitz and I. Scaletsky 1995. Clinical and epidemiological characteristics of acute diarrhea by classical enteropathogenic *Escherichia coli. Rev. Assoc. Med. Bras.* **41**: 259–65.

18 Fagundes-Neto, U. 1996. Enteropathogenic *Escherichia coli* infection in infants: clinical aspects and small bowel morphological alterations. *Rev. Microbiol.* **27**: 117–19.

19 Freestone, P. P., R. D. Haigh, P. H. Williams and M. Lyte 1999. Stimulation of bacterial growth by heat-stable, norepinephrine-induced autoinducers. *FEMS Microbiol. Lett.* **172**: 53–60.

20 Freestone, P. P., M. Lyte, C. P. Neal *et al.* 2000. The mammalian neuroendocrine hormone norepinephrine supplies iron for bacterial growth in the presence of transferrin or lactoferrin. *J. Bacteriol.* **182**: 6091–8.

21 Friedberg, D., T. Umanski, Y. Fang and I. Rosenshine 1999. Hierarchy in the expression of the locus of enterocyte effacement genes of enteropathogenic *Escherichia coli. Molec. Microbiol.* **34**: 941–52.

22 Gillen, K. L. and K. T. Hughes 1991. Molecular characterization of *flgM*, a gene encoding a negative regulator of flagellin synthesis in *Salmonella typhimurium. J. Bacteriol.* **173**: 6453–9.

23 Giron, J. A., A. S. Ho and G. K. Schoolnik 1991. An inducible bundle-forming pilus of enteropathogenic *Escherichia coli. Science* **254**: 710–13.

24 Giron, J. A., A. G. Torres, E. Freer and J. B. Kaper 2002. The flagella of enteropathogenic *Escherichia coli* mediate adherence to epithelial cells. *Molec. Microbiol.* **44**: 361–79.

25 Goldberg, M. D., M. Johnson, J. C. Hinton and P. H. Williams 2001. Role of the nucleoid-associated protein Fis in the regulation of virulence properties of enteropathogenic *Escherichia coli. Molec. Microbiol.* **41**: 549–59.

26 Gomez-Duarte, O. G. and J. B. Kaper 1995. A plasmid-encoded regulatory region activates chromosomal *eaeA* expression in enteropathogenic *Escherichia coli*. *Infect. Immun.* **63**: 1767–76.

27 Grant, A. J., M. Farris, P. Alefounder *et al.* 2003. Co-ordination of pathogenicity island expression by the BipA GTPase in enteropathogenic *Escherichia coli* (EPEC). *Molec. Microbiol.* **48**: 507–21.

28 Gunn, J. S. and S. I. Miller 1996. PhoP-PhoQ activates transcription of *pmrAB*, encoding a two-component regulatory system involved in *Salmonella typhimurium* antimicrobial peptide resistance. *J. Bacteriol.* **178**: 6857–64.

29 Haack, K. R., C. L. Robinson, K. J. Miller, J. W. Fowlkes and J. L. Mellies 2003. Interaction of Ler at the *LEE5* (*tir*) operon of enteropathogenic *Escherichia coli*. *Infect. Immun.* **71**: 384–92.

30 Henderson, B. W. M., R. McNab and A. J. Lax 2000. Prokaryotic and eukaryotic signaling mechanisms. In B. W. M. Henderson, R. McNab and A. J. Lax (eds), *Cellular Microbiology: Bacteria-Host Interactions in Health and Disease*, 1st edn., pp. 89–162, West Sussex, England: John Wiley and Sons.

31 Henke, J. M. and B. L. Bassler 2004. Quorum sensing regulates type III secretion in *Vibrio harveyi* and *Vibrio parahaemolyticus*. *J. Bacteriol.* **186**: 3794–805.

32 Hicks, S., G. Frankel, J. B. Kaper, G. Dougan and A. D. Phillips 1998. Role of intimin and bundle-forming pili in enteropathogenic *Escherichia coli* adhesion to pediatric intestinal tissue *in vitro*. *Infect. Immun.* **66**: 1570–8.

33 Hoffer, S. M., H. V. Westerhoff, K. J. Hellingwerf, P. W. Postma and J. Tommassen 2001. Autoamplification of a two-component regulatory system results in 'learning' behavior. *J. Bacteriol.* **183**: 4914–17.

34 Hooper, L. V. and J. I. Gordon 2001. Commensal host-bacterial relationships in the gut. *Science* **292**: 1115–18.

35 Horger, S., G. Schultheiss and M. Diener 1998. Segment-specific effects of epinephrine on ion transport in the colon of the rat. *Am. J. Physiol.* **275**: G1367–76.

36 Hughes, K. T., K. L. Gillen, M. J. Semon and J. E. Karlinsey 1993. Sensing structural intermediates in bacterial flagellar assembly by export of a negative regulator. *Science* **262**: 1277–80.

37 Jarvis, K. G., J. A. Giron, A. E. Jerse *et al.* 1995. Enteropathogenic *Escherichia coli* contains a putative type III secretion system necessary for the export of proteins involved in attaching and effacing lesion formation. *Proc. Natn. Acad. Sci. USA* **92**: 7996–8000.

38 Jones, K. and S. B. Bradshaw 1996. Biofilm formation by the enterobacteriaceae: a comparison between *Salmonella enteritidis*, *Escherichia coli* and a nitrogen-fixing strain of *Klebsiella pneumoniae*. *J. Appl. Bacteriol.* **80**: 458–64.

39 Kanamaru, K., I. Tatsuno, T. Tobe and C. Sasakawa 2000. SdiA, an *Escherichia coli* homologue of quorum-sensing regulators, controls the expression of virulence factors in enterohaemorrhagic *Escherichia coli* O157:H7. *Molec. Microbiol.* **38**: 805–16.

40 Kenny, B., R. DeVinney, M. Stein *et al.* 1997. Enteropathogenic *E. coli* (EPEC) transfers its receptor for intimate adherence into mammalian cells. *Cell* **91**: 511–20.

41 Kimmitt, P. T., C. R. Harwood and M. R. Barer 1999. Induction of type 2 Shiga toxin synthesis in *Escherichia coli* O157 by 4-quinolones. *Lancet* **353**: 1588–9.

42 Kimmitt, P. T., C. R. Harwood and M. R. Barer 2000. Toxin gene expression by Shiga toxin-producing *Escherichia coli*: the role of antibiotics and the bacterial SOS response. *Emerg. Infect. Dis.* **6**: 458–65.

43 Kinney, K. S., C. E. Austin, D. S. Morton and G. Sonnenfeld 2000. Norepinephrine as a growth stimulating factor in bacteria – mechanistic studies. *Life Sci.* **67**: 3075–85.

44 Kjaergaard, K., M. A. Schembri, H. Hasman and P. Klemm 2000. Antigen 43 from *Escherichia coli* induces inter- and intraspecies cell aggregation and changes in colony morphology of *Pseudomonas fluorescens*. *J. Bacteriol.* **182**: 4789–96.

45 Knutton, S., I. Rosenshine, M. J. Pallen *et al.* 1998. A novel EspA-associated surface organelle of enteropathogenic *Escherichia coli* involved in protein translocation into epithelial cells. *EMBO J.* **17**: 2166–76.

46 Kutsukake, K., Y. Ohya and T. Iino 1990. Transcriptional analysis of the flagellar regulon of *Salmonella typhimurium*. *J. Bacteriol.* **172**: 741–7.

47 Lamont, I. L., P. A. Beare, U. Ochsner, A. I. Vasil and M. L. Vasil 2002. Siderophore-mediated signaling regulates virulence factor production in *Pseudomonas aeruginosa*. *Proc. Natn. Acad. Sci. USA* **99**: 7072–7.

48 Liu, X. and P. Matsumura 1994. The FlhD/FlhC complex, a transcriptional activator of the *Escherichia coli* flagellar class II operons. *J. Bacteriol.* **176**: 7345–51.

49 Lyon, W. R., J. C. Madden, J. C. Levin, J. L. Stein and M. G. Caparon 2001. Mutation of *luxS* affects growth and virulence factor expression in *Streptococcus pyogenes*. *Molec. Microbiol.* **42**: 145–57.

50 Lyte, M., B. P. Arulanandam and C. D. Frank 1996. Production of Shiga-like toxins by *Escherichia coli* O157:H7 can be influenced by the neuroendocrine hormone norepinephrine. *J. Lab. Clin. Med.* **128**: 392–8.

51 Lyte, M., A. K. Erickson, B. P. Arulanandam *et al.* 1997. Norepinephrine-induced expression of the K99 pilus adhesin of enterotoxigenic *Escherichia coli*. *Biochem. Biophys. Res. Commun.* **232**: 682–6.

52 Lyte, M., C. D. Frank and B. T. Green 1996. Production of an autoinducer of growth by norepinephrine cultured *Escherichia coli* O157:H7. *FEMS Microbiol. Lett.* **139**: 155–9.

53 Martinez-Laguna, Y., E. Calve and J. L. Puente 1999. Autoactivation and environmental regulation of *bfpT* expression, the gene coding for the transcriptional activator of *bfpA* in enteropathogenic *Escherichia coli*. *Molec. Microbiol.* **33**: 153–66.

54 McNab, R. M., 1996. Flagella and motility. In F. C. Neidhardt (ed.), *Escherichia coli and* Salmonella, 2nd edn, vol.1, pp. 123–45. Washington, DC: ASM Press.

55 Mellies, J. L., S. J. Elliott, V. Sperandio, M. S. Donnenberg and J. B. Kaper 1999. The Per regulon of enteropathogenic *Escherichia coli*: identification of a regulatory

cascade and a novel transcriptional activator, the locus of enterocyte effacement (LEE)-encoded regulator (Ler). *Molec. Microbiol.* **33**: 296–306.

56   Merkel, T. J., C. Barros and S. Stibitz 1998. Characterization of the *bvgR* locus of *Bordetella pertussis. J. Bacteriol.* **180**: 1682–90.

57   Michael, B., J. N. Smith, S. Swift, F. Heffron and B. M. Ahmer 2001. SdiA of *Salmonella enterica* is a LuxR homolog that detects mixed microbial communities. *J. Bacteriol.* **183**: 5733–42.

58   Miller, S. T., K. B. Xavier, S. R. Campagna *et al.* 2004. *Salmonella typhimurium* recognizes a chemically distinct form of the bacterial quorum-sensing signal AI-2. *Molec. Cell* **15**: 677–87

59   Nataro, J. P. and J. B. Kaper 1998. Diarrheagenic *Escherichia coli. Clin. Microbiol. Rev.* **11**: 142–201.

60   Neely, M. N. and D. I. Friedman 1998. Functional and genetic analysis of regulatory regions of coliphage H-19B: location of Shiga-like toxin and lysis genes suggest a role for phage functions in toxin release. *Molec. Microbiol.* **28**: 1255–67.

61   Nishimura, A. and Y. Hirota 1989. A cell division regulatory mechanism controls the flagellar regulon in *Escherichia coli. Molec. Gen. Genet.* **216**: 340–6.

62   Perna, N. T., G. Plunkett III, V. Butland 2001. Genome sequence of enterohae-morrhagic *Escherichia coli* O157:H7. *Nature* **409**: 529–33.

63   Pratt, L. A. and R. Kolter 1998. Genetic analysis of *Escherichia coli* biofilm formation: roles of flagella, motility, chemotaxis and type I pili. *Molec. Microbiol.* **30**: 285–93.

64   Pruss, B. M., J. W. Campbell, T. K. Van Dyk *et al.* 2003. FlhD/FlhC is a regulator of anaerobic respiration and the Entner-Doudoroff pathway through induction of the methyl-accepting chemotaxis protein Aer. *J. Bacteriol.* **185**: 534–43.

65   Pruss, B. M., D. Markovic and P. Matsumura 1997. The *Escherichia coli* flagellar transcriptional activator *flhD* regulates cell division through induction of the acid response gene *cadA. J. Bacteriol.* **179**: 3818–21.

66   Pruss, B. M. and P. Matsumura 1996. A regulator of the flagellar regulon of *Escherichia coli, flhD,* also affects cell division. *J. Bacteriol.* **178**: 668–74.

67   Raivio, T. L., D. L. Popkin and T. J. Silhavy 1999. The Cpx envelope stress response is controlled by amplification and feedback inhibition. *J. Bacteriol.* **181**: 5263–72.

68   Rothbaum, R., A. J. McAdams, R. Giannella and J. C. Partin 1982. A clinico-pathologic study of enterocyte-adherent *Escherichia coli*: a cause of protracted diarrhea in infants. *Gastroenterology* **83**: 441–54.

69   Roy, C. R., J. F. Miller and S. Falkow 1990. Autogenous regulation of the *Bordetella pertussis bvgABC* operon. *Proc. Natn. Acad. Sci.USA* **87**: 3763–7.

70   Sanchez-SanMartin, C., V. H. Bustamante, E. Calva and J. L. Puente 2001. Transcriptional regulation of the *orf19* gene and the *tir-cesT-eae* operon of entero-pathogenic *Escherichia coli. J. Bacteriol.* **183**: 2823–33.

71   Scaletsky, I. C., M. L. Silva and L. R. Trabulsi 1984. Distinctive patterns of adherence of enteropathogenic *Escherichia coli* to HeLa cells. *Infect. Immun.* **45**: 534–6.

72 Schauder, S. and B. L. Bassler 2001. The languages of bacteria. *Genes Dev.* **15**: 1468–80.

73 Schauder, S., K. Shokat, M. G. Surette and B. L. Bassler 2001. The LuxS family of bacterial autoinducers: biosynthesis of a novel quorum-sensing signal molecule. *Molec. Microbiol.* **41**: 463–76.

74 Schembri, M. A., K. Kjaergaard and P. Klemm 2003. Global gene expression in *Escherichia coli* biofilms. *Molec. Microbiol.* **48**: 253–67.

75 Schembri, M. A. and P. Klemm 2001. Biofilm formation in a hydrodynamic environment by novel *fimH* variants and ramifications for virulence. *Infect. Immun.* **69**: 1322–8.

76 Shin, S., M. P. Castanie-Cornet, J. W. Foster *et al.* 2001. An activator of glutamate decarboxylase genes regulates the expression of enteropathogenic *Escherichia coli* virulence genes through control of the plasmid-encoded regulator, *Per. Molec. Microbiol.* **41**: 1133–50.

77 Shin, S. and C. Park 1995. Modulation of flagellar expression in *Escherichia coli* by acetyl phosphate and the osmoregulator OmpR. *J. Bacteriol.* **177**: 4696–702.

78 Silverman, M. and M. Simon 1974. Characterization of *Escherichia coli* flagellar mutants that are insensitive to catabolite repression. *J. Bacteriol.* **120**: 1196–203.

79 Sircili, M. P., M. Walters, L. Trabulsi and V. Sperandio 2004. Modulation of enteropathogenic *E. coli* (EPEC) virulence by quorum sensing. *Infect. Immun.* **72**: 2329–37.

81 Soncini, F. C. and E. A. Groisman 1996. Two-component regulatory systems can interact to process multiple environmental signals. *J. Bacteriol.* **178**: 6796–801.

82 Soncini, F. C., E. G. Vescovi and E. A. Groisman 1995. Transcriptional autoregulation of the *Salmonella typhimurium phoPQ* operon. *J. Bacteriol.* **177**: 4364–71.

83 Sperandio, V., C. C. Li and J. B. Kaper 2002. Quorum-sensing *Escherichia coli* regulator A (QseA): a regulator of the LysR family involved in the regulation of the LEE pathogenicity island in enterohemorrhagic *Escherichia coli. Infect. Immun.* **70**: 3085–93.

84 Sperandio, V., J. L. Mellies, R. M. Delahay *et al.* 2000. Activation of enteropathogenic *Escherichia coli* (EPEC) *LEE2* and *LEE3* operons by Ler. *Molec. Microbiol.* **38**: 781–93.

85 Sperandio, V., J. L. Mellies, W. Nguyen, S. Shin and J. B. Kaper 1999. Quorum sensing controls expression of the type III secretion gene transcription and protein secretion in enterohemorrhagic and enteropathogenic *Escherichia coli. Proc. Natn. Acad. Sci. USA* **96**: 15196–201.

86 Sperandio, V., A. G. Torres, J. A. Giron and J. B. Kaper 2001. Quorum sensing is a global regulatory mechanism in enterohemorrhagic *Escherichia coli* O157:H7. *J. Bacteriol.* **183**: 5187–97.

87 Sperandio, V., A. G. Torres, B. Jarvis, J. P. Nataro and J. B. Kaper 2003. Bacteria-host communication: the language of hormones. *Proc. Natn. Acad. Sci. USA* **100**: 8951–6.

88 Sperandio, V., A. G. Torres and J. B. Kaper 2002. Quorum sensing *Escherichia coli* regulators B and C (QseBC): a novel two-component regulatory system involved in the regulation of flagella and motility by quorum sensing in *E. coli*. *Molec. Microbiol.* **43**: 809–21.

89 Surette, M. G. and B. L. Bassler 1998. Quorum sensing in *Escherichia coli* and *Salmonella typhimurium*. *Proc. Natn. Acad. Sci.USA* **95**: 7046–50.

90 Surette, M. G., M. B. Miller and B. L. Bassler 1999. Quorum sensing in *Escherichia coli, Salmonella typhimurium*, and *Vibrio harveyi*: a new family of genes responsible for autoinducer production. *Proc. Natn. Acad. Sci.USA* **96**: 1639–44.

91 Swift, S., M. J. Lynch, L. Fish *et al.* 1999. Quorum sensing-dependent regulation and blockade of exoprotease production in *Aeromonas hydrophila*. *Infect. Immun.* **67**: 5192–9.

92 Taga, M. E., S. T. Miller and B. L. Bassler 2003. Lsr-mediated transport and processing of AI-2 in *Salmonella typhimurium*. *Molec. Microbiol.* **50**: 1411–27.

93 Taga, M. E., J. L. Semmelhack and B. L. Bassler 2001. The LuxS-dependent autoinducer AI-2 controls the expression of an ABC transporter that functions in AI-2 uptake in *Salmonella typhimurium*. *Molec. Microbiol.* **42**: 777–93.

94 Tischler, A. D. and A. Camilli 2004. Cyclic diguanylate (c-di-GMP) regulates *Vibrio cholerae* biofilm formation. *Molec. Microbiol.* **53**: 857–69.

95 Tobe, T., G. K. Schoolnik, I. Sohel, V. H. Bustamante and J. L. Puente 1996. Cloning and characterization of *bfpTVW*, genes required for the transcriptional activation of *bfpA* in enteropathogenic *Escherichia coli*. *Molec. Microbiol.* **21**: 963–75.

96 Wang, X. D., P. A. de Boer and L. I. Rothfield 1991. A factor that positively regulates cell division by activating transcription of the major cluster of essential cell division genes of *Escherichia coli*. *EMBO J.* **10**: 3363–72.

97 Watnick, P. I., C. M. Lauriano, K. E. Klose, L. Croal and R. Kolter 2001. The absence of a flagellum leads to altered colony morphology, biofilm development and virulence in *Vibrio cholerae* O139. *Molec. Microbiol.* **39**: 223–35.

98 Winzer, K., K. R. Hardie, N. Burgess *et al.* 2002. LuxS: its role in central metabolism and the in vitro synthesis of 4-hydroxy-5-methyl-3(2H)-furanone. *Microbiology* **148**: 909–22.

99 Wosten, M. M., L. F. Kox, S. Chamnongpol, F. C. Soncini and E. A. Groisman 2000. A signal transduction system that responds to extracellular iron. *Cell* **103**: 113–25.

100 Xavier, K. B. and B. L. Bassler 2003. LuxS quorum sensing: more than just a numbers game. *Curr. Opin. Microbiol.* **6**: 191–7.

101 Yanagihara, S., S. Iyoda, K. Ohnishi, T. Iino and K. Kutsukake 1999. Structure and transcriptional control of the flagellar master operon of *Salmonella typhimurium*. *Genes. Genet. Syst.* **74**: 105–11.

102 Yokota, T. and J. S. Gots 1970. Requirement of adenosine 3', 5'-cyclic phosphate for flagella formation in *Escherichia coli* and *Salmonella typhimurium*. *J. Bacteriol.* **103**: 513–16.

CHAPTER 8

Quorum sensing and cell-to-cell communication in the dental biofilm

Donald R. Demuth
School of Dentistry, University of Louisville, KY, USA

Richard J. Lamont
Department of Oral Biology, University of Florida, Gainesville, FL, USA

175

INTRODUCTION

The microbial community that exists in the oral cavity is perhaps the most accessible, complex and pathogenic of the naturally occurring human biofilms. Over 500 different species of bacteria have been identified in the mature biofilm that forms on tooth surfaces (38). This complex community tenaciously adheres to and develops on the acquired salivary pellicle, a conditioning film of salivary proteins and glycoproteins adsorbed to oral tissue surfaces. The initial colonizers of the salivary pellicle are predominantly Gram-positive facultative anaerobes such as the streptococci; these organisms normally exist in commensal harmony with the host. However, as the oral biofilm matures, there is a change in the microbial composition, with an increasing presence of Gram-negative organisms. The two most common oral diseases in humans, dental caries and periodontal disease, arise from populational shifts in the biofilm in response to a variety of host and/or environmental stimuli. This results in over-representation of pathogenic organisms in the biofilm at afflicted sites in the oral cavity. For example, excessive consumption of dietary sucrose favors the overgrowth of highly fermentative acidophilic organisms such as *Streptococcus mutans*. The acidic local environment generated by these organisms promotes demineralization of the hydroxyapatite matrix of enamel, thus increasing the risk of dental caries. In contrast, periodontal disease is caused by a biofilm that thrives in the subgingival pocket and induces a chronic inflammatory condition that results in the destruction of the connective tissues and bone that support the teeth (23). This subgingival biofilm comprises predominantly of Gram-negative obligate anaerobes, including high numbers of periodontal pathogens such as *Porphyromonas*

Bacterial Cell-to-Cell Communication: Role in Virulence and Pathogenesis, ed. D. R. Demuth and R. J. Lamont. Published by Cambridge University Press. © Cambridge University Press 2005.

gingivalis, Actinobacillus actinomycetemcomitans, Tannerella forsythensis, Treponema spp., and *Prevotella* spp.

Although the composition of the dental biofilm may vary among individuals and from site to site in the oral cavity, its development and maturation follows an ordered progression of events. Temporally distinct patterns of microbial colonization are consistently observed in biofilms on all surfaces in the oral cavity. For example, the initial colonizers of the salivary pellicle on the coronal tooth surface are principally Gram-positive bacteria such as the viridans streptococci, actinomyces and related Gram-positive organisms. Establishment of these organisms facilitates the subsequent colonization of additional actinomyces and related Gram-positive rods, along with Gram-negative species such as *Veillonella* and *Fusobacterium nucleatum*. Further maturation of the oral biofilm is characterized by the colonization of additional Gram-negative anaerobes, including the periodontal pathogens mentioned above (22). Spatially, one of the most dramatic features of the oral biofilm is the presence of column-like microcolonies of early colonizers during its early stages of development (18). In addition, distinct clusters of organisms are often recovered from symptomatic sites of periodontitis patients, suggesting that biofilms comprising defined subsets of bacteria are associated with the initiation and progression of oral disease (41, 42). These reproducible characteristics of biofilm development and the populational shifts that can occur in the mature biofilm have long been taken as *a priori* evidence for the occurrence of intra- and/or interspecies communication among the bacterial residents of dental plaque. However, until recently, the identification of these putative signaling pathways and the outcomes of bacterial communication have been elusive. This chapter will focus on recent work that elucidates two mechanisms of interspecies communication among the oral bacteria, cell-contact-dependent communication and intra- and interspecies signaling that is mediated by a soluble quorum sensing signal related to autoinducer 2, the cyclic borate diester produced by *Vibrio harveyi*.

CONTACT-DEPENDENT SIGNALING IN ORAL BACTERIA

The survival and persistence of bacteria in the human oral cavity requires that the organisms rapidly adhere to a tissue surface and adapt to growth in a biofilm (22). Indeed, recent studies suggest that the act of adhering to the salivary pellicle can alter patterns of gene expression and initiate a response in the bacterium that may facilitate biofilm growth. Du and Kolenbrander (11) showed that the interaction of *Streptococcus gordonii* with saliva induced the expression of several genes, including *sspAB*, and decreased expression of various metabolic genes encoding glucose kinase, dihydropiccolinate synthetase, and an oligopeptide-binding lipoprotein. Interestingly, *sspA* and *sspB* encode streptococcal receptors for a mucin-like glycoprotein, salivary gp-340 (10). The interaction of SspA and SspB with gp-340 promotes colonization by mediating the adherence of streptococci to the salivary pellicle. SspA and SspB also promote coaggregative interactions between *S. gordonii* and actinomyces that may facilitate the subsequent colonization of streptococcal substrates by *Actinomyces* spp., leading to increased complexity in the developing oral biofilm (12). More recently, Zhang *et al.* (54) showed that streptococci that were bound to saliva-coated hydroxyapatite (sHA) beads exhibited increased expression of a two-component system, termed *bfrAB*. Streptococcal cells that were unable to express *bfrAB* were defective in colonizing sHA and did not form biofilms on polystyrene surfaces. These results suggest that signal transduction mediated by the *bfrAB* two-component system may control important aspects of streptococcal adherence and biofilm growth. Although the gene targets that are regulated by *bfrAB* have not yet been identified, the studies above suggest that oral streptococci may utilize specific signal transduction pathways to respond to saliva or other extracellular signals in order to regulate the expression of genes required for adherence to the salivary pellicle and/or for adapting to biofilm growth.

Maturation of the oral biofilm inevitably involves numerous intimate interactions among the various constituent microbial species. Many studies have shown that these interactions are not passive events but can serve specific functions, either synergistic or antagonistic in nature. In some cases, the interbacterial interactions are driven by physiological compatibility between the interacting organisms. Thus, it becomes beneficial for bacteria to assemble into groups that can utilize a complex substrate to maximum efficiency. For example, oral streptococci and actinomyces produce lactate as the end product of fermentation of carbohydrates. In turn, the Gram-negative coccus *Veillonella*, which adheres tightly to oral streptococci, utilizes lactate as a fermentable substrate (35). A similar nutritional cross-feeding occurs through the close association of *Porphyromonas gingivalis* with *Treponema denticola*. Here, *P. gingivalis* utilizes succinate that is produced by *T. denticola*, whereas growth of *T. denticola* is in turn enhanced by isobutyric acids generated by *P. gingivalis*. Although these interactions are well documented, it is unlikely that the cellular responses of the participating organisms involve specific cell-to-cell communication that extends beyond the physiological adaptation to nutrient availability.

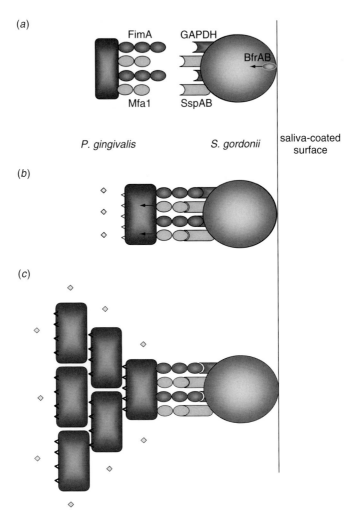

Figure 8.1. Contact-dependent and quorum sensing signaling in the formation of a *Porphyromonas gingivalis*–*Streptococcus gordonii* mixed-species biofilm. (*a*) *S. gordonii* cells adhere to saliva-coated surfaces of oral tissues and the teeth. Contact with saliva induces changes in gene expression within *S. gordonii* that are mediated at least in part by the two-component system BfrAB (54). One of the genes that is induced on streptococcal contact with saliva encodes the Ssp surface adhesin (11). The Ssp polypeptide functions as a ligand for the attachment of *P. gingivalis* (9). Thus, the adherent *S. gordonii* cells serve as a substratum for the subsequent colonization of the developing biofilm by *P. gingivalis* cells. The initial association of *P. gingivalis* with *S. gordonii* (*b*) is complex and requires the interaction of the major fimbrial subunit protein (FimA) with cell-surface glyceraldehyde-3-phosphate dehydrogenase on *S. gordonii* (29). This interaction does not promote the development of *P. gingivalis* microcolonies in the absence of the minor fimbriae.

In other cases, interbacterial cell-to-cell adherence among oral bacteria has been shown to elicit a response that does appear to extend beyond a simple adaptation to nutrient availability (see Figure 8.1). For example, using confocal microscopy, Cook *et al.* (8) showed that *P. gingivalis* adheres to *S. gordonii* cells that had previously colonized a saliva-coated coverglass in a flow chamber. *P. gingivalis* subsequently developed into biofilms consisting of microcolonies approximately 60–80 μm in depth. Such biofilms were not observed when *P. gingivalis* was incubated with the saliva-conditioned surface in the absence of *S. gordonii* or when *P. gingivalis* was exposed to a saliva-coated surface colonized by mutans streptococci (9). These studies indicate that the interaction of *P. gingivalis* with oral streptococci exhibits a high degree of selectivity and suggest that communication occurs between these organisms that may facilitate the adaptation of *P. gingivalis* to a biofilm lifestyle. Indeed, preliminary microarray analyses of the *P. gingivalis* transcriptome has identified over 40 genes that are differentially regulated in *P. gingivalis* upon contact with *S. gordonii* (R. J. Lamont, unpublished data). These genes span several functional groups, indicating that a broad change in the physiology of the organism occurs as it prepares for a biofilm mode of existence. Furthermore, there is no apparent overlap between these contact-regulated genes and the genes controlled by the LuxS-dependent quorum sensing system during biofilm growth (see below). Collectively, these data indicate that the regulation of biofilm-related genes in *P. gingivalis* is complex and can occur at several levels, which may correspond to distinct developmental steps in biofilm initiation and maturation.

Contact-dependent signaling may also be involved in antagonistic interactions that occur between organisms in the oral biofilm, as exemplified by *P. gingivalis* adherence to *Streptococcus cristatus*. Studies in flow chambers similar to those described above suggest that the interaction of *P. gingivalis* and *S. cristatus* is of low affinity and that *P. gingivalis* cells are easily washed off the streptococcal substratum by relatively low shear forces (flow rates). As a result, mixed *P. gingivalis–S. cristatus* biofilms do not develop to the extent observed with *P. gingivalis* and *S. gordonii* (51).

Figure 8.1. (cont.)

Engagement of the *P. gingivalis* minor fimbrial subunit (Mfa1) with the streptococcal Ssp polypeptide necessary for the formation of *P. gingivalis* micro-colonies (*c*). Adherence of *P. gingivalis* and *S. gordonii* (*b*) does not require the quorum sensing signal AI-2, whereas the development of microcolonies (*c*) occurs only in the presence of AI-2 (31). Updated and reproduced with permission from Lamont *et al.* 2002 (24).

A mechanistic basis for these results arises from the observation that the initial contact between *P. gingivalis* and *S. cristatus* results in a significant decrease in the transcription of *fimA*, encoding the major fimbrial adhesin that mediates adherence of *P. gingivalis* to several streptococcal species (23). Downregulation of the FimA adhesin would likely weaken the association between *P. gingivalis* and *S. cristatus*, presumably by reducing the valency of the interaction. Downregulation of *fimA* appears to be independent of the *fimRS* two-component system that has been implicated in the regulation of *fimA* expression (21). Instead, this process is initiated by the interaction of *P. gingivalis* cells with a protein of approximately 60 kDa on the surface of *S. cristatus*. Xie *et al.* (52) found that contact of *P. gingivalis* with *S. cristatus* regulates the expression of genes encoding two outer-membrane lipoproteins, *pg2167* and *pg2131*, as well as *ptpA*, encoding a putative prolyl tripeptidyl peptidase, and *pgo707*, encoding a TonB-dependent outer-membrane protein. However, genes encoding other *P. gingivalis* virulence factors, such as the cysteine proteases, were not affected by interaction with *S. cristatus*. Furthermore, Xie *et al.* (52) showed that *S. cristatus*-dependent regulation of *fimA* expression did not occur in a null mutant of *pg2167*. Indeed, this mutant exhibited constitutively high expression of FimA. This suggests that *pg2167* is involved in a regulatory pathway that may function to repress *fimA*. In contrast, mutation of *pg2131* did not influence the transcription of *fimA* or the contact-dependent regulation of *fimA*. In this mutant, the concentrations of the FimA polypeptide were significantly reduced. Thus, PG2131 appears to be involved in the post-transcriptional processing and/or transport of FimA (52). Interestingly, the primary sequences of PG2167 and PG2131 exhibit significant similarity (e-64) and may be paralogs (www.LANL). Further elucidation of their roles in the contact-dependent regulation of *fimA* will likely reveal much about regulatory networks that control the production of fimbriae in *P. gingivalis* in the context of a multispecies biofilm.

QUORUM-SENSING-DEPENDENT COMMUNICATION AMONG ORAL BACTERIA

Communication that is mediated by secreted diffusible molecules is widespread among bacteria and regulates a number of important physiological and virulence-related properties in these organisms. As highlighted in the first several chapters of this volume, acyl-homoserine lactone (AHL)-dependent quorum sensing systems are essential for aspects of virulence and biofilm growth in a variety of organisms. Some of the first evidence that soluble signal molecules may regulate the growth of oral bacteria arose from the studies of Liljemark *et al.* (26). Measurements of the dynamics of bacterial growth in developing dental biofilms that formed on implanted enamel chips suggested that there was a rapid period of cell division during the first 24 h of biofilm growth and that this period of rapid growth accounted for 90% of the total biomass in the mature biofilm (26). In addition, this period of rapid growth occurred in a cell-density-dependent manner. Cell-free supernatants of medium in contact with biofilm bacteria were shown to increase the incorporation of labeled thymidine on a per cell per time basis when compared with control medium that was not exposed to bacteria. This result suggested that conditioned medium from biofilm cultures contained a soluble molecule (designated START) that influenced the rate of bacterial cell division in the biofilm. Unfortunately, there are no subsequent studies that characterize the START component and determine its relationship to known quorum-sensing signal molecules (e.g. AHLs or AI-2).

More recently, Frias *et al.* (15) examined 33 strains representing 16 different oral bacterial species for the production of quorum-sensing signals, by using sensor 1- or sensor 2-deficient *Vibrio harveyi* reporter organisms (which cannot respond to AHL or AI-2 signals, respectively). None of the oral bacterial strains that were tested induced bioluminescence of *V. harveyi* B886 (sensor 1^+ sensor 2^-), suggesting that AHL-like autoinducers are not produced by these organisms. This result is consistent with the studies of Whittaker *et al.* (47), who utilized three different reporter systems but failed to detect the production of AHL-like autoinducers from Gram-negative oral bacteria. In addition, the recent availability of complete genome sequences of several oral bacteria has facilitated analyses *in silico* to identify genes that are similar to those known to be essential for the production of, detection of, and response to AHL signals in other organisms, such as homologs of LuxR, LuxI, SdiA. These studies have also failed to demonstrate the presence of AHL quorum sensing systems in oral bacteria. Thus, the available data suggest that quorum sensing systems dependent on AHL signals may not be common among organisms that populate the dental biofilm. However, it should be noted that AHL auto-inducers that direct species-specific signaling may not be readily detected by the available reporter strains. Furthermore, genome sequences are currently available for only a small number of the bacterial species that exist in the dental biofilm. Thus, it is possible that AHL-dependent quorum sensing systems that are species-specific may exist in the dental biofilm but cannot be identified by using the current detection methodologies.

QUORUM SENSING BY AUTOINDUCER 2 IN ORAL BACTERIA

In contrast to the results obtained by using the sensor 2-deficient *V. harveyi* reporter, Frias *et al.* (15) showed that conditioned medium obtained from several oral pathogens, e.g. *P. gingivalis, Prevotella intermedia,* and *Fusobacterium nucleatum,* induced bioluminescence of *V. harveyi* BB170 (sensor 1^- sensor 2^+). This suggests that these organisms produce a signal that stimulates the autoinducer 2 (AI-2) quorum sensing system of *V. harveyi.* Surette *et al.* (44) had previously shown that the *luxS* gene of *V. harveyi* encoded the polypeptide that functions to produce AI-2 and that genes related to *luxS* were present in the genomes of *Escherichia coli, Salmonella typhimurium,* and a variety of other Gram-negative and Gram-positive organisms. Subsequent to the functional bioluminescence data of Frias *et al.* (15), genes encoding polypeptides related to *V. harveyi* LuxS were identified from the complete or partial genome sequences of the Gram-negative oral pathogens *P. gingivalis, A. actinomycetemcomitans,* and *F. nucleatum* (4, 7, 13) as well as several species of oral streptococcus (2, 33, 34, 46). Each of these genes, when introduced into the LuxS-deficient *E. coli* strain DH5α, generated a recombinant organism that produced and secreted a soluble signal capable of inducing bioluminscence in *V. harveyi.* Thus, although the oral organisms examined to date appear to lack AHL-based quorum sensing pathways, several oral pathogens clearly produce and respond to AI-2.

The widespread distribution of LuxS has led to speculation that AI-2 is a universal, or species-non-specific, signal. AI-2 has been suggested to promote interspecies communication in microbial communities or may report the total bacterial biomass of a microbial community (36). Clearly, the ability to communicate across species barriers would be beneficial for oral organisms as they thrive in a complex multi-species biofilm; indeed, AI-2 has been shown to be required for the development of *Porphyromonas gingivalis–Streptococcus gordonii* biofilms (see below). However, the overall contribution of LuxS-dependent signaling in the development of the dental biofilm *in vivo* is not fully understood, in part because the LuxS-dependent quorum-sensing regulons of oral bacteria have not been fully defined. In addition, the AI-2 signal transduction cascade(s) utilized by oral organisms such as *Actinobacillus actinomycetemcomitans,* or by *E.coli,* may differ from the well-characterized signal transduction systems that operate in *Vibrio* species (see below and Chapter 7). Thus, it is possible that the physiological role of LuxS-dependent quorum sensing in oral bacteria, and in other organisms that lack AHL-dependent quorum sensing pathways, may be quite different from that of *Vibrio.* Indeed, the wide variety of bacterial functions that appear to be regulated by AI-2, including diverse activities such as protein secretion, iron uptake, and carbohydrate metabolism, suggests that different organisms may utilize the AI-2 signaling pathway for specific physiological requirements. The remainder of this chapter will describe the specific sets of genes and the cellular processes that are regulated by LuxS-dependent quorum sensing in the oral pathogens *A. actinomycetemcomitans* and *P. gingivalis*. In addition, studies to identify the components of the signal transduction cascade that mediates the response of oral bacteria to AI-2 will be discussed.

Is AI-2 a quorum sensing signal in oral bacteria?

In many organisms, the expression of *luxS* does not appear to be transcriptionally regulated (49). However, Chung *et al.* (7) have shown that expression of *luxS* in *P. gingivalis* is environmentally controlled and varies inversely with the osmolarity of the culture medium. Expression was greatest when cells were cultured in medium with an osmolarity of approximately half human physiological levels; it decreased significantly at physiological salt concentration. Thus, the production of LuxS and AI-2 by *P. gingivalis* may be maximal under hypotonic conditions such as occur in saliva. In *A. actinomycetemcomitans*, the amount of AI-2 secreted into the medium is highest at mid-log phase and decreases by approximately 50% when cultures reach stationary phase. This is consistent with the kinetics of AI-2 production reported for many other bacteria, including *E. coli* and *Salmonella typhimurium* (49). One explanation for the decrease in AI-2 concentrations during later stages of growth is that the signal may be degraded or internalized as cells approach and enter stationary phase (45). Indeed, *S. typhimurium* has been shown to possess a LuxS-regulated ABC transport system encoded by the *lsrACDBFGE* operon. This transporter functions to actively import AI-2 into the cell (45). Once internalized, AI-2 is modified by phosphorylation by LsrK (45). An operon corresponding to *lsrACDBFGE* is also present in *A. actinomycetemcomitans*, suggesting that it may also internalize and process AI-2 during the later stages of cell growth in culture. These observations clearly contrast with the general paradigm of quorum sensing, since the extracellular level of AI-2 is not maximal at the highest *A. actinomycetemcomitans* cell density. These results (along with others discussed in Chapter 6) have led to the suggestion that AI-2-dependent signaling may not simply report bacterial cell density but may represent a signal of the overall metabolic activity of a microbial community (36; see also Chapter 6). Another interesting possibility is that different sets of responses and behaviors are induced by extracellular and internalized AI-2 signals.

Physiologic role of LuxS-dependent signaling: regulation of iron acquisition

Analyses of wild-type and LuxS-deficient strains of the oral pathogens *A. actinomycetemcomitans* and *P. gingivalis* suggested that AI-2 may regulate aspects of iron acquisition in both organisms. The growth rates of wild-type and LuxS-deficient strains of *A. actinomycetemcomitans* are indistinguishable when cultured in iron-replete medium. In addition, wild-type *A. actinomycetemcomitans* exhibits normal growth when cultured in the presence of the ferric ion chelator EDDHA, indicating that wild-type cells effectively compete with the chelator for iron in the culture medium. In contrast, the *luxS* mutant fails to divide when cultured in the presence of the chelator, but complementation of the mutant with a functional plasmid borne copy of *luxS* restores normal growth under iron limitation. Lastly, stunted cultures of the mutant strain cultured under iron limitation resume normal growth when inoculated back into iron-replete medium (14). Studies using real-time PCR revealed that the expression of several genes involved in iron acquisition and intracellular storage are significantly altered in the *luxS* mutant. Steady-state levels of *afuA* mRNA, encoding a periplasmic ferric ion transport protein, and *ftnAB*, encoding the intracellular iron storage protein ferritin, are reduced by 8-fold and more than 50-fold, respectively, in the *luxS* mutant. Furthermore, genes encoding putative *A. actinomycetemcomitans* receptors for heme, transferrin, and hemoglobin were also shown to be downregulated in the mutant strain, albeit at more modest levels (2- to 3-fold). Interestingly, the expression of a ferric citrate transport operon and a second operon encoding a putative enterobactin receptor and an ABC-type transporter are upregulated (by 3- and 10-fold, respectively) in the *luxS* mutant (14). This pattern of regulation is intriguing: it suggests that AI-2 may play a role in the adaptation of *A. actinomycetemcomitans* to the biofilm environment in the host by facilitating a switch from the acquisition of ionic iron via chelators, e.g. enterobactin-like siderophores, to the acquisition of iron from host proteins, e.g. transferrin and hemoglobin. Consistent with this possibility, the *Neisseria meningitidis* homolog of AfuA has been shown to represent the periplasmic iron-binding polypeptide involved in the transport of iron extracted from receptor-bound transferrin (16).

P. gingivalis obtains iron almost exclusively through the acquisition of hemin; it utilizes multiple receptors and pathways to accomplish this task. Indeed, the acquisition of hemin is essential for the expression of virulence by *P. gingivalis*. A *P. gingivalis luxS* mutant grows poorly when cultured under hemin-limiting conditions that support growth of the wild-type organism. The mutant also exhibits delayed recovery in growth when transferred back into hemin replete medium (R. J. Lamont, unpublished data). Consistent with these observations, the LuxS-deficient strain exhibits differential regulation (both up and down) of genes involved in various aspects of hemin acquisition and uptake. For example, inactivation of LuxS reduces the expression of the TonB-dependent hemin/hemoglobin receptor HemR/HmuR, and of TonB itself (7). Expression of the major *P. gingivalis* proteases Rgp and Kgp is also altered in the mutant strain (4, 7); these proteases are thought to play a role in the acquisition of hemin by degrading host hemin-sequestering proteins (43). In addition, a reduction in hemagglutination titer has been reported in a LuxS mutant of *P. gingivalis* (4). In contrast, the *P. gingivalis* homolog of HasF, a putative component of the hemophore heme acquisition system, is upregulated in the LuxS-deficient organism. The differential regulation of these hemin acquisition pathways by AI-2 suggests that LuxS-dependent signaling may function to fine-tune hemin uptake in *P. gingivalis* in response to constraints such as the total amount of hemin available and/or the types of host hemin-containing molecule that are present in the local environment.

The studies described above suggest that LuxS-dependent signaling controls aspects of iron acquisition by the oral pathogens *A. actinomycetemcomitans* and *P. gingivalis*. Indeed, increasing evidence suggests that bacterial cell-signaling mechanisms are intimately associated with the pathways utilized in the acquisition of iron by a variety of other organisms as well. For example, in addition to the regulation of bioluminescence, the quorum sensing pathways of *V. harveyi* have been shown to regulate the production of a siderophore via the response regulator LuxO (27). In addition, pyoverdine, a siderophore produced by *Pseudomonas aeruginosa*, may itself function as a signaling molecule that induces the expression of at least three virulence factors by altering the activity of the extracytoplasmic sigma factor PvdS (1). Finally, the human hormone norepinephrine has been previously suggested to possess siderophore activity; Sperandio *et al.* have suggested that norepinephrine influences quorum sensing in enterohemorrhagic *E. coli* (see Chapter 7).

LuxS-dependent regulation of virulence and biofilm development

In many cases, the acquisition of iron is intimately involved in the expression of virulence by pathogenic organisms. However, there are conflicting data on the role that LuxS-dependent signaling may play in regulating virulence. In oral pathogens, this may be due in large part to the lack of a suitable animal model system that accurately reflects the multispecies environment that occurs in the subgingival pocket of the oral cavity. In *A. actinomycetemcomitans*, a leukotoxin of the RTX family of Gram-negative cytolysins is known to be an important virulence factor in early-onset forms of aggressive periodontitis. Indeed, a specific clonal population of *A. actinomycetemcomitans* which is hyperleukotoxic is found almost exclusively in early-onset periodontitis patients of African origin (20). Fong *et al.* (13) showed that the level of leukotoxin protein is decreased by approximately 3-fold in the *luxS* mutant; real-time PCR indicates a concomitant reduction of toxin mRNA in the mutant strain. These results suggest that AI-2 induces the transcription of the leuko-toxin operon. Consistent with these observations, early exponential-phase *A. actinomycetemcomitans* cells exhibit a 2- to 3-fold increase in leukotoxin expression and cytotoxicity after exposure to conditioned medium from mid to late exponential-phase *A. actinomycetemcomitans* cultures (13). Thus, the expression of leukotoxin may increase when *A. actinomycetemcomitans* colonizes and grows to high cell density in the dental biofilm.

The role of AI-2 in modulating the virulence of *Porphyromonas gingivalis* has not been clearly established. One study of *P. gingivalis* virulence using a murine subcutaneous abscess model did not reveal attenuated virulence of a *P. gingivalis* LuxS mutant (4). In other studies, Chung *et al.* (7) showed that the capacity of a LuxS-deficient strain of *P.gingivalis* to invade gingival epithelial cells did not differ appreciably from that of wild-type cells. However, hemin levels are known to influence the expression of a number of genes involved in the growth, survival and pathogenicity of *P. gingivalis*, e.g. proteases, RecA, fimbriae, LPS, and various outer-membrane proteins (3, 5, 23, 28, 40, 50). Thus, it is possible that a link exists between LuxS-dependent signaling and the expression of *P. gingivalis* virulence through its regulation of hemin uptake mechanisms.

Whereas the role of AI-2 in modulating *P. gingivalis* virulence has yet to be confirmed, the involvement of LuxS-dependent signaling in promoting the development of mixed-species biofilms with *P. gingivalis* has been recently established. The adherence of *P. gingivalis* to species of viridans streptococci but not to mutans streptococci may represent one mechanism by which *P. gingivalis* identifies a suitable niche for initial colonization of the dental biofilm. Using flow cells to mimic the open flow environment of the oral cavity *in vitro*, Cook *et al.* (8) showed that *P. gingivalis* adhered to *Streptococcus gordonii* cells that had previously colonized a saliva-coated coverglass and subsequently accumulated in towering columnar micro-colonies. However, inactivation of *luxS* had no effect on *P. gingivalis* adherence or its formation of microcolonies. Subsequently, McNab *et al.* (33) showed that *S. gordonii* also expressed *luxS* and secreted AI-2. Given the widespread distribution of *luxS* and its suggested role as a universal signal, McNab *et al.* (33) suggested that AI-2 produced by *S. gordonii* complemented the *luxS* mutation in *P. gingivalis*. Indeed, when LuxS-deficient strains of both organisms were analyzed for biofilm development in flow cells, *P. gingivalis* did not form towering microcolonies (33). Thus, the development of *P. gingivalis* microcolonies required a signal produced by *luxS* but biofilm development was not strictly dependent upon the cognate *P. gingivalis* signal. Interestingly, the adherence of *P. gingivalis* and *S. gordonii* was not affected by inactivation of *luxS*, suggesting that the adhesion of *P. gingivalis* and its accretion into towering microcolonies are independent processes and that AI-2 dependent signaling is essential only for the latter. This is consistent with the contact-dependent gene expression studies, described earlier in this chapter, which showed that no significant overlap existed between the genes regulated by contact with *S. gordonii* cells and the genes that exhibit differential expression in the LuxS mutant. NcNab *et al.* (33) also showed that LuxS controls aspects of carbohydrate metabolism in *S. gordonii*. However, it remains to be determined whether defective biofilm development occurs as a consequence of the disruption of carbohydrate metabolism in *S. gordonii*, or of hemin uptake in *P. gingivalis*, or via an as yet undefined mechanism.

AI-2 signal transduction in *A. actinomycetemcomitans* and *P. gingivalis*

The signal transduction cascades that mediate the cellular responses to AI-2 have been identified in their entirety only for *V. harveyi* and *V. cholerae* (25). The LuxS protein of *V. harveyi* is well conserved in many organisms; indeed, comparison of the *V. harveyi* and *A. actinomycetemcomitans* LuxS polypeptides indicates that they exhibit greater than 80% amino-acid sequence identity (13). *A. actinomycetemcomitans* also possesses the Hfq protein, an RNA chaperone that functions in *V. harveyi* and *V. cholerae* in conjunction with four small RNAs as the repressor that acts downstream of LuxO (25). However, a bioinformatics approach using the *V. harveyi* LuxP, LuxQ, LuxU and LuxO sequences as probes failed to identify direct homologs of these proteins in the *A. actinomycetemcomitans* genome that exhibit the high level of sequence identity that was observed for LuxS and Hfq. Furthermore, this finding is not unique to *A. actinomycetemcomitans*: similar results were obtained in searches of the *E. coli*, *Salmonella typhimurium*, and *P. gingivalis* genomes. One explanation for these findings is that AI-2 functions as a quorum sensing signal only in *Vibrio* species and may not mediate cell-to-cell communication in other bacteria that lack the complete signal transduction pathway of *Vibrio*. Indeed, Winzer *et al.* (48) suggest that for many organisms AI-2 may simply represent a toxic metabolite that is generated as a by-product of the activated methyl cycle or, alternatively, may be a nutrient that is excreted and subsequently consumed by these bacteria. Evidence for and against this hypothesis is discussed in detail in Chapter 6. However, the response of *A. actinomycetemcomitans* (and other organisms as well) to AI-2 involves genes that would not be expected to play a role in the secretion/excretion, uptake, or utilization of AI-2. Therefore, an alternative explanation is that some organisms, such as *A. actinomycetemcomitans* and *E. coli*, may communicate via AI-2 but do so by responding to a different form of AI-2 than *Vibrio* (see below and Chapter 6) and/or by utilizing sensor kinases and response regulators that differ from the components of the *Vibrio* signal transduction pathway. Several different experimental approaches have recently identified proteins of *A. actinomycetemcomitans*, *E. coli*, and *S. typhimurium* that may represent potential components of the AI-2 signal transduction cascade in these organisms. For example, two different two-component systems, *QseBC* and *QseEF*, have been suggested to regulate gene expression in response to AI-2 in *E. coli* (see Chapter 7). In addition, searching the *A. actinomycetemcomitans* genome at reduced stringency has identified several proteins that exhibit lower, but significant primary sequence identity (*c.*30%–40%) to LuxP and LuxQ. These proteins and their potential roles in AI-2 signal transduction in *A. actinomycetemcomitans* are summarized in Figure 8.2 and discussed below.

The *A. actinomycetemcomitans* AI-2 receptor

Consistent with previous search results of the *E. coli* genome reported by Surette *et al.* (44), the putative ribose-binding periplasmic RbsB polypeptide of *A. actinomycetemcomitans* was found to be most similar to LuxP, the *V. harveyi* AI-2 receptor. Furthermore, our recent studies have shown that *A. actinomycetemcomitans* RbsB, expressed and purified from a LuxS-deficient *E. coli* host, binds to AI-2 in conditioned medium from both *A. actinomycetemcomitans* and *V. harveyi* cultures (D. R. Demuth, unpublished). Indeed, addition of purified RbsB to conditioned medium to a concentration of 7 μM was sufficient to completely deplete AI-2: the treated samples did not induce bioluminescence of the *V. harveyi* BB170 reporter. In contrast, the untreated conditioned medium induced light production 200- to 2000-fold. These findings suggest that RbsB is capable of interacting with both cognate and heterologous AI-2 signals and that RbsB may

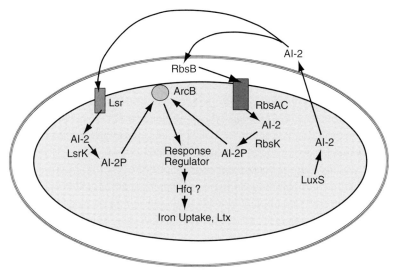

Figure 8.2. The response of *A. actinomycetemcomitans* to autoinducer 2 (AI-2) may require internalization of the signal. *A. actinomycetemcomitans* may internalize AI-2 by two different ABC transport systems. AI-2 may be bound by the periplasmic protein RbsB (D. Demuth *et al.*, unpublished) and is presumably transported into the cytoplasm via the RbsC and RbsA polypeptides. Internalized AI-2 may subsequently be phosphorylated by RbsK. *A. actinomycetemcomitans* also possesses the Lsr ABC transport system, which has been shown to internalize and phosphorylate AI-2 in *S. typhimurium* (44). In addition, genetic studies suggest that the ArcB sensor kinase is involved in the cellular response to AI-2 and may regulate genes involved in iron acquisition and storage (14; see text). The response regulator acted upon by ArcB has not yet been identified.

function as a periplasmic AI-2 receptor in *A. actinomycetemcomitans*. RbsB is part of an operon in *A. actinomycetemcomitans* comprising five genes, *rbsDACBK*, that form a putative ABC-type ribose transport system. This suggests that AI-2 bound by RbsB may be actively transported into the cytoplasm via RbsC and RbsA and subsequently modified by phosphorylation via ribokinase (RbsK). Thus, the *rbsDACBK* operon may be functionally related to the *lsrACDBFGE* operon of *S. typhimurium*. Taga *et al.* (45) demonstrated that *lsrACDBFGE* encodes an ABC-type transporter that actively internalizes AI-2. These genes reside downstream from a second operon in *S. typhimurium*, *lsrRK*, that encodes a regulator of *lsrACDBFGE* expression (*lsrR*) and a kinase (*lsrK*) that phosphorylates AI-2 after it is internalized (45). Interestingly, *A. actinomycetemcomitans* may also be capable of internalizing and modifying AI-2 via an Lsr transport system since it possesses divergently transcribed operons that are highly related to the AI-2 regulated *lsrACDBFGE* and *lsrRK* gene clusters of *S. typhimurium* (45). These observations suggest that *A. actinomycetemcomitans* may be capable of transporting cognate and heterologous AI-2 signals into the cell via two independent ABC transport mechanisms.

From X-ray crystallographic studies, Chen *et al.* (6) determined that the structure of AI-2 bound by LuxP is a borate diester of a hydrated furan diulose. However, Miller *et al.* (37) recently showed that *S. typhimurium* LsrB binds to (2R,4S)-2-methyl-2,3,3,4-tetrahydroxytetrahydrofuran, a form of AI-2 that lacks boron and is structurally distinct from AI-2 of *V. harveyi*. These studies suggest that AI-2 is present in multiple structural forms that may exist in equilibrium (see Chapter 6), and that at least one form of AI-2 may be actively transported into the bacterial cell. The structure of the *A. actinomycetemcomitans* AI-2 signal(s) that is bound by RbsB (or LsrB) is not currently known, although the experiments described above clearly indicate that RbsB interacts with the borate diester form produced by *V. harveyi*. Furthermore, the biological consequences of the interaction of AI-2 with RbsB or LsrB have not yet been determined. Genetic evidence suggests that the sensor kinase ArcB may mediate some aspects of the *A. actinomycetemcomitans* response to AI-2 (see below). However, ArcB does not possess an extracellular domain that could facilitate interactions with a periplasmic receptor and we cannot exclude the possibility that RbsB interacts with a membrane-bound sensor kinase to facilitate the cellular response to signal. Therefore, some important questions remain to be answered. Is internalization of AI-2 required to induce a response in *A. actinomycetemcomitans*? Does the engagement of AI-2 by RbsB or LsrB elicit different cellular responses? Analyses of RbsB- and LsrB-deficient *A. actinomycetemcomitans* mutants are currently being carried out to address these questions.

The *A. actinomycetemcomitans* AI-2 sensor kinase

A low-stringency search of the *A. actinomycetemcomitans* genome indicated that the ArcB sensor kinase was most similar to the *V. harveyi* LuxQ sensor kinase (*c.*30% sequence identity over 500 residues). However, as stated above, ArcB does not possess a large periplasmic domain (8 amino acids) and instead is thought to respond to an intracellular signal rather than interacting with a periplasmic molecule (30). ArcB activity is also influenced by several intracellular allosteric effectors (39). These observations are interesting given the apparent capability of *A. actinomycetemcomitans* to internalize AI-2 (discussed above). To determine whether the ArcB sensor kinase plays a role in the response of

A. actinomycetemcomitans to AI-2, an isogenic *arcB* mutant was constructed and analyzed. In these experiments, it was assumed that if ArcB contributes to the response to AI-2, the knockout mutant should exhibit a phenotype similar to that displayed by the *luxS* knockout strain that cannot produce AI-2. Interestingly, both the *arcB* and *luxS* mutant strains exhibited growth rates under iron-replete conditions that were similar to those of wild-type *A. actinomycetemcomitans*. However, both of the mutants grew poorly and failed to attain high cell density under iron limitation (14). Complementing the mutant strains with a plasmid-borne copy of the appropriate gene (*arcB* or *luxS*) restored normal growth under iron limitation to both mutants. Furthermore, expression of the LuxS-regulated genes *afuA* and *ftnAB* were reduced 16- and 24-fold, respectively, in the *arcB* mutant, consistent with the reduced expression previously reported for these genes in a LuxS-deficient background (14). Thus, the growth and gene expression studies indicate that inactivation of *arcB* and *luxS* have similar effects, suggesting that ArcB may mediate aspects of the response of *A. actinomycetemcomitans* to AI-2.

Several additional characteristics of ArcB are interesting in the context of contributing to the cellular response to AI-2. First, ArcB is a tripartite sensor that contains an integral histidine phosphotransfer domain (32). This function is not present in the LuxQ sensor of *V. harveyi* but is instead carried out by a reversible intermolecular transfer of phosphate between LuxQ and LuxU. AI-2-dependent signal transduction through ArcB would therefore not require an independent phosphotransfer protein related to LuxU. This may explain our inability to detect a homolog of LuxU in the *A. actinomycetemcomitans* genome, even when low stringency searches were performed. ArcB-dependent phosphotransfer reactions to response regulators also appear to be complex and ArcB may crosstalk with multiple signaling pathways. For example, ArcB is capable of activating non-cognate response regulators such as CheY (53) and EnvZ (31) in addition to its cognate ArcA. Thus, ArcB is not stringently coupled to a single response regulator and may regulate the expression of genes through several pathways involving different response regulators. This suggests that ArcB may mediate broad cellular responses to a variety of environmental signals and stimuli. However, very little information is currently available about the specific response regulator(s) that mediates the *A. actinomycetemcomitans* response to AI-2. Identifying these transcriptional factors and demonstrating that they accept the transfer of phosphate from ArcB will be an essential step to confirm the role of this sensor kinase in the AI-2-dependent cellular response.

AI-2 signal transduction in *P. gingivalis*

The components that comprise the AI-2 signal transduction pathway of *P. gingivalis* are largely uncharacterized. *In silico* analysis of the *P. gingivalis* genome identified homologs of LuxQ and LuxO, but not of LuxP or LuxU. Interestingly, the gene encoding the LuxQ homolog, designated *gppX*, has recently been shown to encode a sensor kinase that regulates the expression of *P. gingivalis* proteases (19). Because protease expression is modulated in the *P. gingivalis luxS* mutant, it is possible that GppX represents the sensor kinase that is responsible for AI-2-mediated regulation of the protease genes.

In summary, the AI-2 signal transduction cascade of *Vibrio* species is not wholly conserved in other organisms that produce and respond to a LuxS-dependent signal. Indeed, recent studies suggest that different structural forms of the signal molecule exist. It is clear that additional work will be required to clarify the similarities and differences in the signal transduction pathways contributing to the cellular responses among these diverse organisms.

Crosstalk and implications for AI-2 signal specificity

The widespread distribution and conservation of *luxS* has led to speculation that AI-2 functions as a universal quorum sensing signal that may report total bacterial biomass in a polymicrobial community, or the metabolic status of that community. Support for the universal recognition of AI-2 comes mostly from numerous studies that show that conditioned medium from *luxS*-containing bacteria stimulates bioluminescence of the *V. harveyi* reporter strain BB170. Indeed, *V. harveyi* BB170 is able to respond to AI-2 from many diverse and unrelated organisms. However, is this promiscuity a trait that is unique to *V. harveyi*, or do other organisms also respond to heterologous signals? Our studies have shown that conditioned medium from *A. actinomycetemcomitans* complements a *luxS* mutation in *P. gingivalis*, suggesting that *P. gingivalis* is able to recognize and respond to the *A. actinomycetemcomitans* signal (13). Furthermore, McNab *et al.* (33) showed that a *P. gingivalis luxS* mutant was complemented *in trans* by AI-2 produced by *Streptococcus gordonii*. Normal biofilms formed when the mutant organism adhered to wild-type *S. gordonii* cells, whereas no biofilms developed when a LuxS-deficient *P. gingivalis* adhered to a *luxS* mutant of *S. gordonii* (33). Thus, the ability of AI-2 to communicate between bacterial species is not limited solely to the *Vibrio* bioluminescence response. Although these data are suggestive that AI-2 may function as a universally recognized signal, one cannot yet exclude the possibility that

LuxS-dependent signaling may display some degree of specificity. The AI-2 regulon has not been fully characterized for any single bacterial species and it is not currently known whether all of the genes that respond to the cognate signal of a given organism also exhibit a comparable response to AI-2 from other bacteria. Thus it is possible that some genes may be promiscuous in their response to signal (i.e. respond to both cognate and heterologous signals) whereas other genes may respond preferentially to the cognate signal. Furthermore, from an ecological perspective, the desirability of a universally detected signal is not obvious. In complex mixed-species communities such as the dental biofilm, constituent organisms (many, if not all, expressing *luxS*) will be present at different cell densities, at different metabolic states, and in different local environments. Given the intricate and complex interrelationships that occur among the different organisms in mature multispecies biofilms, it would be beneficial if these organisms could sense not only how many other bacteria were present, but sense specific other organisms that are cohabiting the local environment. In this scenario, LuxS-dependent signaling could serve to define specific niches in complex microbial communities.

Several mechanisms can be envisioned that would support specificity of LuxS-dependent cell-to-cell communication. One possibility is that the fine structure of AI-2 varies between organisms. This possibility has already been acknowledged since the precursor of AI-2, 4,5-dihydroxy-2, 3- pentanedione, is unstable and can potentially rearrange to form different products (49). It is also possible that different bacteria may uniquely modify the core structure of AI-2, which in turn may impart specificity to the signaling process. The best evidence that supports the existence of structural variation of AI-2 signals is the recent finding that AI-2 from *V. harveyi* and *Salmonella typhimurium* are structurally distinct (6, 37). Thus, it will be essential to identify the structure of AI-2 from additional bacteria in order to clarify the extent of structural variation that exists with AI-2, and to determine whether structural variants of the signal induce differential responses from organisms.

Specificity of LuxS-dependent signaling might also occur at the level of signal detection and/or signal transduction. Multiple sensor kinases and/or AI-2 receptors, or the presence of a hierarchy of response elements, can define specific responses of bacteria to AI-2 and contribute to organism-specific responses. In this respect, it is interesting that multiple two-component systems have been shown to control LuxS-regulated genes in *E. coli* (see Chapter 7). Elucidation of these signal transduction pathways will be necessary to determine the overall distribution and the degree of conservation of the AI-2 signal transduction cascades. Only after this information is obtained will we have a more accurate picture of the distribution, specificity, and outcomes of LuxS-dependent cell-to-cell communication.

ACKNOWLEDGEMENTS

Research was supported by Public Health Service Grant DE14605 from the National Institute of Dental and Craniofacial Research.

REFERENCES

1   Beare, P. A., R. J. For, L. W. Martin and I. L. Lamont 2003. Siderophore-mediated cell signaling in *Pseudomonas aeruginosa*: divergent pathways regulate virulence factor production and siderophore receptor synthesis. *Molec. Microbiol.* **47**: 195–207.

2   Blehert, D. S., R. J. Palmer, J. B. Xavier, J. S. Almeida and P. E. Kolenbrander 2003. Autoinducer 2 production by *Streptococcus gordonii* DL1 and the biofilm phenotype of a *luxS* mutant are influenced by nutritional conditions. *J. Bacteriol.* **185**: 4851–60.

3   Bramanti, T. E., S. C. Holt, J. L. Ebersole and T. Van Dyke 1993. Regulation of *Porphyromonas gingivalis* virulence: hemin limitation effects on the outer membrane protein (OMP) expression and biological activity. *J. Periodont. Res.* **28**: 464–6.

4   Burgess, N. A., D. F. Kirke, P. Williams *et al.* 2002. LuxS-dependent quorum sensing in *Porphyromonas gingivalis* modulates protease and haemagglutinin activities but is not essential for virulence. *Microbiology.* **148**: 763–72.

5   Champagne, C. M., S. C. Holt, T. E. van Dyke, B. J. Gordon and L. Shapira 1996. Lipopolysaccharide isolated from *Porphyromonas gingivalis* grown in hemin-limited chemostat conditions has a reduced capacity for human neutrophil priming. *Oral Microbiol. Immunol.* **5**: 319–25.

6   Chen, X., S. Schauder, N. Potier *et al.* 2002. Structural identification of a bacterial quorum sensing signal containing boron. *Nature* **415**: 545–9.

7   Chung, W. O., Y. Park, R. J. Lamont *et al.* 2001. Signaling system in *Porphyromonas gingivalis* based on a LuxS protein. *J. Bacteriol.* **183**: 3903–9.

8   Cook, G. S., Costeton, J. W. and J. J. Lamont 1998. Biofilm formation by *Porphyromonas gingivalis* and *Streptococcus gordonii*. *J. Periodont. Res.* **33**: 323–7.

9   Demuth, D. R., D. C. Irvine, J. W. Costerton, G. S. Cook and R. J. Lamont 2001. Discrete protein determinant directs the species-specific adherence of *Porphyromonas gingivalis* to oral streptococci. *Infect. Immun.* **69**: 5736–41.

10 Demuth, D. R. and H. F. Jenkinson 1997. Structure, function and antigenicity of strepotococcal antigen I/II polypeptides. *Molec. Microbiol.* **23**: 183–90.

11 Du, L. D. and P. E. Kolenbrander 2000. Identification of saliva-regulated genes of *Streptococcus gordonii* DL1 by differential display using random arbitrarily primed PCR. *Infect. Immun.* **68**: 4834–7.

12 Egland, P. G., L. D. Du and P. E. Kolenbrander 2001. Identification of independent *Streptococcus gordonii* SspA and SspB functions in coaggregation with *Actinomyces naeslundii*. *Infect. Immun.* **69**: 7512–16.

13 Fong, K. P., W. O. Chung, R. J. Lamont and D. R. Demuth 2001. Intra- and interspecies regulation of gene expression by *Actinobacillus actinomycetemcomitans* LuxS. *Infect. Immun.* **69**: 7625–34.

14 Fong, K. P., L. Gao and D. R. Demuth 2003. *luxS* and *arcB* control aerobic growth of *Actinobacillus actinomycetemcomitans* under iron limitation. *Infect. Immun.* **71**: 298–308.

15 Frias, J., E. Olle and M. Alsina 2001. Periodontal pathogens produce quorum sensing signal molecules. *Infect. Immun.* **69**: 3431–4.

16 Gomez, J. A., M. T. Criado and C. M. Ferreiros 1998. Cooperation between the components of meningococcal transferrin receptor, TbpA and TbpB, in the uptake of transferrin iron by the 37kDa ferric binding protein (FbpA). *Res. Microbiol.* **149**: 381–7.

17 Grenier, D. and D. Mayrand 1986. Nutritional relationships between oral bacteria. *Infect. Immun.* **53**: 616–20.

18 Guggenheim, M., S. Shapiro, R. Gmür and B. Guggenheim 2001. Spatial arrangements and associative behavior of species in an in vitro oral biofilm model. *Appl. Environ. Microbiol.* **67**: 1343–50.

19 Hasegawa, Y., S. Nishiyama, K. Nishikawa *et al.* 2003. A novel type of two-component regulatory system affecting gingipains in *Porphyromonas gingivalis*. *Microbiol. Immunol.* **47**: 849–58.

20 Haubek, D., O. K. Ennibi, K. Poulsen, N. Benzarti and V. Baelum 2004. The highly leukotoxic JP2 clone of *Actinobacillus actinomycetemcomitans* and progression of periodontal attachment loss. *J. Dent. Res.* **83**: 767–70.

21 Hayashi, J., K. Nishikawa, R. Hirano, T. Noguchi and F. Yoshimura 2000. Identification of a two-component signal transduction system involved in fimbriation of *Porphyromonas gingivalis*. *Microbiol. Immunol.* **44**: 279–82.

22 Kolenbrander, P. E. 2000. Oral microbial communities: biofilms, interactions, and genetic systems. *A. Rev. Microbiol.* **54**: 413–37.

23 Lamont, R. J. and H. F. Jenkinson 1998. Life below the gum line: pathogenic mechanisms of *Porphyromonas gingivalis*. *Microbiol. Molec. Biol. Rev.* **62**: 1244–63.

24 Lamont, R. J., A. El-Sabaeny, Y. Park *et al.* 2002. Role of the *Streptococcus gordonii*. SspB protein in the development of *Porphyromonas gingivalis* biofilms on streptococcal substrates. *Microbiology* **148**: 1627–36.

25 Lenz, D. H., K. C. Mok, B. N. Lilley *et al.* 2004. The small RNA chaperone Hfq and multiple small RNAs control quorum sensing in *Vibrio harveyi* and *Vibrio chloerae*. *Cell* **118**: 69–82.

26 Liljemark, W. F., C. G. Bloomquist, B. E. Reilly *et al.* 1997. Growth dynamics in a natural biofilm and its impact on oral disease management. *Adv. Dent. Res.* **11**: 14–23.

27 Lilley, B. N. and B. L. Bassler 2000. Regulation of quorum sensing in *Vibrio harveyi* by LuxO and sigma-54. *Molec. Microbiol.* **36**: 940–54.

28 Liu, Y. and H. M. Fletcher 2001. Environmental regulation of *recA* gene expression in *Porphyromonas gingivalis*. *Oral Microbiol. Immunol.* **16**: 136–43.

29 Maeda, K., H. Nagata, Y. Yamamoto *et al.* 2004. Glyceraldehyde-3-phosphate dehydrogenase of *Streptococcus oralis* functions as a coadhesin for *Porphyromonas gingivalis* major fimbriae. *Infect. Immun.* **72**: 1341–48.

30 Malpica, R., B. Franco, C. Rodriguez, O. Kwon and D. Georgellis 2004. Identification of a quinine-sensitive redox switch in the ArcB kinase. *Proc. Natn. Acad. Sci. USA* **101**: 13318–23.

31 Matsubara, M., S. I. Kitaoka, S. I. Takeda and T. Mizuno 2000. Tuning of porin expression under anaerobic growth conditions by his-to-asp cross phosphorelay through both the EnvZ-osmosensor and ArcB anaerosensor in *Escherichia coli*. *Genes Cells* **5**: 555–69.

32 Matsushika, A. and T. Mizuno 2000. Characterization of three putative subdomains in the signal input domain of the ArcB hybrid sensor in *Escherichia coli*. *J. Biochem.* **127**: 855–60.

33 McNab, R., S. K. Ford, A. El-Sabaeny *et al.* 2003. LuxS-based signaling in *Streptococcus gordonii*: autoinducer 2 controls carbohydrate metabolism and biofilm formation with *Porphyromonas gingivalis*. *J. Bacteriol.* **185**: 274–84.

34 Merritt, J., F. Qi, S. D. Goodman, M. H. Anderson and W. Shi 2003. Mutation of luxS affects biofilm formation in *Streptococcus mutans*. *Infect. Immun.* **71**: 1972–9.

35 Mikx, F. H. and J. S. van der Hoeven 1975. Symbiosis of *Streptococcus mutans* and *Veillonella alcalescens* in mixed continuous cultures. *Arch. Oral Biol.* **20**: 407–10.

36 Miller, M. B. and B. L. Bassler 2001. Quorum sensing in bacteria. *A. Rev. Microbiol.* **55**: 165–99.

37 Miller, S. T., K. B. Xavier, S. R. Campagna *et al.* 2004. *Salmonella typhimurium* recognizes a chemically distinct form of the bacterial quorum sensing signal AI-2. *Molec. Cell.* **15**: 677–87.

38 Paster, B. J., S. K. Boches, J. L. Galvin *et al.* 2001. Bacterial diversity in human subgingival plaque. *J. Bacteriol.* **183**: 3770–83.

39 Rodreiguez, C., O. Kwon and D. Georgellis 2004. Effect of D-lactate on the physiological activity of the ArcB sensor kinase in *Escherichia coli*. *J. Bacteriol.* **186**: 2085–90.

40 Smalley, J. W., A. J. Birss, A. S. McKee and P. D. Marsh 1991. Haemin-restriction influences haemin-binding, haemagglutination and protease activity of cells and extracellular membrane vesicles of *Porphyromonas gingivalis* W50. *FEMS Microbiol. Lett.* **61**: 63–7.

41 Socransky, S. S. and A. D. Haffajee 1992. The bacterial etiology of destructive periodontal disease: current concepts. *J. Periodontol.* **63**: 322–31.

42 Socransky, S. S., A. D. Haffajee, M. A. Cugini, C. Smith and R. L. Kent 1998. Microbial complexes in subgingival plaque. *J. Clin. Periodontol.* **25**: 134–44.

43 Sroka, A., M. Sztukowska, J. Potempa, J. Travis and C. A. Genco 2001. Degradation of host heme proteins by lysine- and arginine-specific cysteine proteinases (gingipains) of *Porphyromonas gingivalis*. *J. Bacteriol.* **183**: 5609–16.

44 Surette, M. G., M. B. Miller and B. L. Bassler 1999. Quorum sensing in *Escherichia coli, Salmonella typhimurium* and *Vibrio harveyi*: a new family of genes responsible for auto-inducer production. *Proc. Natn. Acad. Sci. USA* **96**: 1639–44.

45 Taga, M. E., S. T. Miller and B. L. Bassler 2003. Lsr-mediated transport and processing of AI-2 in *Salmonella typhimurium*. *Molec. Microbiol.* **50**: 1411–27.

46 Wen, Z. T. and R. A. Burne 2004. LuxS-mediated signaling in *Streptococcus mutans* is involved in regulation of acid and oxidative stress tolerance and biofilm formation. *J. Bacteriol.* **186**: 2682–91.

47 Whittaker, C. J., C. M. Klier and P. E. Kolenbrander 1996. Mechanisms of adhesion by oral bacteria. *A. Rev. Microbiol.* **50**: 513–50.

48 Winzer, K., K. R. Hardie and P. Williams 2002. Bacterial cell-to-cell communication: sorry, can't talk now – gone to lunch. *Curr. Opin. Microbiol.* **5**: 2215–22.

49 Xavier, K. B. and B. L. Bassler 2003. LuxS quorum sensing: more than just a numbers game. *Curr. Opin. Microbiol.* **6**: 191–7.

50 Xie, H., S. Cai and R. J. Lamont 1997. Environmental regulation of fimbrial gene expression in *Porphyromonas gingivalis*. *Infect. Immun.* **65**: 2265–71.

51 Xie, H., G. Cook, J. W. Costerton *et al.* 2000. Intergeneric communication in dental plaque biofilms. *J. Bacteriol.* **182**: 7067–9.

52 Xie, H., N. Kozlova and R. J. Lamont 2004. *Porphyromonas gingivalis* genes involved in *fimA* regulation. *Infect. Immun.* **72**: 651–8.

53 Yaku, H., M. Kato, T. Hakoshima, M. Tsuzuki and T. Mizuno 1997. Interaction between CheY response regulator and the histidine containing phosphotransfer (HPt) domain of the ArcB sensor kinase in *Escherichia coli*. *FEBS Lett.* **408**: 337–40.

54 Zhang, Y., Y. Lei, A. Khammanivong and M. C. Herzberg 2004. Identification of a novel two-component system in *Streptococcus gordonii* V288 involved in biofilm formation. *Infect. Immun.* **72**: 3489–94.

CHAPTER 9

Quorum-sensing-dependent regulation of staphylococcal virulence and biofilm development

Jeremy M. Yarwood

Department of Microbiology, Carver College of Medicine, University of Iowa, IA, USA

(199)

INTRODUCTION

Staphylococci are a genus of bacteria remarkably adept at causing a variety of human and animal diseases. These range from relatively benign skin infections, such as impetigo, to much more serious ones, including endocarditis, osteomyelitis, toxic shock syndrome, and those associated with implanted medical devices. In fact, the staphylococci are a leading cause of nosocomial infections worldwide, and the continuing emergence of highly drug-resistant strains has created an immediate need for the development of new antimicrobial therapies and strategies. Since the identification of the accessory gene regulator (Agr) quorum sensing system in *Staphylococcus aureus*, and subsequently in other staphylococcal species, it has been assigned a central role in the regulation of staphylococcal virulence. As such, it has attracted substantial attention as a potential target for controlling staphylococcal disease.

Although recent studies have shown that virulence-gene regulation by Agr is considerably more complex *in vivo* than initially understood from studies *in vitro*, it remains clear that expression of Agr, or even lack thereof, is an important determinant in staphylococcal disease development. *agr* mutants have been shown to be attenuated for virulence in some animal models of infection, including a murine arthritis model, an osteomyelitis model, and a skin abscess model (reviewed in (34)). It has also been shown that expression of Agr, and of Agr-regulated exotoxins, facilitates escape of *S. aureus* internalized by epithelial cells (49). Furthermore, evidence is beginning to emerge that the Agr system may play a major role in biofilm formation, development, and behavior, with important implications for human disease.

Bacterial Cell-to-Cell Communication: Role in Virulence and Pathogenesis, ed. D. R. Demuth and R. J. Lamont. Published by Cambridge University Press. © Cambridge University Press 2005.

This chapter will discuss the nature of the *agr* locus, potential interaction of Agr with other virulence gene regulators, the emergence of Agr variants, the impact of Agr expression on biofilm formation, and the viability of therapeutic strategies targeting the Agr system. It will focus primarily on the Agr system of *S. aureus*, in which Agr, and virulence in general, has been best studied, and to a lesser extent on that of *Staphylococcus epidermidis*. Agr homologs have been identified in many additional staphylococcal species, but little or no investigation has been conducted into regulation of virulence by Agr in these staphylococci.

THE Agr QUORUM SENSING SYSTEM

The *agr* locus consists of two divergent operons (Figure 9.1) (reviewed in (34)). The P2 operon (*agrACDB*) encodes the proteins necessary for signal synthesis, processing, secretion, and recognition, whereas the transcript of the P3 operon, RNAIII, mediates the regulatory effects of Agr expression. The autoinducing peptide (AIP) signal, which can consist of seven to nine residues, is formed by cleavage and processing of the AgrD protein. A characteristic thiolactone ring is formed between a generally conserved central cysteine and the peptide's C-terminal carboxyl group; this cyclical structure is generally required for the activity of AIP. Both the cleavage of AgrD and the secretion of AIP are thought to be mediated by the membrane protein AgrB, although other proteins may be involved as well. In a comparison of AIP sequences from multiple *Staphylococcus* species, only the central cysteine and the five-membered thiolactone ring are generally conserved (reviewed in (34, 38)). One interesting exception is found in some strains of *Staphylococcus intermedius*, which produce an active nonapeptide AIP, which contains a serine instead of a cysteine group and forms a cyclic lactone. The length of the N-terminal tail varies from two to four amino acids; this results in pheromones with seven to nine residues in all. Lengthening or shortening the tail of an *S. epidermidis* AIP by a single amino acid residue results in a lack of biological activity in this bacterium, suggesting that AIP length is important for biological activity.

In flask cultures, the amount of AIP in the medium generally increases in correlation with increasing cell density. Upon reaching sufficient AIP concentration (this has been reported to be during the mid-log phase, although the timing may vary from strain to strain) signaling via the AgrA–AgrC system leads to increased transcription of both the P2 and P3 operons. AgrA and AgrC form the response regulator and histidine kinase receptor, respectively, of a two-component regulatory system that responds

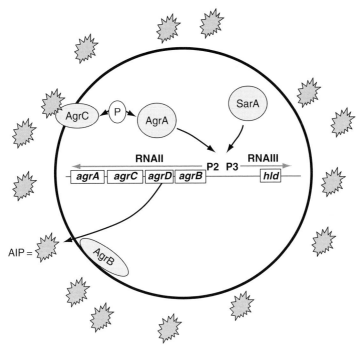

Figure 9.1. Model of the Agr system (see text for additional description). The *agrD* product is processed by AgrB into AIP and secreted into the extracellular environment. Recognition of AIP by AgrC triggers transfer of a phosphate group between AgrC and AgrA. AgrA, together with the transcriptional regulator SarA, acts to increase transcription of both the P2 and P3 operons. The transcript of the P3 operon, RNAIII, is the effector molecule of the *agr* locus and encodes δ-toxin (*hld*) as well.

to the secreted AIP. AgrA is thought to be constitutively phosphorylated, thus its activation may require dephosphorylation. Binding of AgrA to the *agr* promoters has not been demonstrated; other regulatory proteins such as staphylococcal accessory regulator A (SarA) likely are a critical component of Agr autoinduction and response to the AIP.

The transcript of the P3 operon, RNAIII, is considered to be the effector of the *agr* locus in mediating the repression or induction of quorum-controlled genes. Levels and timing of RNAIII transcription vary from strain to strain and can be correlated with the relative levels of Agr-regulated secreted or surface factors (29). δ-Toxin (*hld*) is also translated from the RNAIII molecule, although disruption of δ-toxin translation does not appear to impair the regulatory capabilities of RNAIII. Although the RNAIII nucleotide sequence is not well conserved, the secondary structure is, including several stem–loop motifs. When RNAIII from *S. epidermidis,*

S. simulans, and *S. warneri* were expressed in *S. aureus*, they completely repressed expression of the Protein A (*spa*) gene, similar to native *S. aureus* RNAIII, and they stimulated expression of α-toxin (*hla*) and serine protease, suggesting conservation of some important regulatory function among species (51). However, the ability to stimulate α-toxin and serine protease was impaired compared with *S. aureus* RNAIII, indicating that repression and activation functions might be independently encoded within the RNAIII molecule.

Much remains unknown of how RNAIII exerts its regulatory effects, although RNAIII is capable of regulation both at the transcriptional and translational levels (reviewed in (34)). For instance, the transcription termination loop of RNAIII is necessary for repression of *spa* transcription, whereas the 3′ end of RNAIII is complementary to the translation initiation site of *spa* mRNA and reportedly blocks its translation. In regulation of *hla* translation, the 5′ region of RNAIII is complementary to the *hla* mRNA leader sequence. The *hla* mRNA leader folds into an untranslatable form unless prevented from doing so by RNAIII. Benito *et al.* (2) proposed that several structurally different populations of RNAIII might coexist *in vivo*, and that RNAIII undergoes conformational changes necessary for specific functions. For example, two functions of RNAIII, translation of *hld* and translation regulation of *hla*, are thought to be mutually exclusive from a structural standpoint. In the *in vitro* structure of RNAIII, part of the *hla* mRNA binding sequence is hidden, whereas the RBS sequence and initiation codon of *hld* remain accessible. Presumably, then, some conformational change must occur for the *hla* mRNA–RNAIII interaction to occur. There is also some functional redundancy in the *S. aureus* RNAIII molecule, as non-overlapping 5′ and 3′ regions of the molecule are independently active in stimulating *hla* transcription. This may be related to the presence of nearly identical sequences in two distant stem-loops, which are complementary to the canonical Shine–Delgarno sequence. It has been proposed that RNAIII could thus act by interfering with translation of other regulatory proteins via these anti-S-D sequences or that the C-rich loops could serve as decoy binding sites for regulatory proteins, potentially titrating them out of circulation and/or altering their DNA-binding characteristics (34).

In fact, many of the regulatory effects exerted by RNAIII are likely to be indirect. Recent analysis of Agr regulation of *sed* expression suggests that the late log growth phase increase in *sed* transcription occurs via the Agr-mediated reduction in Rot (repressor of toxin) activity rather than via a direct effect of Agr (54). This was supported by the observation that the *sed* promoter is not regulated by the Agr system in a *rot* mutant background. Agr does not appear to affect *rot* transcription. However, it has been shown to downregulate Rot activity by an as yet undefined mechanism. One explanation is that Rot might be an RNAIII-binding protein, and activation of *agr* results in titration of Rot from its gene targets (32).

Agr-REGULATED GENES

Virulence determinants regulated by Agr are summarized in Table 9.1. In general, *agr* expression in batch cultures leads to the increased expression of secreted virulence factors (exoenzymes and exotoxins) and the decreased expression of several surface-associated adhesins and virulence factors. For most strains this occurs during the transition from late-log growth to early stationary phase, also known as postexponential phase. A model for how this might affect pathogenesis *in vivo* has been proposed

Table 9.1. *Virulence factors regulated by Agr*

Virulence Factor	Agr effect
aureolysin (metalloprotease)	+
TSST-1	+
enterotoxin B	+
enterotoxin C	+
α-, β-, δ-, γ-toxins	+
exfoliatins A and B	+
fatty-acid-modifying enzyme (FAME)	+
hyaluronate lyase	+
lipase	+
phospholipase C	+
proteases (Spl A, B, D, F; V8 protease)	+
staphylokinase	+
cell-surface associated capsular polysaccharide (type 5)	+
collagen-binding protein	+ / −
coagulase	−
fibronectin-binding proteins A and B	−
protein A	−
vitronectin-binding protein	−

Source: Adapted from (5, 41).

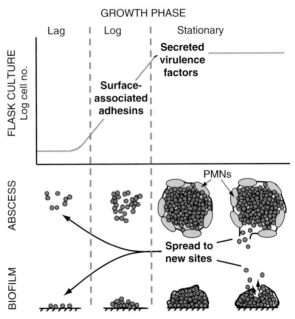

Figure 9.2. Model of Agr regulation of virulence genes in vitro and in vivo. Expression of cell-surface-associated adhesins enhances colonization and proliferation of staphylococci at the site of infection. Growth and maturation of the staphylococcal colony leads to an abscess, a mature biofilm, or both. Expression of extracellular enzymes and toxins facilitates escape of staphylococci from the localized infection and subsequent colonization of secondary sites or hosts. PMNs: polymorphonuclear cells. Adapted from (41).

for abscess formation; similar phenomena may be important in biofilms as well (Figure 9.2). During early growth, expression of surface adhesins and immuno-evasive factors is important for colonization. However, as bacterial cell density becomes high and cells exhaust local nutrient sources, expression of tissue-degrading exoenzymes and immunostimulatory toxins facilitates the spread of staphylococci from this localized infection to new potential sites of infection. Exoenzyme activity may also provide nutrient release from the surrounding tissues.

With the exception of capsular polysaccharides, factors upregulated by the Agr system with particular significance for virulence can generally be divided into two classes, exoenzymes and exotoxins (reviewed in (12)). Exoenzymes, such as lipase and the proteases, contribute to host tissue degradation, perhaps to create a source of nutrients or to facilitate escape from the localized infection. Proteases also degrade host proteins important for the immune response, such as the neutrophil defensins, platelet microbiocidal proteins, and antibodies, thus providing some protection from host immunity. Regulation of lipase production itself may be due to the upregulation by Agr of proteolytic activity, enhancing conversion of the pro-form of the lipase to mature lipase, rather than regulation at the transcriptional level. Fatty-acid modifying enzyme (FAME) is also strongly activated by Agr. FAME can inactivate bactericidal lipids often found in staphylococcal abscesses through esterification of the lipids into alcohols. These lipids are frequently released from glycerides in the abscess, perhaps by the action of staphylococcal lipase. Accordingly, most strains that produce lipase also produce FAME.

Exotoxins upregulated by Agr include the hemolysins (α-, β-, δ-, γ-toxin), which have general lytic activity against a broad range of host cells, and the pyrogenic toxin superantigens (SAgs), which have broad immunostimulatory activity. The α- and δ-toxins are both thought to be pore-forming agents strongly regulated by Agr. α-Toxin is positively regulated at both the transcriptional and translational levels by RNAIII, whereas δ-toxin is directly encoded via RNAIII. α-Toxin is a highly potent toxin, killing erythrocytes, mononuclear immune cells, and epithelial and endothelial cells; δ-toxin is a small, surfactive protein and is active against many types of membranes; β-toxin is likely a sphingomyelinase, causing hydrolysis of sphingomyelin in the membrane outer leaflet patches of erythrocytes and eventual collapse of the lipid bilayer; γ-toxin is a member of a family of bi-component toxins of *S. aureus* in which the pore-forming activity is mediated by two synergistically acting proteins, and is strongly hemolytic but much less leukotoxic.

SAgs generally exert their effects through non-antigen-specific binding of professional antigen-presenting cells and T-cells. Typical antigens might stimulate one out of 10,000 T-cells that specifically recognize that particular antigen; superantigens may stimulate and cause polyclonal proliferation of 20% or more of all circulating T-cells. This results in the release of high levels of cytokines, leading to the symptoms of toxic shock such as vascular dilation, loss of blood pressure, and subsequent organ damage and failure. In general, most of the SAgs are activated by the Agr system, although there are some exceptions. Staphylococcal enteroxin A (SEA), for instance, is produced throughout growth in an Agr-independent manner. It is not clear what the significance of this Agr-independent expression and the relative contribution of Agr-independent toxins versus Agr-activated toxins in a particular infection type might be. Many of the SAgs are also enterotoxins, exhibiting emetic activity, which is separable within the protein structure from their superantigenic activity, and are responsible for the dominance of *S. aureus* as a leading cause of food poisoning.

Several capsular serotypes have been identified in *S. aureus*; more than 90% of clinical isolates produce capsular polysaccharide (reviewed in (12)). Agr has been described as a positive regulator of serotype 5 capsule polysaccharide production, both *in vitro* and *in vivo* in a rabbit endocarditis model (63). Although there are some conflicting data, the preponderance of evidence suggests that capsule production enhances virulence and may resist phagocytosis and clearance by host immune cells. Immunization with certain capsule serotypes also is protective against challenge by *S. aureus* in animal models of infection.

There are numerous factors important for virulence that are down-regulated by Agr as well. These include several surface-associated adhesins such as protein A, fibronectin-binding proteins (*fnbA, fnbB*), vitronectin-binding protein, and coagulase (reviewed in (12)). Many of these proteins can also be found in substantial quantities in the growth medium, suggesting that their release from the cell may have importance as well. Protein A was the first staphylococcal surface protein to be characterized and is noted for its ability to bind the Fc region of mammalian IgG. By binding IgG, Protein A may interfere with the phagocytosis of opsonized bacteria. Protein A can also mediate staphylococcal adherence to von Willebrand factor, a host extracellular matrix protein, suggesting that Protein A may influence several aspects of the colonization and infectious processes. Two structurally similar proteins, FnbpA and FnbpB, have been shown to mediate *S. aureus* binding to fibronectin. Fibronectin is a ubiquitous protein found in the extracellular matrix of most tissues, as well as in soluble form in many body fluids, and is necessary for the adhesion of almost all cell types. Fibronectin is one of the host proteins that rapidly coat foreign objects, such as an intravascular catheter, thus facilitating adherence of staphylococci to this *de facto* biological surface. The fibronectin-binding proteins may also play a role in invasion of host cells by binding soluble fibronectin, which is then recognized by integrins on the host cell. This results in phagocytosis of the host-protein-coated bacteria. Vitronectin is an adhesive glycoprotein found in circulation at several extracellular matrix sites, particular during tissue or vascular remodeling. Similar to fibro-nectin- or fibrinogen-binding proteins, vitronectin binding likely facilitates colonization by staphylococci of host tissues or host-protein-coated implanted devices. Coagulase production is the primary criterion used to distinguish *S. aureus* from other staphylococcal species in a clinical micro-biology setting. Coagulase binds soluble fibrinogen and also binds human prothrombin to form a complex which converts soluble fibrinogen to insoluble fibrin. Coagulase is cell-wall-associated, although it does not have a cell-wall-anchoring sequence. The role of coagulase in staphylococcal pathogenesis is not well understood. It could be that fibrin clotting around infection foci protects the bacteria from elements of host immunity.

Using a *S. aureus* gene array, Dunman *et al.* (11) identified 104 genes induced and 34 genes repressed in an Agr-dependent manner. This study supports in general the idea that extracellular virulence factors are activated by Agr and surface adhesins repressed. However, the majority of genes identified as being Agr-regulated were in fact not known virulence factors, but were instead involved in such cellular processes as amino-acid metabolism and nutrient transport. Considering this evidence, as well as the identification of Agr homologs in other, less virulent staphylococci, one can speculate that the Agr system may not have evolved originally to facilitate virulence, but perhaps for the coordination of more basic biological functions.

QUORUM SENSING IN OTHER STAPHYLOCOCCI

Staphylococcus epidermidis

In general, the Agr system in *S. epidermidis* appears to be highly similar to that in *S. aureus* (55, 60). It is growth-phase-dependent and, with a few exceptions, upregulates exoprotein production while downregulating several surface-associated proteins. In particular, both lipase and protease activity are greatly downregulated in a *S. epidermidis agr* mutant. Overall, homology of the *agr* loci between *S. aureus* and *S. epidermidis* is 68%. The δ-toxin presumably encoded by the *S. epidermidis* RNAIII molecule differs in three amino acids from that produced by *S. aureus*, and is upregulated in post-exponential phase, as is RNAIII. δ-Toxin activity was found in 21 of 23 *S. epidermidis* strains tested. Agr was also shown to be indirectly involved in production of the antibiotic epidermin by *S. epidermidis* via regulation of EpiP, a protease involved in the formation of mature epidermin (27).

Most infections caused by *S. epidermidis* are chronic and generally less severe than those caused by *S. aureus* (reviewed in (12)) owing to the absence in *S. epidermidis* of most of the immunostimulatory toxins found in *S. aureus*. However, severe disease can still result from colonization by *S. epidermidis*, particularly in immunocompromised patients; *S. epidermidis* is perhaps the most frequent contaminant of implanted medical devices. Potential consequences of *S. epidermidis* infection include abscess formation, endocarditis, peritonitis, ventriculitis, and sepsis. Much of the development of these diseases is dependent in large part upon the response of the human innate immune system to specific microbial products, or immune "modulins." The lipopolysaccharide (LPS) of Gram-negative bacteria is an example of a particularly potent immune modulin. The peptidoglycan and lipoteichoic acid (LTA) of Gram-positive bacteria are also immune modulators, but to a much lesser extent than LPS. However, *S. epidermidis* was also found to produce the phenol-soluble modulins (PSM, reviewed in (58)). These modulins are a pro-inflammatory complex of peptides that have been shown to induce cytokine production by monocytes, cause degranulation, and inhibit spontaneous apoptosis in human neutrophils, and serve as a chemoattractant for neutrophils and monocytes. *S. epidermidis* has been shown to produce at least three of these peptides, PSMα, PSMβ, and PSMγ. PSMγ is in fact δ-toxin and encoded by the *agr* locus, PSMα is similar to δ-toxin, and PSMβ is similar to the "SLUSH" peptides from *S. lugdunensis* and the gonococcal growth inhibitor peptides from *S. haemolyticus*. Recently, it was found that *agr* regulates the production of these modulins in *S. epidermidis* (58). This study was the first to ascribe a major role to quorum sensing in host–pathogen interaction during *S. epidermidis* infection. PSM production was completely ablated in an *agr* mutant; the *agr* mutant did not induce production of TNFα by human myeloid cells nor did it induce chemotaxis of neutrophils to the extent that a wild-type strain did. The authors propose that quorum-sensing regulation allows PSM production only at the appropriate stage of infection, enabling *S. epidermidis* growth and survival in the host. For instance, production of immunostimulatory factors early in infection when only a few bacterial cells were present would stimulate neutrophil activity and attraction would likely result in clearance from the host. Alternatively, production of PSM and extracellular enzymes at the later stages of infection when bacterial cell density is high would result in tissue degradation associated with a strong immune response and the enzymatic activity, both providing nutrients and facilitating escape from the localized abscess (similar to the model of quorum sensing in *S. aureus* infections).

Staphylococcus lugdunensis

An *agr*-related sequence was originally described in this organism by Vandenesch *et al.* (57). A transcript with homology to RNAIII was identified in some of the strains, and these same strains had hemolysin activity similar to that caused by *S. aureus* δ-toxin, but the RNAIII molecule from *S. lugdunensis* does not encode any peptide homologous to δ-toxin. More recently, it was found that this hemolytic activity was actually mediated by three closely related "SLUSH" peptides encoded by a non-*agr* locus (9).

Staphylococcus saprophyticus

Recently, an *agr* homolog was cloned from *S. saprophyticus* (44), a cause of urinary tract infections. Although no genes have yet been confirmed to be regulated by Agr in this organism, RNAIII levels do appear to increase with time and the *agr* locus was found in all clinical isolates tested. The *S. saprophyticus* RNAIII does not appear to encode any peptide with homology to δ-toxin, consistent with the observation that *S. saprophyticus* is not hemolytic.

Agr homologs have also been identified by PCR amplification and sequencing in several additional staphylococcal species by Dufour *et al.* (10) in a study described below.

Agr SPECIFICITY GROUPS

Four distinct Agr groups, based on the identity of the AIP produced, have been described in *S. aureus* (reviewed in (34)). AIP produced by one group generally inhibits signaling by staphylococci from a different Agr group by competitive binding to the AgrC receptor. In addition, one AIP produced by *S. epidermidis* inhibits signaling by three of the four *S. aureus* *agr* groups. Conversely, *S. aureus* Agr group IV is the only one capable of inhibiting *S. epidermidis* signaling.

In a survey of numerous staphylococcal species, Dufour *et al.* (10) found the presence of *agr* homologs in 12 other staphylococcal species besides those already described in *S. aureus* {4}, *S. epidermidis* {3}, and *S. lugdunensis* {2} (numbers in brackets represent the number of Agr groups identified for each species). These included *S. auricularis* {2}, *S. capitis* subsp. *capitis* {2}, *S. caprae* {2}, *S. simulans* {2}, *S. arlettae* {1}, *S. carnosus* {1}, *S. cohnii* subsp. *cohnii* {1}, *S. cohnii* subsp. *urealyticum* {1}, *S. gallinarum* {1}, *S. intermedius* {1}, and *S. xylosus* {1}. The authors found extreme divergence in the *agr* locus sequence among species, with only 10% of the *agrBCD* nucleotides being absolutely conserved. The most extensive divergence appears to be within the regions thought to be important for signal synthesis (*agrD*), processing (C-terminal portion of *agrB*), and recognition (portion of the N-terminal half of *agrC*). AgrA, which does not interact directly with the AIP, shows much less sequence divergence. This study also found significant covariance between these genes, as one might expect in order for a strain to properly secrete and recognize a unique AIP. In general, phylogenetic relationships established by using 16 S rDNA strongly resembled relationships determined by using *agrC* or *agrB*.

However, phylogenicity established by using *agrD* was very different from that determined by using 16 S, suggesting that *agrD* sequences are the most divergent of all. Thus, the authors propose that the driving force in Agr evolution might be mutation of the *agrD* sequence coding for the AIP. Under the appropriate environment, the strain carrying the mutated *agrD* might accumulate compensatory mutations in *agrB* and *agrC* to allow recovery of a functional Agr quorum sensing system. A model for Agr evolution in the context of a biofilm is discussed later in this chapter.

There is some correspondence as well between phylogenetic trees, Agr groups, and disease type (24). For instance, most menstrual TSS strains belong to Agr group III, whereas most exfoliatin-producing strains belong to Agr group IV. Interestingly, a strong correlation was found between vancomycin treatment failure and infection due to MRSA encoding Agr group II. It may simply be that this Agr group is associated with select *S. aureus* clones, known as glycopeptide intermediate-resistance *S. aureus* (GISA), that have intrinsic advantages under vancomycin selection (33). Li *et al.* (28) did find that genetic polymorphism of Agr was linked to pathogenicity of *S. epidermidis* in China. Group I isolates were predominant in pathogenic isolates, whereas isolates from healthy patients tended to be group II. It was also shown that some Agr alleles were more highly associated with infection of a specific host (human vs. bovine) (17).

It has been proposed that Agr polymorphism might contribute to exclusion of other strains of the same species from the colonization or infection site or to isolate certain staphylococcal populations (25, 34), but there is no clear evidence for this phenomenon. From an epidemiological standpoint, it is difficult to untangle whether Agr plays an active or a passive role in colonization and disease, owing to the covariance of many other virulence-associated genes among Agr groups. Further complicating the studies are questions about whether or not the *agr* system is even expressed in many environments *in vivo*, particularly in areas of low cell density. Indeed, RNAIII expression was frequently not detected in the lungs of cystic fibrosis (CF) patients (20). Nevertheless, Goerke *et al.* (22) addressed this issue of Agr interference in a study of cystic fibrosis patients and healthy controls. They found that strain replacement in CF patients was accompanied by a change in the Agr group 80% of the time. However, in 10 cases where more than one strain of *S. aureus* was detected in the patients, 6 consisted of co-colonization by strains from interfering Agr groups, suggesting that Agr-group-based interference was not an active phenomenon in these cases. This was a much more difficult issue to assess in healthy individuals. Although the authors found several cases where healthy carriers became sequentially colonized by strains belonging to an interfering Agr group and were rarely co-colonized by strains from interfering Agr groups, they were unable to rule out the possibility of a time gap between the loss of one strain and colonization by a second. However, in three cases where a person was colonized by more than one strain, two cases were of strains belonging to the same Agr group and the third consisted of isolates from Agr groups I and IV, which are only partly inhibitory for each other. Kahl *et al.* (26) found that the success of isolates from CF patients, as measured by their prevalence, persistence, or ability to co-colonize with other *S. aureus* strains, was independent of Agr group specificity. In addition, the Agr specificity group distribution did not differ significantly when compared with isolates from other infection types and from healthy carriers. Lina *et al.* (30) did show that, among healthy volunteers, *S. aureus* colonization rates were negatively correlated with the rate of colonization by *Corynebacterium* spp. and *S. epidermidis*. Only one type of *S. aureus agr* allele was detected in each carrier, but the results did not generally support in vitro Agr-specific cross-inhibition experiments. For example, the relatively frequent co-isolation of Agr groups not mutually inhibitory, such as Agr groups I and IV, should have been observed if cross-inhibition was an important determinant in colonization, but this was not the case. Jarraud *et al.* (24) concluded that the data do not necessarily support a direct role for Agr group in the type of human disease caused, but may instead reflect an ancient evolutionary division of *S. aureus* in which other determinants developed in subsequent staphylococcal lineages facilitate pathogenesis and disease type. Regardless, additional studies are needed to address the potential functions of various Agr groups in staphylococcal colonization or disease pathogenesis, as the studies conducted so far have been limited to very few disease types.

It has also been proposed that the ability of *S. epidermidis* to generally inhibit quorum sensing by most *S. aureus* strains may help to explain why *S. epidermidis* predominates on the skin and in infections of indwelling medical devices (38). This assumes that the *agr* locus (i) is expressed in these loci, and (ii) facilitates colonization, ideas for which little either supporting or conflicting evidence exists.

Agr REGULATION OF VIRULENCE IN THE CONTEXT OF OTHER REGULATORS AND THE ENVIRONMENT IN VIVO

Unarguably, *agr* expression must occur within the context of the extracellular environment, growth phase, and overall cell metabolism. Early on in the study of Agr regulation, it was recognized that there is a temporal program of exotoxin expression that is unaltered by the timing of RNAIII expression. For instance, Vandenesch *et al.* (56) found that *hla* transcription could be delayed as much as six hours after RNAIII reached its maximum level. In the case of Agr-activated toxins, several other appropriate environmental conditions are required for full expression of these proteins, at least in vitro. These conditions include neutral pH, sufficient oxygen and carbon dioxide concentrations, and elevated protein. Thus, interaction of Agr with other regulatory systems that respond to diverse signal inputs is essential for a complete understanding of the staphylococcal quorum response.

SrrAB

Yarwood *et al.* (69) proposed one potential link between cellular metabolism and quorum sensing: the two-component system SrrAB (also described independently by Throup *et al.* (52)). This homolog of the *Bacillus subtilis* ResDE two-component system regulates certain virulence factors in response to oxygen concentration; *srrAB* mutants are growth-defective in the absence of oxygen, presumably owing to the role of *srrAB* in activating enzymes necessary for synthesis of alternative electron acceptors. It is possible that SrrAB responds to an intermediate of the electron-transport pathway in staphylococci, such as reduced menaquinone, thus providing a direct link to cellular energy metabolism.

The *sae* locus

A second regulatory locus, *sae*, encoding a two-component regulatory system, SaeRS, and two additional genes (*saeP* and *saeQ*) was identified with effects on multiple staphylococcal virulence factors (18, 19, 34, 35). Nuclease and coagulase in particular were repressed at the transcriptional level in an *saeRS* mutant. SaeRS expression does not appear to affect RNAIII production and instead is thought to act together with or downstream of Agr. Coincident with the onset of RNAIII synthesis, one early *sae* locus transcript disappears and three new ones appear. This switch is blocked in an *agr* mutant as well as by diverse environmental conditions, including NaCl, pH < 6, and subinhibitory levels of clindamycin, conditions that were previously shown to affect exoprotein synthesis. Thus, it has been proposed that *sae* coordinates quorum sensing with some environmental signals (35).

ArlRS

ArlRS is a two-component regulatory system that regulates autolysis and expression of the NorA multidrug efflux pump (13). Increased expression of *agr* was observed in an *arl* mutant (14). Correspondingly, the expression of several Agr-activated genes were increased in *arl* mutants, such as *a*-toxin, lipase, and serine protease. Interestingly, *spa* (Protein A) expression was also increased in *arl* mutants, an effect apparently mediated in part by SarA (discussed later in this chapter) as well as by *agr*. Expression of *arlRS* was itself reduced in an *agr* mutant, so ArlRS and Agr apparently form an autorepression circuit in that *arlRS* counters *agr* induction and *vice versa*.

The σ^B factor

The staphylococcal alternative sigma factor, σ^B, is activated by environmental stresses (such as ethanol treatment) and energy depletion; σ^B appears to act generally opposite of Agr as it represses many secreted virulence factors while upregulating some surface adhesins (3). In certain conditions, σ^B may contribute to biofilm development by staphylococci. However, despite its apparent role in regulation of virulence factors and biofilm formation, a clear role for σ^B in virulence has not yet been established. Perhaps this is due to the limited number of studies addressing σ^B-controlled virulence, particularly in models of long-term, chronic infections. No direct link between σ^B and *agr* expression has been identified, although lower levels of the RNAIII transcript have been found in strains with an intact *sigB* operon (4, 23).

ClpX and ClpP

Clp proteolytic complexes have been shown to be important for virulence and survival by several pathogenic bacteria under stress conditions. A ClpX or ClpP mutant of *S. aureus* is attenuated in virulence, and both ClpX and ClpP are important for growth under oxidative stress or at low temperature (16). Interestingly, absence of either ClpX or ClpP reduces the transcription of RNAIII and AIP activity. Correspondingly, expression of several extracellular proteins, including α-toxin, are reduced in Clp mutants. No mechanism has yet been identified whereby the Clp proteases exert this regulation on the Agr system, though it may serve as an important regulatory pathway for virulence regulation under certain conditions.

The SarA family

No consensus sequence has been identified as a regulatory target for Agr among Agr-regulated genes. Presumably, then, RNAIII must frequently exert its regulatory activity via intermediary proteins (37). Some of these intermediary regulators likely belong to the SarA family, an expanding group of homologous winged helix–turn–helix transcription factors (reviewed in (5)). This family includes SarA, SarR, SarT, SarS, SarU, SarY, and Rot. Many of these proteins are an integral part of virulence gene regulation, often influencing expression of the *agr* operon itself and sometimes even opposing Agr regulation of certain genes. SarA, the most extensively studied member of this family, is thought to regulate *agr* expression directly via binding to the *agr* promoter region. The *agr* transcripts RNAII and RNAIII are both diminished in *sarA* mutants; a 29 bp sequence located between the *agr* P2 and P3 promoters, identified as a SarA binding site, is required for the transcription of RNAII in *S. aureus* (6). In addition, hemolysin production can be restored to wild-type levels in *sarA* mutants by induction of RNAIII supplied *in trans*. However, SarA also mediates virulence gene regulation independently of Agr, activating some genes also activated by Agr (e.g. hemolysins) while repressing others also activated by Agr (e.g. proteases). Many of these effects are likely to be direct regulatory actions, as several SarA-regulated genes have a consensus SarA binding site upstream. The relationship to Agr of other members of the SarA family is diverse, and likely helps to fine-tune the quorum response in response to other signals. SarR downregulates SarA expression through binding of the *sarA* promoter, in effect repressing Agr expression, whereas SarU activates RNAIII transcription. SarS activates protein A synthesis; SarT represses α-toxin production. Experimental evidence suggests that Agr likely represses *spa* by suppressing *sarS* transcription. In contrast, Agr does not appear to upregulate *hla* expression via repression of *sarT*. The global regulator of virulence, Rot, generally acts to counter Agr activity in repressing some secreted proteins and activating surface-associated virulence genes (43).

REGULATION BY Agr IN VIVO

The regulation of virulence genes by Agr appears to be considerably more complex in vivo than was understood from experimentation in vitro. The contribution of Agr to virulence may depend heavily upon the site of infection; and one animal model of infection will imply a role for Agr very different from that found by using another animal model. For instance, growth and virulence gene regulation in the kidney, where both pH and osmolarity are high, is vastly different from growth in an endocarditis vegetation. Even within an endocarditis vegetation, exposure of cells to nutrients is much greater on the surface of the vegetation than in the depths of the vegetation.

When *agr* expression has been examined in vivo, a complex picture of Agr regulation has emerged. Goerke *et al.* (20) found that RNAIII expression by staphylococci in the sputum of cystic fibrosis patients was highly variable and did not correlate with expression of *hla*, which is strongly regulated by Agr in vitro. The same group also determined that mutation of *agr* did not affect *hla* expression in a guinea pig model of device-related infection (21). They found that, in the two strains examined, RNAIII expression was significantly lower in vivo than during growth in vitro and negatively correlated with bacterial densities. Yarwood *et al.* (68) used microarray analysis of virulence gene expression in vivo in a rabbit model of TSS. Surprisingly, despite the repression of *agr* expression in serum cultures and in vivo, several secreted virulence factors were upregulated, sufficient to cause acute disease in the animals. There was also little difference between an *agr* wild type and an *agr* mutant in expression of exotoxins and causation of disease. This lack of correlation between *agr* expression and toxin expression in vivo suggests the presence of other regulatory mechanisms that can up- or downregulate many virulence factors independently of Agr activity.

Xiong *et al.* (65) did confirm several aspects of the in vitro model of the Agr regulatory circuit in an experimental endocarditis model, as well as identifying an intriguing exception. As might be expected, maximal RNAII activation in vegetations occurred early, followed by increasing RNAIII expression. This correlated with increased bacterial densities within the vegetations, which were higher in the vegetations than in kidney or spleen tissues. The authors concluded that RNAIII activation in vivo is time- and cell-density-dependent, and perhaps also tissue-specific. Surprisingly, RNAIII activation was also observed in vegetations formed by using Agr signaling mutants (though to a lesser extent than in the wild type), suggesting that an RNAII-independent mechanism of RNAIII activation may exist *in vivo*. In addition, there was no correlation between RNAII promoter activity and vegetation densities.

Rothfork *et al.* (42) hypothesized that bacterial clumping mediated by fibrinogen might create a microenvironment within the host that promotes density-dependent activation of *agr* expression independently of overall bacterial burden. The plasma protein fibrinogen is an important component of the acute inflammatory response. It helps to promote neutrophil migration and adhesion, induction of cytokine synthesis, coating of foreign bodies, walling off of infection sites, and initiation of wound healing. *S. aureus* possesses several cell-associated and secreted factors that directly interact with fibrinogen or its soluble precursor, fibrin. Rothfork *et al.* showed that, in a murine abscess model, transient depletion of the animal of fibrinogen significantly reduced the bacterial burden and overall morbidity and mortality in the animals. This was not observed in infection by an *agr* mutant. Fibrinogen depletion also inhibited activation in vivo of RNAIII transcription, as well as expression of the quorum-activated virulence factors α-toxin and capsule. The data suggest that fibrinogen-mediated clumping is sufficient to concentrate the autoinducer and promote quorum sensing. The same effects could also be mediated by fibronectin. This study provides an important mechanistic link between the innate immune response and pathogenesis of *S. aureus*, as well as insight into regulation of *agr* expression in vivo.

Agr AND STAPHYLOCOCCAL BIOFILMS

The concept of biofilm formation and its relevance to human disease has been well reviewed elsewhere (8, 39). In general, biofilms are a complex community of bacteria enclosed in a matrix that is usually self-produced. They are of particular clinical relevance, as biofilm-associated bacteria are more resistant to, or tolerant of, antimicrobial treatment and are also resistant to clearance by the host immune system. Many staphylococcal infections appear to be associated with, or resemble, biofilms: these include endocarditis, osteomyelitis, and even some skin infections. However, the most common staphylococcal biofilm-associated infections are associated with implanted medical devices, such as intravascular catheters. These infections are most commonly caused by *S. epidermidis*; in fact, the ability to form biofilms is considered the primary virulence factor of *S. epidermidis*. Quorum sensing in staphylococcal biofilms is an important, emerging area of investigation, as many factors thought to be important in adherence of staphylococci to biological surfaces are under the control of the Agr system. Furthermore, emerging evidence suggests that the Agr phenotype of staphylococci might influence the development and outcome of long-term, chronic staphylococcal infections.

There are at least three important stages in staphylococcal biofilm development and behavior. The first is the initial attachment of cells to a biotic or abiotic surface, usually mediated by surface adhesins. The second, or maturation, stage involves the accumulation of cells into multilayered clusters enclosed in a self-produced matrix. The matrix usually consists of the polysaccharide intercellular adhesion (PIA). In the host milieu, however, it is not entirely clear whether PIA is in fact required to form a biofilm-like community. One might imagine that, through intracellular binding mediated by host cell matrix components, a biofilm-like structure could be achieved together with the important characteristics of a biofilm (nutritional gradients, protection from host immune factors and predation, etc.) without the presence of PIA. The third stage of biofilm development involves detachment of cells from the biofilm; this may facilitate the colonization of sites distant from the original infection site. The factors contributing to detachment are both external and internal to the biofilm. Physical factors such as shear and physical disruption of the biofilm induce large-scale detachment; emerging evidence suggests that biofilm-associated bacteria may also actively promote their own detachment.

There are potential mechanisms whereby Agr expression might impact each of these stages of biofilm development, based on a very limited number of studies in vitro.

Initial attachment

There appear to be two general mechanisms by which staphylococci attach to a surface, as illustrated by colonization of an intravascular catheter. During insertion of the catheter, attachment to the naked polymer surface occurs through non-specific, physiochemical interactions, such as hydrophobic interactions. Subsequent to implantation, the catheter surface becomes coated with components of the host matrix, such as fibrinogen, fibronectin, and collagen. This facilitates more specific interactions between the staphylococci and what is now a biological surface mediated by specific receptors on the staphylococci, such as the fibrinogen- and fibronectin-binding proteins. Several of these specific staphylococcal receptors are negatively regulated by Agr. In some staphylococcal species large proteins that might mediate non-specific, hydrophobic interactions with the uncoated polymer surface are also regulated by Agr (e.g. the autolysin AtlE in *S. epidermidis*).

Maturation

Little evidence exists for or against a contribution of the Agr system to the maturation of biofilms. In particular, the expression of PIA is not regulated by Agr (59). However, it is conceivable that accumulation in the host of staphylococcal cells in a biofilm, particularly through mutual binding by adjacent bacteria of host factors, would be enhanced by the continued expression of surface adhesins.

Detachment

Vuong *et al.* (62) have proposed that expression of δ-toxin, a protein with surfactant properties and encoded by the *agr* locus, might contribute to detachment of cells from a biofilm. Thus, in combination with the downregulation of surface adhesins, Agr may well play an important role in facilitating release of staphylococcal cells from the biofilm. Indeed, we have observed enhanced detachment of *agr*-expressing cells from a biofilm (67), but have not yet been able to confirm the contribution of Agr to this phenomenon. Large-scale detachment events would also be expected to influence mature biofilm structure, at least temporarily, thus potentially influencing the maturation stage of biofilm development as well.

At first glance, the various existing literature regarding the role of the Agr system in staphylococcal biofilm formation and behavior appear somewhat inconsistent. Several investigators have grown different strains under different growth conditions and, not surprisingly, have obtained different results. In a survey of 105 *S. aureus* strains, Vuong *et al.* (62) found a strong correlation between lack of Agr activity (as measured by lack of δ-toxin production) and ability to adhere to polystyrene. The authors attributed this, at least in part, to the surfactant properties of δ-toxin, as addition of increasing concentrations of δ-toxin decreased attachment of *S. aureus* to polystyrene. In two studies with somewhat conflicting results, Shenkman *et al.* first found that *agr* mutants showed (compared with wild-type *S. aureus*) increased adherence to immobilized fibrinogen, increased induction of platelet aggregation, and had little impact on adherence to immobilized fibronectin, von Willebrand factors, bovine corneal extracellular matrix, and endothelial cells (47). The difference in adherence properties developed primarily under flow conditions, suggesting different adhesion mechanisms under static and flow conditions. In the second study, it was concluded that RNAIII downregulated *S. aureus* adherence to fibrinogen under static conditions while upregulating *S. aureus* adherence to fibronectin and endothelial cells under both static and flow conditions (48). The authors also found that the contribution of activated platelets in *S. aureus* adherence to endothelial cells was downregulated by RNAIII, likely due to decreased adherence to fibrinogen, a plasma protein thought to bridge

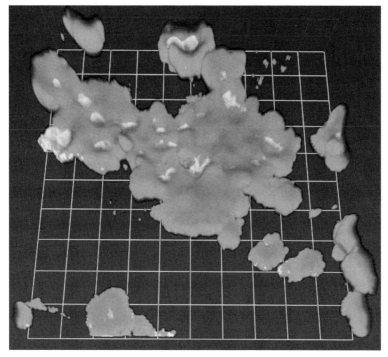

Figure 9.3. Three-dimensional reconstruction of a *Staphylococcus aureus* biofilm. Cells expressing a quorum-controlled green fluorescent protein reporter are green; the remaining biofilm is red from staining with propidium iodide. Each side of the grid represents about 600 µm. (See also color plate section.)

S. aureus, platelets and endothelial cells. Pratten *et al.* (40) showed pleotropic effects of both the *agr* and *sar* operons on expression of surface molecules responsible for binding to substrata.

To address whether the variable results found in the literature were the result of different strains or different growth conditions, Yarwood *et al.* (67) grew the same isogenic pair (wild-type versus *agr* mutant) in biofilms under several conditions. In this study, the contribution of Agr to biofilm development was found to be dependent on growth conditions. In some cases, *agr* expression decreased bacterial attachment and biofilm formation. Under other conditions, it enhanced biofilm formation or, in the case of flow-cell biofilms, appeared to have no effect at all, even when clearly expressed (Figure 9.3; for additional images see (67)). Not surprisingly, the nature of the growth surface appears to be especially important in detecting a contribution of Agr to biofilm development, for example, whether or not the surface is coated with biologically relevant proteins with which staphylococci can interact and receptors for which might be regulated by *agr*.

In the first study of its kind to address directly the biofilm-forming capabilities of *agr* mutants in vivo, Vuong *et al.* (61) found that a *S. epidermidis agr* mutant showed increased binding to epithelial cells and a higher colonization rate in a rabbit model of an indwelling medical-device-related infection. They also confirmed that deletion of *agr* or inhibition of Agr activity led to thicker biofilms *in vitro*. These results were consistent with those of a study conducted earlier by the same laboratory group, in which a *S. epidermidis agr* mutant showed increased primary attachment and biofilm formation, as well as expression of the cell-surface-associated autolysin AtlE (59). (Repetitive sequences in AtlE are thought to interact hydrophobically with abiotic surfaces.) As in *S. aureus*, production of PIA by the *S. epidermidis agr* mutant was similar to that of the wild type. As expected, the *agr* mutant lacked δ-hemolysin production. Addition of increasing concentrations of δ-toxin resulted in decreased attachment of *S. epidermidis* cells to polystyrene, where 10 mg ml^{-1} δ-toxin was sufficient to reduce biofilm formation of the *agr* mutant strain to the same levels found in the *agr* wild-type strain.

Interestingly, there may also be some role for *agr* expression in the resistance of staphylococcal biofilms to antibiotic exposure. Under conditions where an *agr* mutant formed a smaller biofilm than its wild-type parent did, the mutant was also more sensitive to rifampicin treatment, but not to oxacillin (67). The basis for this variation in sensitivity is unknown, although there is precedent for the regulation of other antibiotic resistance mechanisms by Agr. Regulation of NorA, a multidrug efflux pump involved in resistance to quinolones, by the DNA-binding protein NorR, was found to require an intact Agr system (53).

Thus, the role of Agr in biofilm development and behavior varies from species to species and from one environment to another, although some consistent themes are emerging. It will be critical in future investigations to identify those growth conditions that best mimic the environment *in vivo* in order to most effectively study Agr regulation of virulence factors in biofilms. Even more desirable are studies examining the contribution of Agr to biofilm formation in vivo, the first of which is described in the next section.

Agr VARIANTS AND THEIR ROLE IN STAPHYLOCOCCAL PATHOGENESIS

Agr variants (cells either overexpressing or underexpressing Agr compared with the parental strain) have been frequently isolated from cultures *in vitro*, suggesting that staphylococci maintain some capacity to alter their Agr phenotype or maintain Agr-negative subpopulations. Somerville *et al.* (50) found that repeated passage of *S. aureus* in vitro resulted in the loss of Agr function in a large percentage of the population, along with corresponding hemolytic and aconistase activity. The authors hypothesized that frequent mutations of *agr* create a mixed population of bacteria, with some cells expressing colonization factors while others tend to express secreted exotoxins. Under a particular environment with specific ecological and/or immunological selection, the Agr variant best able to adapt would emerge.

Agr mutants are frequently found among clinical isolates. Vuong *et al.* (61) showed that the percentage of strains with defective quorum-sensing systems was significantly higher among isolates from patients with infections of joint prostheses than among isolates from the skin of healthy controls (36% and 5%, respectively). This same group also found that 26% of 105 *S. aureus* isolates failed to produce δ-toxin, indicating that they were deficient in quorum-sensing-mediated regulation (62). When Goerke *et al.* (20) isolated staphylococci from the lungs of cystic fibrosis patients, not only did the strains generally express low levels of RNAIII, but several isolates were also found to be Agr-negative. Fowler *et al.* (15) showed that the percentage of *S. aureus* isolates recovered from patients with persistent bacteremia with defective δ-toxin production (a consistent indicator of Agr activity) was higher than in isolates from patients with resolving bacteremia (71% and 39%, respectively). The authors postulated that lack of *agr* expression might contribute to persistent bacteremia through the increased expression of the *S. aureus* surface adhesion gene, *fnbA*, in these mutants. The fibronectin-binding protein encoded by *fnbA* has been shown to enhance *S. aureus* adhesion to, invasion of, and persistence within endothelial cells. Intracellular invasion may contribute then to resistance to antibiotics, as vancomycin penetrates poorly into endothelial cells. Thus, lack of *agr* expression may facilitate a protected intracellular reservoir for *S. aureus*. Indeed, an *agr* mutant was incapable of escape from the endosome (49) or of inducing apoptosis (64), suggesting a prominent role for Agr in invasion of and persistence in host cells.

One area of particular concern in staphylococcal pathogenesis is the emergence of staphylococci with intermediate resistance to glycopeptide antibiotics (GISA). Interestingly, GISA are frequently isolated from biomedical-device-related infections, which are also likely to be biofilm-associated; these same GISA have been shown to be predominantly Agr-negative (45). The same study also suggested that loss of Agr function might in fact contribute to the development of vancomycin tolerance, an intriguing idea yet to be confirmed.

Schwan *et al.* (46) examined the behavior of mixed populations of hyperhemolytic, hemolytic, and non-hemolytic variants in a murine abscess model of infection. They found that the percentage of non-hemolytic variants, likely representing Agr-negative bacteria, recovered from the wound increased over time, whereas the number of hyperhemolytic variants (Agr overexpressors) decreased dramatically over the same time period. A wound infection model demonstrated the same trend, though to a lesser degree. In contrast, in a model of systemic infection, hemolytic variants seemed to be favored in isolates recovered from murine livers and spleens. Thus, Agr activity likely facilitates survival and pathogenesis in some host environments, but not in others. Additional studies will be of great importance in monitoring Agr phenotypes of clinical isolates from various infection types (preferably multiple isolates from each patient) and corresponding this to disease progression and outcome.

Although there is only scattered evidence for this idea, one can imagine that Agr-negative variants are better suited to biofilm formation and long-term, chronic infection as (i) they tend to express the surface adhesins that mediate cell-to-cell and cell-to-surface interactions, while downregulating factors that may facilitate detachment, such as δ-toxin, and (ii) they tend to express more immuno-evasive factors, such as protein A, than immuno-stimulatory ones (such as the superantigens). In addressing the first idea, our laboratory has found that Agr-negative variants become the predominant form in biofilms grown in a serum-based medium (66). It is not yet clear whether this is due to a selective pressure against Agr-positive cells, increased generation of Agr variants in the biofilm, active detachment of cells expressing Agr, or some combination of all three. Preliminary data also indicate that the Agr-positive population is not completely lost from the biofilm (J. M. Yarwood, unpublished data), suggesting that there may be some mechanism to retain the capability to express invasive factors at an appropriate stage of infection. Indeed, we have detected the frequent detachment of cells expressing *agr* from the biofilm (67). This may have important clinical implications, as detaching cells expressing *agr* are also likely to be expressing extracellular virulence factors important in causing acute infection.

One potential model of *S. aureus* Agr evolution in the context of a chronically infected host is presented in Figure 9.4. Upon establishment of infection, mutations (usually point mutations) accumulate in the *agr* loci of *S. aureus* cells. These mutations result in the conversion of a significant

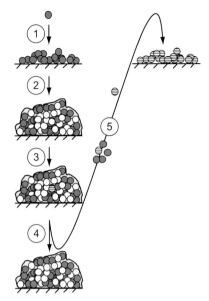

Figure 9.4. Model of Agr variant generation in *Staphylococcus* (see text for additional description). Colonization of a surface by an Agr-positive strain (gray cells) leads to microcolony development (step 1). Agr-negative variants (white cells) arise in the biofilm (step 2) through point mutation, genetic rearrangement, transposon insertion, etc. Occasionally, an Agr-negative variant acquires compensatory mutations to recognize a variant AIP and regains a functional Agr system (banded cells), albeit one distinct from the original strain (step 3). This new Agr specificity group strain detaches, along with other Agr-positive cells (step 4). Ecological pressures are then exerted over an extended time period (step 5). These pressures may select for certain virulence characteristics of the new strain, which is likely also divergent from the original strain in other virulence factors, and the strain emerges as a major Agr specificity group.

part of the population to a quorum-sensing negative phenotype. The Agr-negative phenotype confers some protection on the staphylococcal population as a whole, due to increased expression of immuno-evasive factors, and facilitates attachment and accumulation through increased expression of host protein-binding factors. This protected environment is conducive to the continued growth of staphylococci and additional accumulation of mutations in the *agr* locus. On very rare occasions, appropriate mutations are acquired by the *agr*-negative variant to return functionality of the *agr* locus, such as alteration of the AgrC receptor to recognize an AIP variant. In an extremely rare case, it is also possible that mutations in all appropriate areas of the *agr* locus are gained simultaneously to create a functionally unique Agr specificity group. Along with other Agr-positive cells, these new Agr specificity-group cells detach from the biofilm through expression of invasive virulence factors or production of δ-toxin and establish infection elsewhere in the host or, alternatively, colonize a secondary host. In some cases, appropriate ecological pressures are present to allow emergence of this new Agr specificity group from among the established groups. The combined rarity of these events – accumulation of several, eventually positive mutations, and selection for any emergent Agr specificity group – would only give rise to a major new *S. aureus* Agr group very infrequently. This would be consistent with the identification of only four distinct *S. aureus agr* groups thus far, despite frequent mutation of the *agr* locus. The driving force behind this cycle is, in part, the advantage conferred by maintaining a mixed population of cells, where Agr-negative variants prevent recognition by immune surveillance and cells expressing *agr* provide additional nutrient sources through host tissue degradation or facilitate escape from the localized infection at appropriate times. Thus, the emergence of distinct Agr groups may be a by-product of this mode of Agr evolution in which the generation of variants itself is important. However, it is noteworthy that this Agr-negative phenotype is often generated through non-reversible mutation of the *agr* locus, rather than by a reversible, conditional switching of *agr* expression on and off. This is consistent with some evolutionary advantage for generation of distinct Agr specificity groups.

INHIBITION OF STAPHYLOCOCCAL QUORUM SENSING AS A THERAPEUTIC TOOL

The use of inhibitors of quorum sensing has been proposed as one mechanism for controlling staphylococcal infections (25; see also Chapter 4). In a skin abscess model it was shown that co-administration of the synthetic Agr group II AIP together with the bacterial inoculation significantly attenuated an infection caused by an Agr group I strain (31). However, it has also been found that cross-inhibiting pheromones mimic *agr* mutations in both *S. aureus* and *S. epidermidis* and enhance biofilm formation (59, 61, 62). This result calls for extreme caution in the use of quorum-sensing inhibitors, as it is conceivable that such treatments, while mitigating the acute phase of infections, might facilitate chronic, biofilm-associated infections.

RIP/RAP

With some degree of controversy, a second quorum-sensing system has been described in *S. aureus* that is proposed to regulate Agr activity. This system (see (1, 7, 34) and other studies by N. Balaban and colleagues) consists of the autoinducer RNAIII-activating protein (RAP) and its target molecule TRAP. RAP is described as an ortholog of the ribosomal protein L2 and as being synthesized early in growth. Immunization against RAP was shown to mitigate pathology in a murine cutaneous *S. aureus* infection model. Reportedly, when RAP reaches a threshold concentration, it induces the histidine phosphorylation of the membrane protein TRAP. This event leads to the upregulation of *agr* transcription through an undescribed mechanism. Once AIP is made, it has been reported to lead to the downregulation of TRAP phosphorylation. A protein produced by *S. xylosus* and resembling, in the N-terminal sequence of RAP, RNAIII inhibiting peptide (RIP) has been reported to act as an agonist of TRAP, inhibiting its phosphorylation and, consequently, *agr* expression. Treatment with synthetic RIP inhibits several types of *S. aureus* and *S. epidermidis* infections, including those that are biofilm-related or caused by multiple drug-resistant staphylococci. Apparently, these therapeutic effects can be observed when RIP is applied locally or systemically.

The RAP/TRAP/RIP story thus far is somewhat unsatisfying. Little has been described regarding the properties of RIP, and even its amino-acid sequence and structure remain questionable. It is also conceivable that, when used at high concentrations, RIP has sufficient amphipathic, or surfactant, properties to prevent bacterial attachment, not unlike many other proteins. All of the experiments in vivo have administered RIP before or coincident with bacterial challenge and it is clear that, in these cases, RIP inhibits bacterial adherence to surfaces. However, there is no evidence that RIP has any effect against established biofilm, and few data are available as to the overall effect of RIP on bacterial physiology or virulence. It is also not clear whether a RIP-impregnated intravascular catheter would continue to inhibit staphylococcal adhesion against the more or less continual "challenge" that likely occurs in vivo, from either the epithelium or transient bacteremias. Implanted devices are soon coated with host matrix proteins, a fact that limits the efficacy of many device surface treatments as the now biotic surface provides several protein-receptor-specific targets for staphylococci to bind to. Furthermore, very little is known as to the regulatory targets of this system. Besides being reported to inhibit Agr activation, no other gene targets have been conclusively identified. This is particularly unsatisfying, as inhibition of Agr activity would be expected to increase bacterial adhesion, yet, in fact, the RIP-treated cells are less likely to adhere; this result suggests that RIP may in fact simply be acting as a surfactant-like molecule. Finally, other laboratory groups have been unable to detect the RNAIII-activating activity in supernatants of *agr*-null strains, despite the presumed presence of RAP, contributing to the controversy as to the true nature of this molecule (34, 36).

CONCLUSION

As more studies regarding quorum sensing emerge, it is becoming clear that the role and expression of Agr is intimately tied to environmental conditions, infection type, and the input of other virulence regulators. As with many biological processes, there are few rules regarding quorum sensing to which there are not important exceptions. However, the data also suggest that, under certain conditions, quorum sensing may have an important part to play in everything from influencing the chronic nature of disease to staphylococcal metabolism. Thus, additional studies are sorely needed to flesh out the role of Agr in staphylococcal virulence, particularly through animal models of infection and epidemiological monitoring of Agr variants. Even given the caveats associated with the use of Agr inhibitors in staphylococcal infections, interference with quorum signaling may yet prove to be beneficial in treatment of certain staphylococcal infections, particularly those in which inhibition of toxin production would be of immediate benefit.

ACKNOWLEDGEMENTS

J. M. Y. was supported by a Ruth L. Kirschstein National Research Service Award from the National Institute of General Medical Sciences and a gift from the Procter & Gamble Company. E. Peter Greenberg is gratefully acknowledged for helpful discussions.

REFERENCES

1 Balaban, N., T. Goldkorn, R. T. Nhan *et al.* 1998. Autoinducer of virulence as a target for vaccine and therapy against *Staphylococcus aureus*. *Science* **280**: 438–40.
2 Benito, Y., F. A. Kolb, P. Romby *et al.* 2000. Probing the structure of RNAIII, the *Staphylococcus aureus agr* regulatory RNA, and identification of the RNA domain involved in repression of protein A expression. *RNA* **6**: 668–79.
3 Bischoff, M., P. Dunman, J. Kormanec, *et al.* 2004. Microarray-based analysis of the *Staphylococcus aureus* σ^B regulon. *J. Bacteriol.* **186**: 4085–99.
4 Bischoff, M., J. M. Entenza and P. Giachino 2001. Influence of a functional *sigB* operon on the global regulators *sar* and *agr* in *Staphylococcus aureus*. *J. Bacteriol.* **183**: 5171–9.

5 Cheung, A. L. and G. Zhang 2002. Global regulation of virulence determinants in *Staphylococcus aureus* by the SarA protein family. *Front. Biosci.* **7**: 1825–42.

6 Chien, Y., A. C. Manna, S. J. Projan and A. L. Cheung 1999. SarA, a global regulator of virulence determinants in *Staphylococcus aureus*, binds to a conserved motif essential for *sar*-dependent gene regulation. *J. Biol. Chem.* **274**: 37169–76.

7 Dell'Acqua, G., A. Giacometti, O. Cirioni *et al.* 2004. Suppression of drug-resistant staphylococcal infections by the quorum-sensing inhibitor RNAIII-inhibiting peptide. *J. Infect. Dis.* **190**: 318–20.

8 Donlan, R. M. and J. W. Costerton 2002. Biofilms: survival mechanisms of clinically relevant microorganisms. *Clin. Microbiol. Rev.* **15**: 167–93.

9 Donvito, B., J. Etienne, L. Denoroy *et al.* 1997. Synergistic hemolytic activity of *Staphylococcus lugdunensis* is mediated by three peptides encoded by a non-*agr* genetic locus. *Infect. Immun.* **65**: 95–100.

10 Dufour, P., S. Jarraud, F. Vandenesch *et al.* 2002. High genetic variability of the *agr* locus in *Staphylococcus* species. *J. Bacteriol.* **184**: 1180–6.

11 Dunman, P. M., E. Murphy, S. Haney *et al.* 2001. Transcription profiling-based identification of *Staphylococcus aureus* genes regulated by the *agr* and/or *sarA* loci. *J. Bacteriol.* **183**: 7341–53.

12 Fischetti, V. A., R. P. Novick, J. J. Ferretti, D. A. Portnoy and E. J. I. Rood 2000. *Gram-Positive Pathogens.* Washington, DC: ASM Press.

13 Fournier, B., R. Aras and D. C. Hooper 2000. Expression of the multidrug resistance transporter NorA from *Staphylococcus aureus* is modified by a two-component regulatory system. *J. Bacteriol.* **182**: 664–71.

14 Fournier, B., A. Klier and G. Rapoport 2001. The two-component system ArlS-ArlR is a regulator of virulence gene expression in *Staphylococcus aureus*. *Mol. Microbiol.* **41**: 247–61.

15 Fowler, V. G., Jr., G. Sakoulas, L. M. McIntyre *et al.* 2004. Persistent bacteremia due to methicillin-resistant *Staphylococcus aureus* infection is associated with *agr* dysfunction and low-level in vitro resistance to thrombin-induced platelet microbiocidal protein. *J. Infect. Dis.* **190**: 1140–9.

16 Frees, D., S. N. Qazi, P. J. Hill and H. Ingmer 2003. Alternative roles of ClpX and ClpP in *Staphylococcus aureus* stress tolerance and virulence. *Mol. Microbiol.* **48**: 1565–78.

17 Gilot, P. and W. van Leeuwen 2004. Comparative analysis of *agr* locus diversification and overall genetic variability among bovine and human *Staphylococcus aureus* isolates. *J. Clin. Microbiol.* **42**: 1265–9.

18 Giraudo, A. T., A. Calzolari, A. A. Cataldi, C. Bogni and R. Nagel 1999. The *sae* locus of *Staphylococcus aureus* encodes a two-component regulatory system. *FEMS Microbiol. Lett.* **177**: 15–22.

19 Giraudo, A. T., A. L. Cheung and R. Nagel 1997. The *sae* locus of *Staphylococcus aureus* controls exoprotein synthesis at the transcriptional level. *Arch. Microbiol.* **168**: 53–8.

20 Goerke, C., S. Campana, M. G. Bayer *et al.* 2000. Direct quantitative transcript analysis of the *agr* regulon of *Staphylococcus aureus* during human infection in comparison to the expression profile *in vitro*. *Infect. Immun.* **68**: 1304–11.

21 Goerke, C., U. Fluckiger, A. Steinhuber, W. Zimmerli and C. Wolz 2001. Impact of the regulatory loci *agr*, *sarA* and *sae* of *Staphylococcus aureus* on the induction of alpha-toxin during device-related infection resolved by direct quantitative transcript analysis. *Mol. Microbiol.* **40**: 1439–47.

22 Goerke, C., M. Kummel, K. Dietz and C. Wolz 2003. Evaluation of intraspecies interference due to *agr* polymorphism in *Staphylococcus aureus* during infection and colonization. *J. Infect. Dis.* **188**: 250–6.

23 Horsburgh, M. J., J. L. Aish, I. J. White *et al.* 2002. σ^B modulates virulence determinant expression and stress resistance: characterization of a functional *rsbU* strain derived from *Staphylococcus aureus* 8325–4. *J. Bacteriol.* **184**: 5457–67.

24 Jarraud, S., C. Mougel, J. Thioulouse *et al.* 2002. Relationships between *Staphylococcus aureus* genetic background, virulence factors, *agr* groups (alleles), and human disease. *Infect. Immun.* **70**: 631–41.

25 Ji, G., R. Beavis and R. P. Novick 1997. Bacterial interference caused by auto-inducing peptide variants. *Science* **276**: 2027–30.

26 Kahl, B. C., K. Becker, A. W. Friedrich *et al.* 2003. *agr*-dependent bacterial interference has no impact on long-term colonization of *Staphylococcus aureus* during persistent airway infection of cystic fibrosis patients. *J. Clin. Microbiol.* **41**: 5199–201.

27 Kies, S., C. Vuong, M. Hille *et al.* 2003. Control of antimicrobial peptide synthesis by the *agr* quorum sensing system in *Staphylococcus epidermidis*: activity of the antibiotic epidermin is regulated at the level of precursor peptide processing. *Peptides* **24**: 329–38.

28 Li, M., M. Guan, X. F. Jiang *et al.* 2004. Genetic polymorphism of the accessory gene regulator (*agr*) locus in *Staphylococcus epidermidis* and its association with pathogenicity. *J. Med. Microbiol.* **53**: 545–9.

29 Li, S., S. Arvidson and R. Mollby 1997. Variation in the *agr*-dependent expression of alpha-toxin and protein A among clinical isolates of *Staphylococcus aureus* from patients with septicaemia. *FEMS Microbiol. Lett.* **152**: 155–61.

30 Lina, G., F. Boutite, A. Tristan *et al.* 2003. Bacterial competition for human nasal cavity colonization: role of staphylococcal *agr* alleles. *Appl. Environ. Microbiol.* **69**: 18–23.

31 Mayville, P., G. Ji, R. Beavis *et al.* 1999. Structure-activity analysis of synthetic autoinducing thiolactone peptides from *Staphylococcus aureus* responsible for virulence. *Proc. Natn. Acad. Sci. USA* **96**: 1218–23.

32 McNamara, P. J., K. C. Milligan-Monroe, S. Khalili and R. A. Proctor 2000. Identification, cloning, and initial characterization of *rot*, a locus encoding a regulator of virulence factor expression in *Staphylococcus aureus*. *J. Bacteriol.* **182**: 3197–203.

33   Moise-Broder, P. A., G. Sakoulas, G. M. Eliopoulos *et al.* 2004. Accessory gene regulator group II polymorphism in methicillin-resistant *Staphylococcus aureus* is predictive of failure of vancomycin therapy. *Clin. Infect. Dis.* **38**: 1700–5.

34   Novick, R. P. 2003. Autoinduction and signal transduction in the regulation of staphylococcal virulence. *Mol. Microbiol.* **48**: 1429–49.

35   Novick, R. P. and D. Jiang 2003. The staphylococcal *saeRS* system coordinates environmental signals with *agr* quorum sensing. *Microbiology* **149**: 2709–17.

36   Novick, R. P., H. F. Ross, A. M. N. S. Figueiredo *et al.* 2000. Activation and inhibition of the staphylococcal AGR system. *Science* **287**: 391.

37   Novick, R. P., H. F. Ross, S. J. Projan *et al.* 1993. Synthesis of staphylococcal virulence factors is controlled by a regulatory RNA molecule. *EMBO J.* **12**: 3967–75.

38   Otto, M. 2001. *Staphylococcus aureus* and *Staphylococcus epidermidis* peptide pheromones produced by the accessory gene regulator system. *Peptides* **22**: 1603–8.

39   Parsek, M. R. and P. K. Singh 2003. Bacterial biofilms: an emerging link to disease pathogenesis. *Annu. Rev. Microbiol.* **57**: 677–701.

40   Pratten, J., S. J. Foster, P. F. Chan, M. Wilson and S. P. Nair 2001. *Staphylococcus aureus* accessory regulators: expression within biofilms and effect on adhesion. *Microbes Infect.* **3**: 633–7.

41   Projan, S. J. and R. P. Novick 1997. The molecular basis of pathogenicity. In K. B. Crossley and G. L. Archer (eds.), *The Staphylococci in Human Disease*, pp. 55–81. New York: Churchill Livingstone.

42   Rothfork, J. M., S. Dessus-Babus, W. J. Van Wamel, A. L. Cheung and H. D. Gresham 2003. Fibrinogen depletion attenuates *Staphyloccocus aureus* infection by preventing density-dependent virulence gene up-regulation. *J. Immunol.* **171**: 5389–95.

43   Said-Salim, B., P. M. Dunman, F. M. McAleese *et al.* 2003. Global regulation of *Staphylococcus aureus* genes by Rot. *J. Bacteriol.* **185**: 610–19.

44   Sakinc, T., P. Kulczak, K. Henne and S. G. Gatermann 2004. Cloning of an *agr* homologue of *Staphylococcus saprophyticus*. *FEMS Microbiol. Lett.* **237**: 157–61.

45   Sakoulas, G., G. M. Eliopoulos, R. C. Moellering, Jr. *et al.* 2002. Accessory gene regulator (*agr*) locus in geographically diverse *Staphylococcus aureus* isolates with reduced susceptibility to vancomycin. *Antimicrob. Agents Chemother.* **46**: 1492–502.

46   Schwan, W. R., M. H. Langhorne, H. D. Ritchie and C. K. Stover 2003. Loss of hemolysin expression in *Staphylococcus aureus agr* mutants correlates with selective survival during mixed infections in murine abscesses and wounds. *FEMS Immunol. Med. Microbiol.* **38**: 23–8.

47   Shenkman, B., E. Rubinstein, A. L. Cheung *et al.* 2001. Adherence properties of *Staphylococcus aureus* under static and flow conditions: roles of *agr* and *sar* loci, platelets, and plasma ligands. *Infect. Immun.* **69**: 4473–8.

STAPHYLOCOCCAL VIRULENCE AND BIOFILMS

48 Shenkman, B., D. Varon, I. Tamarin *et al.* 2002. Role of *agr* (RNAIII) in *Staphylococcus aureus* adherence to fibrinogen, fibronectin, platelets and endothelial cells under static and flow conditions. *J. Med. Microbiol.* **51**: 747–54.

49 Shompole, S., K. T. Henon, L. E. Liou *et al.* 2003. Biphasic intracellular expression of *Staphylococcus aureus* virulence factors and evidence for Agr-mediated diffusion sensing. *Mol. Microbiol.* **49**: 919–27.

50 Somerville, G. A., S. B. Beres, J. R. Fitzgerald *et al.* 2002. In vitro serial passage of *Staphylococcus aureus*: changes in physiology, virulence factor production, and *agr* nucleotide sequence. *J. Bacteriol.* **184**: 1430–7.

51 Tegmark, K., E. Morfeldt and S. Arvidson 1998. Regulation of *agr*-dependent virulence genes in *Staphylococcus aureus* by RNAIII from coagulase-negative staphylococci. *J. Bacteriol.* **180**: 3181–6.

52 Throup, J. P., F. Zappacosta, R. D. Lunsford *et al.* 2001. The *srhSR* gene pair from *Staphylococcus aureus*: genomic and proteomic approaches to the identification and characterization of gene function. *Biochemistry* **40**: 10392–401.

53 Truong-Bolduc, Q. C., X. Zhang and D. C. Hooper 2003. Characterization of NorR protein, a multifunctional regulator of *norA* expression in *Staphylococcus aureus*. *J. Bacteriol.* **185**: 3127–38.

54 Tseng, C. W., S. Zhang and G. C. Stewart 2004. Accessory gene regulator control of staphylococcal enterotoxin d gene expression. *J. Bacteriol.* **186**: 1793–801.

55 Van Wamel, W. J., G. van Rossum, J. Verhoef, C. M. Vandenbroucke-Grauls and A. C. Fluit 1998. Cloning and characterization of an accessory gene regulator (*agr*)-like locus from *Staphylococcus epidermidis*. *FEMS Microbiol. Lett.* **163**: 1–9.

56 Vandenesch, F., J. Kornblum and R. P. Novick 1991. A temporal signal, independent of *agr*, is required for *hla* but not *spa* transcription in *Staphylococcus aureus*. *J. Bacteriol.* **173**: 6313–20.

57 Vandenesch, F., S. J. Projan, B. Kreiswirth, J. Etienne and R. P. Novick 1993. Agr-related sequences in *Staphylococcus lugdunensis*. *FEMS Microbiol. Lett.* **111**: 115–22.

58 Vuong, C., M. Durr, A. B. Carmody *et al.* 2004. Regulated expression of pathogen-associated molecular pattern molecules in *Staphylococcus epidermidis*: quorum-sensing determines pro-inflammatory capacity and production of phenol-soluble modulins. *Cell. Microbiol.* **6**: 753–9.

59 Vuong, C., C. Gerke, G. A. Somerville, E. R. Fischer and M. Otto 2003. Quorum-sensing control of biofilm factors in *Staphylococcus epidermidis*. *J. Infect. Dis.* **188**: 706–18.

60 Vuong, C., F. Gotz and M. Otto 2000. Construction and characterization of an *agr* deletion mutant of *Staphylococcus epidermidis*. *Infect. Immun.* **68**: 1048–53.

61 Vuong, C., S. Kocianova, Y. Yao, A. B. Carmody and M. Otto 2004. Increased colonization of indwelling medical devices by quorum-sensing mutants of *Staphylococcus epidermidis* in vivo. *J. Infect. Dis.* **190**: 1498–1505.

62 Vuong, C., H. L. Saenz, F. Gotz and M. Otto 2000. Impact of the *agr* quorum-sensing system on adherence to polystyrene in *Staphylococcus aureus*. *J. Infect. Dis.* **182**: 1688–93.

63 Wamel, W. V., Y.-Q. Xiong, A. Bayer *et al.* 2002. Regulation of *Staphylococcus aureus* type 5 capsular polysaccharides by *agr* and *sarA* in vitro and in an experimental endocarditis model. *Microb. Pathogen.* **33**: 73–9.

64 Wesson, C. A., L. E. Liou, K. M. Todd *et al.* 1998. *Staphylococcus aureus* Agr and Sar global regulators influence internalization and induction of apoptosis. *Infect. Immun.* **66**: 5238–43.

65 Xiong, Y. Q., W. Van Wamel, C. C. Nast *et al.* 2002. Activation and transcriptional interaction between *agr* RNAII and RNAIII in *Staphylococcus aureus* in vitro and in an experimental endocarditis model. *J. Infect. Dis.* **186**: 668–77.

66 Yarwood, J. M. 2004. Quorum sensing in *Staphylococcus aureus* biofilms. Presented at the 11th International Symposium on Staphylococci and Staphylococcal Infections, Charleston, SC, October 2004.

67 Yarwood, J. M., D. J. Bartels, E. M. Volper and E. P. Greenberg 2004. Quorum sensing in *Staphylococcus aureus* biofilms. *J. Bacteriol.* **186**: 1838–50.

68 Yarwood, J. M., J. K. McCormick, M. L. Paustian, V. Kapur and P. M. Schlievert 2002. Repression of the *Staphylococcus aureus* accessory gene regulator in serum and in vivo. *J. Bacteriol.* **184**: 1095–101.

69 Yarwood, J. M., J. K. McCormick and P. M. Schlievert 2001. Identification of a novel two-component regulatory system that acts in global regulation of virulence factors of *Staphylococcus aureus*. *J. Bacteriol.* **183**: 1113–23.

CHAPTER 10

Cell-density-dependent regulation of streptococcal competence

M. Dilani Senadheera, Celine Levesque and Dennis G. Cvitkovitch

Dental Research Institute, University of Toronto,
Toronto, Canada

INTRODUCTION

A brief history

In the 1920s Frederick Griffith, a medical officer at the Ministry of Health in Britain, made a significant discovery regarding *Streptococcus pneumoniae*, a bacterium that caused a pneumonia epidemic in London. While examining the strain variability within different groups of pneumococci, Griffith noted that an avirulent strain of the bacterium could revert to the virulent type or remain unchanged following subculture (37). Because this phenomenon enabled the bacterium to acquire a novel heritable phenotype, Griffith coined the term "transformation principle" to describe the phenotypic changes he observed.

In his classic experiment, Griffith studied a highly infective, encapsulated S strain that formed smooth colonies, and an avirulent R strain, which had no capsule and formed rough colonies when grown on blood agar (37). When healthy mice were injected with the S strain, they died of septicemia, whereas separate admission of the R strain or the heat-killed S strain appeared to be harmless. However, when the live R strain and the heat-killed S strain were injected simultaneously, the mice died. Surprisingly, when blood samples drawn from these dead animals were analyzed, both R and S live strains were detected. Based on these results, Griffith concluded that a "transforming factor", present in the heat-killed S strain, was able to "transform" an avirulent R strain into a capsulated, virulent S strain.

Over the next few decades, Griffith's inspiring work on transformation was followed up by a number of scientists. In 1944, by isolating various

Bacterial Cell-to-Cell Communication: Role in Virulence and Pathogenesis, ed. D. R. Demuth and R. J. Lamont. Published by Cambridge University Press. © Cambridge University Press 2005.

components of a dead encapsulated virulent pneumococcal strain and testing each component for its ability to transform, Oswald T. Avery, Colin M. MacLeod, and Maclyn J. McCarty concluded that a nucleic acid of the deoxyribonucleic type was the factor involved in transformation (6). However, it was not until 1952 that Alfred Hershey and Martha Chase conclusively demonstrated that DNA and not proteins are the transforming factor by using bacteriophage infecting *Escherichia coli* cells (46). They observed that, when bacteriophage containing ^{32}P-labeled DNA and ^{35}S-labeled proteins were used to infect *E. coli*, only ^{32}P-DNA entered the cell, whereas the ^{35}S-proteins stayed outside. Hence, they concluded that DNA carried the hereditary information that coded for the replication of the bacteriophage.

These pioneering experiments were aimed at elucidating the mechanism of genetic exchange, responsible for horizontal gene transfer. We now know that bacteria can acquire foreign DNA through three distinct mechanisms: transduction, which requires an intermediate bacteriophage; conjugation, which involves cell-to-cell contact via an organelle called a pilus; and by transformation, which is simply the uptake and integration of free foreign DNA into the cell's chromosome. Understanding these mechanisms of genetic exchange, responsible for horizontal gene transfer, is indispensable to understanding the genetic plasticity of bacteria and to re-evaluating some of our therapeutic strategies against bacterial infections.

In this chapter we examine how natural competence for genetic transformation in streptococci is controlled by a cell-density-dependent quorum-sensing system. Genetic competence is a transient physiological state that enables a bacterium to recognize, process, and integrate foreign DNA into its genome, leading to transformation (16, 23, 39, 69). We will begin by focusing on *S. pneumoniae*, whose transformation process has been studied for decades, and which has one of the best-characterized systems among Gram-positive cocci. Consequently, we will also discuss the control of gene transfer in oral streptococci, whose transformation mechanisms have also been extensively studied. Interestingly, the oral cavity harbours over 400 different species, many of which can act as DNA donors and provide access to a "communal gene pool" (26). As organisms colonizing the primary portals of entry to the human body, the ability of oral microbes to uptake DNA and transform is potentially of great significance to human health. *S. pneumoniae*, whose ecological niche is the human nasopharynx, has been previously shown to acquire B-lactam (e.g. penicillin) resistance from oral streptococci such as *Streptococcus mitis* and *S. oralis* via transformation

(21, 22). With this in mind, we will discuss how genetic competence is regulated in response to cell population density in streptococci.

Cell-density-dependent competence development

Bacterial transfer of genes via natural transformation depends on a number of factors. The recipient cells have to be in a metabolically active, competent state; it is also essential that free DNA (chromosomal or plasmid) be available for active uptake (16, 23, 39, 69). Because a cell utilizes a considerable amount of its resources and energy for the processing, uptake and incorporation of such DNA, it makes intuitive sense that competence is synchronized with cell population density, which would be an indication of the free homologous DNA that is available for uptake.

In the 1960s Pakula and Walczak (89) and Tomasz and Hotchkiss (116) discovered that in streptococcal cultures the induction of genetic competence took place only at a certain cell population density. In their report, Tomasz and Hotchkiss describe two phases of competence: an initial lag phase yielding a low number of competent cells, and a second phase displaying the fast onset and explosive spread of the competent state throughout the majority of cells in the culture. To test whether this latter phase was activated by an external cell product, they conditioned a non-competent cell culture with cell-free, filtered supernatant that was derived from a competent culture. Interestingly, an immediate induction of competence was observed in the treated, initially non-competent, culture. Further testing revealed that a protein-like macromolecular compound present in the supernatant of competent cultures acted as an "activator" to induce competence in otherwise non-competent cells (89, 114–116).

Following these pioneering experiments, the competence-activating peptide molecule was successfully purified from *S. pneumoniae* strain CP1200 (40, 41). Sequence analysis of this secreted molecule revealed a 17-residue cationic peptide that retained its full biological activity when chemically synthesized (40). Because this synthetic heptadecapeptide could stimulate competence in streptococcal cultures, it was designated CSP, for competence stimulating peptide (reviewed below).

The ability to control cell physiology in concert with cell population density, a widespread phenomenon termed quorum sensing, is observed in both Gram-negative and Gram-positive organisms (8, 55, 112, 122). It is well established that, by density-dependent cell-to-cell communication, bacteria are able to respond to changes in the environment by altering gene expression. Such coordinated expression of genes facilitates group behavior among cells that is typically observed in multicellular organisms. In addition to inducing competence, quorum sensing is known to control a variety of physiological activities in bacteria, including antibiotic production, sporulation, biofilm differentiation, conjugation, and bioluminescence (18, 30, 40, 64, 76, 90, 107, 111). In Gram-positive bacteria, quorum-sensing systems often consist of three components: a signal peptide messenger molecule and a two-component signal transduction system (TCSTS) comprising a histidine kinase (HK) and a response regulator (RR) (8, 112). As bacteria multiply, corresponding secreted peptide molecules accumulate proportionately in the surrounding environment. These molecules are then detected by a membrane-associated HK protein once a minimal stimulatory concentration is reached (110). As a result, the sensor kinase is autophosphorylated on a conserved histidine residue, and subsequently the phosphate group is transferred to a cognate response regulator (RR) protein. Phosphorylation of the RR results in its activation, thereby controlling the transcription of various gene(s) by altering the binding affinity of RNA polymerase to one or more of the promoter regions belonging to the target genes.

Among streptococcal species, *S. pneumoniae* has one of the best-characterized density-dependent signal transduction systems. In this bacterium, two independent quorum-sensing mechanisms control competence and bacteriocin production. In addition, the quorum-sensing systems of *S. mutans* and *S. gordonii* have also been well studied. In the proceeding sections, we will discuss the competence-inducing quorum-sensing system of *S. pneumoniae* and how it triggers the expression of genes involved in DNA uptake.

GENETIC COMPETENCE REGULATION IN *S. PNEUMONIAE*

The ComCDE competence regulon in *S. pneumoniae*

In *S. pneumoniae*, competence for genetic transformation is regulated by a quorum-sensing system encoded by two genetic loci, *comCDE* and *comAB* (13, 45). The *comC*, *comD*, and *comE* genes encode the CSP peptide-precursor, the histidine kinase (HK), and the response regulator (RR), respectively (12, 40, 42, 90). The secretion apparatus necessary for CSP maturation and export is encoded by the *comA* and *comB* genes. When the CSP reaches a critical extracellular concentration it is detected by the ComD receptor, which undergoes autophosphorylation at a conserved histidine

```
S. gordonii (Challis)   MKKKNKQNLLPKELQQFEILTERKLEQVTGGDVRSNKIRLWWENIFFNKK--
S. gordonii (7865)      MKKKNKQNLLPKELQQFEILTERKLEQVTGGDIRHRINNSIWRDIFLKRK--
S. pneumoniae (R6)      MKNTVK-------LEQFVALKEKDLQKIKGGEMR--LSKFFRDFILQRKK--
S. pneumoniae (TIGR4)   MKNTVK-------LEQFVALKEKDLQKIKGGEMR--ISRIILDFLFLRKK--
S. mutans (UA159)       MKKTLS------LKNDFKEIKTDELEIIIGGSGSLSTFFRLFNRSFTQALGK
                        MK      QNLLPK    F   .   L   .  GG                GK
```

Figure 10.1. Multiple sequence alignment of the deduced amino acid sequences of the CSP precursors from different strains belonging to *Streptococcus gordonii*, *S. pneumoniae* and *S. mutans*. A vertical box indicates the processing site that is followed by the Gly–Gly motif in each sequence.

residue. The phosphorylated ComD then transfers its phosphate group to ComE, which is activated and, in turn, controls the transcription of its own production, as well as that of the so-called late genes that are essential for DNA processing, uptake, and recombination. The induction of this latter pathway, which induces competence-related genes, is dependent upon the activation of an alternative sigma factor designated ComX (60). In streptococcal cultures, the accumulation of CSP and the development of competence is not passive, but a tightly regulated property. In the following sections, the ComABCDE regulon and the induction of the early and late competence genes will be discussed.

The competence-stimulating peptide (CSP)

The discovery of CSP was significant, not only as a valuable research ingredient in pneumococcal genetics, but also as an important tool that has been utilized to enhance our understanding of the quorum-sensing mechanisms of Gram-positive bacteria in general. The gene, *comC*, that encodes CSP was identified by searching the *S. pneumoniae* genome for the reverse-translated amino acid sequence that was derived from CSP (40, 41). Sequence analysis of *comC* revealed that its deduced primary translational product (the precursor peptide of CSP) was a 41 amino-acid peptide, of which 17 amino acids of the C-terminus formed the biologically active CSP that was secreted. Activation of CSP was found to be dependent on cleavage of the pro-peptide at a Gly–Gly residue present in the −1 and −2 location (relative to the processing site) in its 24 amino-acid N-terminal leader sequence during post-translational modification (Figure 10.1) (41, 90). Further analysis of this N-terminal sequence revealed that it belonged to the double-glycine-type leader peptide family and was unrelated to the N-terminal signal sequences that direct proteins across the cytoplasmic membrane via the sec-dependent pathway (44). Analysis of leader peptides revealed the following consensus: hydrophobic residue (-15), Leu (-12), Ser (-1), Glu (-8), Leu (-7), Ile (-4), Gly (-2), Gly (-1). The residues in -4, -7, -12 and -15 positions almost always occupied hydrophobic residues, but only the Gly at -2 was universally conserved among these peptides.

Gly–Gly leader peptides were first discovered in the precursors of lactococcin A, a bacteriocin secreted by *Lactobacillus lactis* (48). Bacteriocins are ribosomally synthesized, proteinaceous compounds that have antimicrobial activity against closely related bacteria and are found in both Gram-negative and Gram-positive bacteria (102). In addition to bacteriocins, these double-glycine-type leaders were later discovered in pheromone peptide precursors (such as in the CSP precursor) of Gram-positive bacteria that are involved in cell-to-cell communication (40). Very often, their precursor peptides are encoded by genes that are cotranscribed with one or more neighboring genes whose products act as ATP-binding cassette (ABC) transporters and their accessory proteins. It has been shown that disruption of the structural genes that code for ABC transporters will prevent the secretion of their substrate peptides (49, 86), thereby providing evidence that the export of these activators requires a dedicated ABC transporter (41) (Figure 10.2).

The *comAB* locus in *S. pneumoniae* encodes the CSP secretion apparatus, which includes an ABC-transporter (ComA) and its accessory protein (ComB) (49, 50). The indispensable role of *comA* in the development of genetic competence in *S. pneumoniae* was discovered by virtue of an insertion–duplication mutant with a disrupted *comA*, which was non-transformable and unable to produce the competence factor (86). The mutant cells were, however, capable of becoming competent when supplemented with exogenous competence factor. It was therefore hypothesized that the defect in competence resulted from interruption of the synthesis, release, or activity of the competence factor. It was later shown that ComAB was the secretion apparatus of CSP (41). In ComA, the N-terminal domain has two conserved sequence motifs associated with proteolytic activity that are responsible for cleaving the leader peptide from the precursor CSP at the Gly–Gly residues (13). Interestingly, ComA falls into a unique family of ABC transporters that carry out proteolytic processing of their substrates concomitant with export, and are dedicated to exporting bacteriocins (35, 81). As in the case of CSP, bacteriocins exported by these ABC-transporters comprise a double-glycine leader in their N-terminal leader sequence. Because CSP has no apparent bacteriocin-like activity, it is probable that these transporters may serve other functions as well.

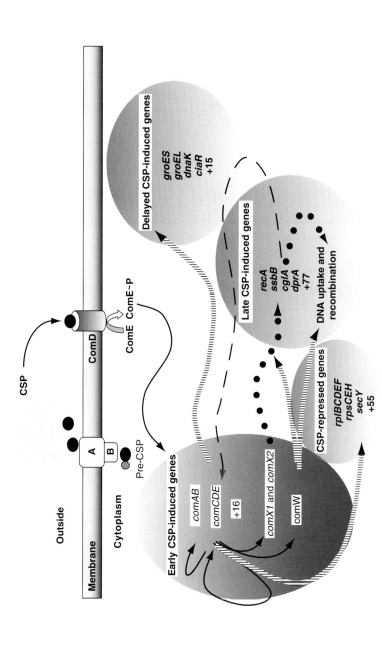

Figure 10.2. The dynamic network of genes regulated by the CSP-dependent quorum sensing mechanism that regulates competence in *Streptococcus pneumoniae*. This figure is modified from that published by Peterson *et al.* (95) based on results from their microarray analysis of CSP-modulated genes. Black solid arrows: regulation of the autocatalytic quorum-sensing circuit. Black dotted arrows: known links to late genes and DNA processing proteins. Black striped arrows: hypothetical links of early gene products to late, repressed, or delayed genes. Black dashed arrow: hypothetical retro-regulation (95).

The ComD, CSP receptor

During quorum sensing, when the cell population density reaches a critical threshold, CSP is detected by a membrane-associated HK protein. The CSP receptor, ComD, was first identified by using two *Streptococcus gordonii* strains, Challis and NCTC 7865 (42). It had been observed that the streptococcal competence pheromones exhibit strain specificity such that a signaling peptide from one strain may or may not be able to induce competence in another non-competent strain. In *S. gordonii*, the Challis and NCTC 7865 strains represent two distinct pherotypes or "transformation groups" that only respond to their own CSPs. For instance, the competence factor present in the supernatant of a Challis culture is unable to induce competence in NCTC 7865 and *vice versa*. On the other hand, Challis is able to induce competence in a non-competent *S. gordonii* strain (Wicky) that belongs to the same pherotype. To identify the competence peptide receptor, this strain-specificity of CSP was exploited. Specifically, the *comD* genes belonging to Challis and NCTC 7865 were expressed in strain Wicky by using a Campbell-type mutagenesis strategy (42). Subsequently, the ability of the recombinant strain to induce competence when supplemented with CSP derived from Challis and NCTC 7865 strains was assayed. In contrast to the parent Challis strain that responded only to its own CSP, the recombinant Wicky strain became competent in the presence of CSP derived from NCTC 7865, thereby proving that ComD is the receptor for CSP.

The ComD receptor consists of an N-terminal membrane-spanning domain and a C-terminal cytoplasmic kinase domain, which contains the histidine residue that is phosphorylated (100, 110). By conducting multiple sequence alignment of 348 kinase domains, more than 90% of the proteins belonging to the histidine kinase superfamily were divided into distinct subfamilies (36). As a result, most of the peptide pheromone-activated sensor kinases, including ComD in *S. pneumoniae*, were incorporated in an HPK10 subfamily. Members of this HPK10 subfamily belong to a group of proteins called the orthodox kinases, which comprise a membrane-spanning N-terminal domain and a C-terminal cytoplasmic kinase domain (36, 110). Interestingly, this group differs from other surface-located histidine kinases in their regions of homology and in their membrane-spanning domains. These "homology boxes" are characteristic highly conserved (fingerprint) amino-acid sequences that are believed to play crucial roles in substrate binding, catalysis, and/or structure (e.g. the H-, N-, D-, F-, and G-boxes). Members of the HPK10 subfamily have no proline residue in the

H-box that acts as the site of histidine phosphorylation, and contain a characteristic tyrosine that is located two residues downstream from the conserved histidine residue. In addition, HPK10 HKs lack a D-box, which is part of the nucleotide-binding domain, and their N-box contains only one asparagine residue. In addition, in contrast to most prokaryotic histidine kinases, which consist of two such transmembrane segments, orthodox kinases usually possess six or seven transmembrane segments in the N-terminal membrane-associated domain (20, 42, 67). Sequence analysis of ComD of *Streptococcus pneumoniae* Rx and *S. pneumoniae* A66 revealed a high level of sequence similarity in the receptor domain of these kinases (90). Despite this similarity, the corresponding peptide pheromones were not able to cross-activate these kinases (96), thereby highlighting the importance of this region in the ComD receptor for CSP specificity. With reference to peptide molecules that activate the HPK10 subfamily of kinases, almost all peptide pheromones that have been characterized are unmodified peptides (54). As with CSP in *S. pneumoniae*, these signaling molecules are synthesized as precursor peptides containing a double glycine leader, which serves as a secretion signal that is cleaved concomitant with export.

The ComE response regulator

The ComE comprises the RR belonging to the ComDE TCSTS that is activated when CSP reaches a critical concentration. Previously, it was shown that the disruption of *comE* abolished two processes: response to synthetic CSP and endogenous competence induction (90). Like other response regulators, ComE consists of an N-terminal regulatory domain and a C-terminal DNA-binding effector domain (110). Following the phosphorylation of ComD, the phosphate group is transferred to a conserved aspartate residue present in the regulatory domain in ComE. The phosphorylation of ComE, designated ComE~P, enhances its binding to various promoter regions throughout the genome, thereby regulating gene expression at the transcriptional level. ComE~P-regulated genes belong to the early phase of the competence development pathway, which sets the stage for the synthesis of proteins that are directly involved in the transport and integration of extracellular DNA.

Notably, it was demonstrated that, after exposure to CSP, the transcription rates of the *comCDE* and *comAB* operons were significantly increased during competence development; this was consequently followed by a burst in CSP secretion (2, 90, 119). It was also observed that this response did not occur in ComE-deficient mutants (90), thereby suggesting a

ComE-mediated role in the transcriptional activation of *comCDE*. Later, by using electrophoretic mobility shift assays, it was demonstrated that ComE specifically binds to the *comC* and *comA* promoters, strongly supporting its role as a transcriptional factor that activates the *comAB* and *comCDE* promoters (119).

The ComE binding-site was found to possess two 9 bp imperfect direct repeats separated by 12 bp (119). More specifically, it was identified as the following sequence: aCAtTTct(a/g)G————12 bp————ACA(t/g)TtgAG, where the most conserved bases are capitalized. The presence of the ComE binding-site in the *comA* and *comC* promoters and their resultant increase in expression generates an autocatalytic regulatory loop representative of a positive feedback mechanism (i.e. during competence induction, *comC* can stimulate its own synthesis and secretion; phosphorylation of ComE increases its response to CSP and also amplifies the production and secretion rate of this signal molecule) (45). This abrupt rise in CSP probably ensures that the entire population of cells reaches its competence state simultaneously. However, this self-inducing mechanism is transient. Based on Northern hybridization studies and *lacZ* reporter gene assays, the *comAB* and *comCDE* transcripts are highest only 5 min after CSP addition, and then decrease and disappear almost entirely at 20 min (3, 119). These genes whose transcripts appear the earliest when cells are exposed to CSP are designated the early competence genes and include at least 8 genes (3, 11, 60, 73, 91, 94, 95, 90).

ComX: the link to late competence genes

After the ComE binding-site had been identified, it was compared with a common regulatory sequence motif in the promoters belonging to coordinately expressed late competence genes previously identified by Campbell *et al.* (11) and Pestova and Morrison (91). Because these sequences did not share a common sequence motif, it was believed that ComE activated the late genes via one or more intermediate factors, later identified as ComX (60). ComX, also called Sigma X (σ^x), is synthesized by duplicate genes, *comX1* and *comX2*, and constitutes an alternative sigma factor that can be derived by co-purification with RNA polymerase isolated from competent streptococcal cultures (60, 75). The ComX shows high homology to the σ^H in *Bacillus subtilis*, a member of the σ^{70} RNA polymerase sigma subunit family, and is essential for the transcription of the late genes involved in the processing, binding, uptake, and recombination of DNA during transformation (73, 94, 95, 98). ComX acts by directing an

RNA polymerase to a recognition sequence, TACGAATA, referred to as the cin-box or com-box, that is present in the late-gene promoters (59). Interestingly, the regulation of ComX is dependent on CSP, the phosphorylation of ComE, and the activity of both *comX1* and *comX2* to obtain a normal level of competence (60). It was shown that, in a mutant lacking both *comX* genes, transformation ability was abolished altogether. More recently, Luo *et al.* identified a second early gene product, ComW, shown to be important for the optimal activity of ComX (74). Although its mode of action is yet to be characterized, the ectopic coexpression of *comX* and *comW* demonstrated that, together, just the products of these early genes were sufficient to induce competence.

Before discussing the late competence-induced genes, it is important to describe the actual transformation process. Transformation is initiated by the binding of exogenous double-stranded DNA molecules in linear or circular form (plasmids) to the bacterial cell surface. Consequently, these double-stranded molecules are split into single strands by the activity of EndA (125). One strand is actively transported into the cell in a 3' to 5' direction, whereas the complementary strand is degraded by nucleases (84). Depending on the presence of homologous sequences, the newly translocated DNA strands can be fully or partly integrated into the host genome via recombination. In a competent state, this recombination event is favored by the presence of single-stranded gaps throughout the host chromosome (23). In plasmid DNA the re-circularization process is less efficient because a second strand, complementary to the first, must be present.

With the recent advances in molecular technology that includes the use of high-density DNA microarrays, researchers have been able to monitor the global changes in gene expression during competence development (94, 95, 98). Based on these studies, it has been demonstrated that the transcriptional kinetics of at least 188 genes are transiently affected during this process. Most recently, in a DNA microarray study conducted by Peterson *et al.* (95), CSP-induced genes exhibited four temporally distinct expression waves: early, late, and delayed gene induction, and gene repression. Notably, the majority of the CSP-responsive genes belonged to the late phase, whose expression was optimal at approximately 10 min following CSP addition. These included 81 genes in 21 clusters, whose products have a direct role in DNA uptake and homologous recombination during transformation. These include *ssbB* (*cilA*), *dalA, ccl, cglABCD, celAB,* and *cflAB* (11, 14, 60, 91). Once imported into the cell, the DNA molecules are integrated into the host chromosome via homologous recombination assisted by RecA (79, 87). Interestingly, the *recA*, *cinA*, *dinF*, and *lytA* genes that comprise the recA operon were induced in the late phase by CSP (94, 95, 98). The genes that displayed a delayed expression pattern showed the accumulation of mRNA starting from the first minute after exposure to CSP until after the optimal peak was obtained for the late genes. Although the expression of certain delayed genes appeared to be unaffected by the inactivation of ComX (95), the majority of these genes were associated with bacterial stress response. For instance, the entire *dnaK* heat-shock operon, which included *hrcA*, *grpE*, *dnaK*, and *dnaJ* (103) was induced in response to CSP.

Competence-induced cell lysis

Any discussion of competence would naturally evoke the question regarding the source of donor DNA that is used for transformation. Previously, it was believed that these DNA molecules were derived from dead bacteria that had lysed by natural causes. In contrast to this traditional view, emerging investigations have demonstrated that, in pneumococci, the release of donor DNA is triggered by the addition of CSP (88, 108, 109). More specifically, during co-cultivation, competent streptococci can actively acquire DNA by killing 5%–20% of their non-competent neighbors of the same strain in a planktonic population when exposed to CSP (108). Moreover, it was observed that cell lysis did not occur in a ComE-deficient mutant, thereby demonstrating that the ComDE signal transduction system was involved in DNA release as well as in its uptake. At the present, the mechanism of lysis and how it is related to competence induction remains to be investigated. However, based on current evidence, DNA release is accomplished by heterolysis (i.e. lysis of one bacterium that is caused by another) coordinated by proteins such as autolytic amidase, LytA, and an autolytic lysozyme, LytC (88, 108, 109). From an evolutionary perspective, the discovery of competence-induced lysis is of enormous significance in increasing the genetic plasticity of pneumococci. Previously, Steinmoen *et al.* (109) indicated that, in *S. pneumoniae*, cell lysis and DNA uptake were coordinated to ensure the presence of a sufficient amount of homologous DNA during competence development, which would increase the chances for homologous recombination leading to the emergence of novel genotypes. In contrast to this "DNA release-and-uptake" model, a more recent investigation into CSP-induced lysis by Moscoso and Claverys (88) provided evidence that would question the validity of cell lysis as a method evolved for maximized genetic exchange. They showed that, although competence decreased after its maximum at 20 min after CSP addition, the amount of liberated DNA continued to increase and reached a maximum in the stationary phase, when cells were no longer capable of DNA uptake. Supporting their view is the appearance of nuclease activity (EndA and at least one other nuclease) when competence is triggered. Hence, Moscoso and Claverys suggest that competence-induced DNA release serves a role different from that previously thought and may possibly have a role in nutrient acquisition, biofilm formation, or the release of toxins (e.g. pneumolysin, teichoic and lipoteichoic acids).

Competence shutoff

In *S. pneumoniae*, genetic competence is a transient phenomenon that takes place at a certain cell population density. Based on emerging evidence, this process is tightly regulated; this makes sense, because constitutive uptake of foreign DNA would generate havoc in the cell by overloading the DNA repair systems, causing damage to the host chromosome and also by generating lethal mutations. Despite the importance of competence shutdown for cell viability, there has been little insight into its mechanism in the past. However, several hypotheses describing the mechanism of competence shutoff have been described.

One such hypothesis, proposed by Lee and Morrison, states that the complete shutoff of competence followed by its continued suppression is ComX-mediated (60). Consistent with this is the rapid degradation of ComX as competence declines (60, 75). This is likely caused by a proteolytic event that prevents perpetual competence. In addition to degradation of ComX, it was assumed that an additional negative control mechanism was acting to shut off competence, because a rapid decrease of late competent gene transcripts was observed even when the ComX protein was still present. In addition, Lee and Morrison (60) suggested that competence shutdown was likely mediated by the inhibitory activity of a putative late competent gene product, which was designated ComI. To identify such inhibitory candidates, Peterson *et al.* (95) performed targeted gene deletions to screen previously uncharacterized late competence genes, which were identified by global genome analysis, whose temporal profile would significantly alter competence decay. Based on their results, none of the candidate genes selected for mutagenesis was involved in the competence shutdown mechanism.

Previously, Alloing *et al.* (3) had suggested that a ComE-specific phosphatase, CEP, could manipulate competence activation as well as its shutdown by altering the unphosphorylated to phosphorylation ratio of ComE (ComE : ComE~P). More specifically, optimal transcription of *comCDE* would occur in a partly phosphorylated ComE/ComE~P state, whereas a fully phosphorylated ComE would increase the transcription of late competence genes, including the gene responsible for CEP production. Increased production of CEP would then deplete ComE~P, which, in turn, would reduce the expression of late competence genes, thereby shutting down competence. However, there was no evidence of a putative CEP, based on microarrays conducted by different researchers (95, 98), thereby warranting the requirement for more studies to elucidate the molecular pathway(s) that lead to competence shutdown.

The CiaRH and VicRK two-component signal transduction systems (TCSTS)

The *S. pneumoniae* CiaRH TCSTS was linked to competence by Guenzi *et al.* (38) based on an observation that an amino-acid change in CiaH (*ciaH*^{T230P}) caused complete inhibition of competence. Because supplementing these *ciaH*^{T230P} mutant cultures with exogenous competence factor did not restore competence, these authors argued that the CiaRH signal transduction system was involved in the early steps of competence regulation. A few years later, it was observed that a mutant deficient in the *ciaR* gene became competent under conditions that inhibited the development of competence in the wild-type strain (33). However, by examining *ciaR* mutants that were derived from spontaneous revertants of *ciaH*^{T230P}, that had a restored ability to transform, it became apparent that competence was derepressed in these *ciaR* mutants. A link between the CiaRH and ComCDE signal transduction systems was first obtained by Echenique *et al.* (24). In their studies, inactivation of *ciaRH* resulted in the overexpression of *comCDE*, and it was proposed that CiaR negatively regulated the competence regulon. The work by Martin *et al.* (80) provided supporting evidence by using transposon mutagenesis to identify *comCDE*-upregulated mutants (called ComCDE^{UP} or CUP) that included independent insertions in the *ciaR* gene. Additional evidence for the CiaR-mediated negative regulation of the competence regulon was provided recently based on two independent transcriptome studies (82, 104). Despite accumulating knowledge suggesting some type of cross-regulation between these signal transduction systems, it remains to be established whether CiaR is directly or indirectly involved in *comCDE* repression. With reference to this, the *ciaRH* and *comCDE* systems exhibited 83% identity over a fragment

24 bp long that included their −10 promoter regions (80). Hence, it is possible that *comCDE* would be targeted for regulation by harboring a CiaR binding site in this segment.

In addition to CiaRH, the VicRK TCSTS (58) has also been associated with competence development (25, 117). The *S. pneumoniae vicRK*, also called TCS02 (58), 492hk/rr (113), and *micAB* (25), is the ortholog of the essential *yycFG* signal transduction system in *B. subtilis* (27). In contrast to *B. subtilis*, which requires both the HK and the RR pair for its viability, only the RR is essential for survival in *S. pneumoniae* (58). Based on the findings of Echenique and Trombe (25), the phosphorylated VicR acts upstream of ComDE to repress competence when oxygen is limited. Interestingly, VicK is an atypical kinase that harbours a PAS domain. Usually, PAS domains are involved in monitoring intracellular signals such as redox potential (124), suggesting that the response to oxygen limitation might be sensed intracellularly. Interestingly, in a VicK-deficient strain, transformability was decreased by three orders of magnitude.

Based on accumulating evidence, competence development in *S. pneumoniae* involves many dynamic interactions within the cell that not only involve a mechanism that monitors the population density, but also seem to connect DNA uptake to other important physiological processes. For instance, competence is influenced by oxygen availability, temperature, pH, and Ca^{2+} and Mg^{2+} concentrations (45). Also importantly, the *comCDE* competence regulon is located near the putative origin of chromosome replication, *ori* (31). Hence, the *com–ori* co-location enables the bacterium to sense the rate of replication (which is dependent on nutrient availability) and develop competence in concert with the environmental conditions (15). Previously, Alloing *et al.* (3) discovered that quorum sensing is affected in an *obl*-deficient mutant (*obl* for oligopeptide-binding lipoprotein) suggesting that transformation is affected by an oligopeptide permease in response to nutrient availability. However, more studies are needed to elucidate the molecular and genetic mechanisms that connect these signals to the competence regulon.

The Blp quorum-sensing system

The BlpABCSRH regulon forms a second quorum-sensing system in *S. pneumoniae* that shares common features with the ComABCDE competence regulon (19). The Blp regulon (*blp* for bacteriocin-like peptide) controls the production of class II bacteriocins and their immunity proteins in response to a threshold cell population density (19, 97). In this system, the *blpC* encodes the BIP (bacteriocin-inducing peptide) pheromone, whereas the *blpRH* encodes the histidine kinase sensor (BlpH) and response regulator (BlpR), respectively. Although the function of BlpS is unknown, it shares high similarity with the C-terminal DNA-binding domain of BlpR, but lacks the N-terminal domain. Similar to CSP, BlpC contains a characteristic Gly–Gly motif that is believed to be cleaved by the *blpAB*-encoded secretion apparatus, prior to export of the mature BIP signaling peptide. Sequence analysis of BlpC from different strains demonstrated allelic variation resulting in at least four different BIPs, thereby leading to strain-specific activity of this signaling molecule reminiscent of the pherotype variation for CSP.

Previously, based on microarray analyses, it was believed that cross-communication was unlikely between the Com and Blp quorum-sensing systems at the pheromone or histidine kinase level (15). It was argued that genes affected by crosstalk would possibly have been detected during global genome analyses conducted to identify genes regulated by either CSP or BIP addition. More recently, the addition of BIP and CSP signaling peptides were demonstrated to activate the transcription of an operon, encoding an ABC transporter (QsrAB) of unknown function in *S. pneumoniae* Rx strain (56). Moreover, cross-induction was achieved by a hybrid-direct-repeat motif present in the target promoter that responded to both ComE and BlpR. Because homology searches suggested a putative role for QsrAB as a sodium pump, it is possible that it protects the bacterium against osmotic stress (56).

GENETIC COMPETENCE IN ORAL STREPTOCOCCI

In the oral cavity, there is a diverse range of habitats including both soft and hard tissues. Consequently, to be able to survive and prosper, many oral bacteria have evolved highly specialized mechanisms to overcome fluctuations in the composition of nutrient supply, local oxygen availability, pH, shear forces due to saliva flow and mastication, and a range of host defense mechanisms (77). Oral streptococci, originally called viridans streptococci, are the species that predominantly inhabit the mouth and upper respiratory tracts of humans as commensal bacteria and constitute approximately 20% of the normal human oral flora (57). Oral streptococci can cause opportunistic infections at oral or non-oral sites when environmental conditions favor an overgrowth of particular species or when they escape their normal ecologic niches and establish at another site, usually as a consequence of some breakdown in normal host defenses (101).

Table 10.1. *Naturally competent human oral streptococci*

Group	Species
Anginosus	S. anginosus
	S. constellatus
	S. intermedius
	S. mitis
Mitis	S. cristatus
	S. oralis
Mutans	S. mutans
	S. ratti
Sanguinis	S. sanguinis
	S. gordonii

Source: Adapted from (16).

The classification of oral streptococci has undergone significant changes in the past several years. A comprehensive review (28) divided the oral streptococci into five distinct species groups based on phenotypic characteristics, DNA–DNA re-association, and 16 S rRNA gene sequencing: (i) the *Streptococcus anginosus* group, (ii) the *Streptococcus mitis* group, (iii) the *Streptococcus mutans* group, (iv) the *Streptococcus salivarius* group, and (v) the *Streptococcus sanguinis* group.

Oral streptococci respond to signals resulting from the proximity, density, and identity of microbial neighbors. Through the process of quorum sensing, bacteria can indirectly determine population density by sensing concentration of a signal molecule (8). As discussed earlier, the work on pneumococcal competence had pioneered a research area on quorum-sensing regulation of streptococcal competence. Also importantly, this gene-exchange phenomenon has long been described as a characteristic of species within the mitis group (62). To map the incidence of natural competence in the genus *Streptococcus* (43) a number of streptococcal strains were screened by PCR for the presence of the competence operon (*comCDE*). By using primers complementary to the Arg- and Glu-tRNA genes that flank the *comCDE* operon in *S. pneumoniae*, it was observed that most streptococcal species belonging to the mitis and anginosus groups possessed the *comCDE* operon. Recently, natural genetic competence has been reported in other oral streptococcal species belonging to the sanguinis and mutans groups (Table 10.1) (for review see 16).

Horizontal gene transfer exerts a strong selective force on a bacterial population, leading to the evolution of prokaryotic genomes (10). As natural competence for genetic transformation is involved in horizontal gene transfer, it plays a significant role in gene acquisition/loss and strain heterogeneity, thereby leading to overall evolution (29). The potential for gene transfer is high among streptococci in oral biofilms; emerging evidence suggests that quorum sensing may be coupled with expression of competence for acquisition of foreign DNA via transformation. In the following sections, specific examples of quorum-sensing regulation of natural competence in two important oral streptococci, namely *Streptococcus gordonii* and *S. mutans*, will be discussed.

Transformation in *Streptococcus gordonii*

Genetics of competence development

S. gordonii is a human commensal bacterium classified into the sanguinis group (28). It participates in the formation of the dental biofilm as a primary colonizer of the tooth pellicle, creating a template for the subsequent attachment of other bacteria and leading to the establishment of the complex oral biofilm (121). By producing hydrogen peroxide, this species may also play a protective role in relation to periodontal diseases, as hydrogen peroxide inhibits the growth of Gram-negative periodontopathogens (47). Although *S. gordonii* is considered to be relatively non-pathogenic, this bacterium is among the most common oral bacteria associated with infective endocarditis (32).

S. gordonii Challis (formerly *Streptococcus sanguis* Challis) was the first oral streptococcus shown to be capable of transformation (62). Competence in *S. gordonii* develops in the early to mid exponential growth phase and is modulated by environmental stimuli that trigger the production of an extracellular competence factor (CF). Although all of the stimuli are still unknown, it has been demonstrated that a slightly basic pH and the presence of albumin or heat-inactivated serum can trigger competence development in *S. gordonii* (70). CF acts in the same manner as *S. pneumoniae* CSP by functioning as a population density indicator for competence induction. Two CFs have been isolated from *S. gordonii* species; both are 19 amino-acid peptides that are processed from a 50 amino-acid precursor peptide containing the characteristic N-terminal double-glycine type leader, as seen in *S. pneumoniae* (42, 72). Amino-acid sequence alignment showed that these two peptides have 74% amino-acid identity. The *comC* gene encoding CF is located immediately upstream of the *comD* and *comE* genes which encode the HK and RR of a TCSTS, respectively (42). As with *S. pneumoniae*, the *S. gordonii comCDE* operon is located between the Arg-tRNA and Glu-tRNA genes (42, 43).

At a different genomic locus, there is a *comAB* operon encoding the ComAB transporter, a member of the ABC-type transporters present in *S. pneumoniae*. Amino-acid sequences of the ComA proteins from *S. pneumoniae* and *S. gordonii* displayed 82% identity (71); the ComB sequences revealed that the peptides were 55% identical. Analysis of the *S. gordonii* genome located 3' proximal to *comB* revealed the presence of an additional locus, called *orfX*, that had no similarity to known sequences (71). The *S. gordonii orfX* locus contains three small putative open reading frames, designated *comX*, *orfM*, and *orfO*. The *comX* gene encodes a putative 52 amino-acid basic peptide with a predicted isoelectric point (pI) of 10.2. Complementation experiments demonstrated that ComX is able to complement transformation deficiency in the non-transformable *S. gordonii* strain Wicky, a non-strain that had lost the ability to synthesize the CF. As ComX is responsible for competence-stimulating activity in S. gordonii Wicky, it has been proposed that CF may not be the only signal molecule; consequently, a CF independent pathway for competence induction may exist in *S. gordonii* (96, 98).

Regulation of competence

Experimental evidence suggested that the mechanisms involved in cell-density-dependent regulation of competence development in *S. pneumoniae* and *S. gordonii* were fundamentally similar. A model depicting the pathway involved in the regulation of competence in *S. gordonii* has been proposed by Jenkinson (51) based on the data obtained for *S. pneumoniae* (Figure 10.2). Competence develops through a quorum-sensing mechanism in which CF serves as the signal molecule. The CF precursor is most likely processed and exported outside the cell by the ComAB transporter system. When the extracellular CF reaches a critical threshold concentration, it is sensed by the N-terminal domain of the membrane-bound ComD receptor (70). This interaction results in the autophosphorylation of ComD and the subsequent activation of the response regulator ComE by phosphorylation (ComE~P). In general, phosphorylation enhances the affinity of response regulators for their DNA binding sites, leading to increased expression of target genes.

In addition to controlling the expression of the *comCDE* operon, ComE~P also regulates transcription of late competence genes (*comY* operon, *cipA*, and *recA*) via a regulatory protein (Figure 10.3). The *comY* operon contains four genes, designated *comYA*, *comYB*, *comYC*, and *comYD*, and encodes a putative dedicated secretion system thought to be involved in DNA uptake. It has been demonstrated that a *S. gordonii comYA*-defective mutant was completely deficient in transformation and exhibited decreased levels of DNA binding and hydrolysis (72). Interestingly, the *comYA* gene product shares significant similarity (57% homology) to the ComGA transporter of *Bacillus subtilis* (72). The *S. gordonii comYA* gene also represents the first competence gene conserved between the transformable streptococci and the well-studied transformable *B. subtilis* (70, 72). The *cipA* and *recA* genes encode proteins necessary for DNA recombination. The product encoded by the *cipA* (competence-inducible peptide) gene is predicted to be a putative 51 kDa protein believed to be involved in chromosomal integration and plasmid reassembly. As expected, *S. gordonii cipA* mutants are defective in chromosomal transformation (70).

Another ABC-type transporter, designated Hpp, has been shown to be involved in competence development by *S. gordonii* (Figure 10.3). Hpp is a lipoprotein-dependent oligopeptide permease that preferentially transports hexa- and heptapeptides (52). The Hpp permease is composed of cytoplasmic membrane-bound lipoproteins that are highly similar (60% amino acid identity) to the substrate-binding components of the *S. pneumoniae* Ami oligopeptide permease (4). It has been demonstrated that mutations in genes encoding components of the *S. gordonii* Hpp oligopeptide permease transporter affect competence development by reducing the efficiency of transformation (52). A possible explanation is that the Hpp-transported peptides may modulate the expression or activity of a ComE-specific phosphatase, which, in turn, controls the level of ComE~P (51). The activity of Hpp also affects the expression of *cshA*, a gene encoding a high-molecular-mass cell-wall-anchored polypeptide essential for colonization of the oral cavity by *S. gordonii* (83). CshA mediates the binding of *S. gordonii* to human fibronectin, *Actinomyces naeslundii*, *Streptococcus oralis*, and *Candida albicans* (53).

In Gram-positive bacteria, it has been demonstrated that peptide transport systems are not only essential for the uptake of peptides as a source of nutrients, but also involved in affecting other cellular functions and properties, such as the development of competence, sporulation, and bacterial adherence. The Hpp system may influence the development of *S. gordonii* genetic competence by functioning in sensing environmental changes during growth. As proposed by Jenkinson (51), the utility of this would be to provide an additional level of control on the ComE regulatory circuit and

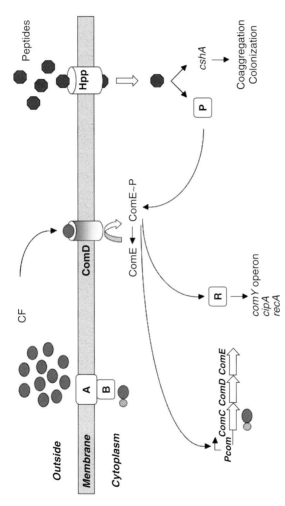

Figure 10.3. Schematic representation of quorum-sensing regulation of genetic competence in *Streptococcus gordonii*. Model proposed by Jenkinson (51). See text for details. R: regulatory protein. P: phosphatase.

a means by which alternative environmental signals might be integrated into the *S. gordonii* quorum-sensing-dependent competence pathway.

Transformation in *Streptococcus mutans*

Genetic regulation of competence development

S. mutans is considered a major causative agent of dental caries, one of the most common human infectious diseases. In the United States, billions of dollars are spent annually to treat dental caries or caries-related infections. *S. mutans* initiates dental caries while living in the biofilm environment of dental plaque, which comprises a diverse community of microorganisms found on tooth surfaces and embedded in an extracellular matrix of bacterial and host origin (78). The main virulence factors associated with *S. mutans* cariogenicity are its ability to adhere and coaggregate on teeth, its acidogenicity, and its aciduricity (7, 85). Consequently, the tooth surface is an indispensable natural habitat for *S. mutans*. As demonstrated for *S. gordonii*, *S. mutans* was also found to possess a quorum sensing system that closely resembles the *com* regulon of *S. pneumoniae*. One genomic locus contains the *comC*, *comD*, and *comE* genes encoding the precursor to the competence-stimulating peptide (CSP), a transmembrane HK, and an RR, respectively, with the *comD*-encoded protein being the CSP receptor (65). In contrast to what was found in *S. pneumoniae* and other oral streptococci, in *S. mutans* the *comC* gene is encoded divergently proximal to the *comDE* genes. Moreover, in *S. mutans*, the *comCDE* region is not flanked by the Arg- and Glu-tRNA genes (64) and therefore cannot be detected by PCR using primers complementary to the tRNA-encoding genes (43). The *S. mutans* CSP was found to be a 21 amino-acid peptide derived from a 46 amino-acid peptide precursor. The peptide precursor is believed to be produced constitutively and cleaved during its export via the ComAB transporter system (93). As seen in *S. pneumoniae* and *S. gordonii*, the *S. mutans comCDE* region is not closely linked to the *comAB* operon (1). Interestingly, deletion mutations in any of the *S. mutans com* genes resulted in mutants with a residual level of transformation (64). This is in contrast to the transformability of *S. pneumoniae*, in which the *com* mutants had no detectable transformation (12). This is one clue that the mechanisms of the quorum-sensing-dependent competence pathway may be regulated differently among various streptococcal species. Indeed, another fundamental difference between *S. pneumoniae* and *S. mutans* is the optimal level of competence achieved under saturating CSP concentrations. In this condition, nearly all of the *S. pneumoniae* cells become competent, whereas this percentage is typically between 0.1% and 1% in an *S. mutans* population. While the reasons for this are unclear, it is probable that, in *S. mutans*, alternate signals and/or shutoff mechanisms may hinder the spread of competence throughout the whole population.

A model describing the general pathway of quorum-sensing-mediated regulation of genetic competence in *S. mutans* is illustrated in Figure 10.4. When the mature CSP reaches a critical threshold concentration at cell densities typical of the early to mid exponential growth phase, it interacts with the membrane-associated HK, ComD. This interaction results in the phosphorylation of ComD, which in turn stimulates the phosphorylation of its cognate RR, ComE. The phosphorylated form of ComE acts at promoter sites (*Pcom*) for genes whose expression is upregulated during the development of competence. Based on the work done in pneumococci, the following consensus sequence for the ComE-binding site has been proposed: aCAtTTca/gG-N_{12}-ACAt/gTTgAG (119). In *S. mutans*, at least six genes are known to be upregulated by the ComCDE two-component signal transduction system, *comC*, *comDE*, *comAB*, and *comX* (66). By using a transcriptional fusion of the *comX* promoter to a promoterless reporter gene, Aspiras *et al.* (5) demonstrated a positive correlation between the *pcomX* transcription in the presence of CSP and competence development. Similar to *S. pneumoniae*, it is believed that *comX* encodes an alternative sigma factor that directs the transcription of RNA polymerase by altering its binding affinity to the promoters of several late competence-specific genes involved in DNA uptake and processing (73). Indeed, bioinformatic analysis of the *S. mutans* UA159 genome revealed that many putative open reading frames, including late-competence gene homologs, possess a sequence in their promoter regions (TTTTT-N_9-TACGAATA, where N_9 represents a 9 bp linker region) that resembles the *com* box of *S. pneumoniae* (11). Interestingly, several of these genes are unrelated to genetic competence. These findings suggest a complex network of genes that are regulated by the quorum-sensing system and that control other cellular functions in addition to competence induction in *S. mutans*.

Study of competence in model biofilms

Dental biofilm formation is initiated by the interactions between planktonic cells and tooth surfaces in response to appropriate environmental signals. Bacteria in a biofilm mode of growth are distinct from planktonic cells in their gene expression and cellular physiology (106). Interestingly, it has been demonstrated that *S. mutans* cells growing in biofilms can be transformed more efficiently (at a rate 10- to 600-fold higher) than their

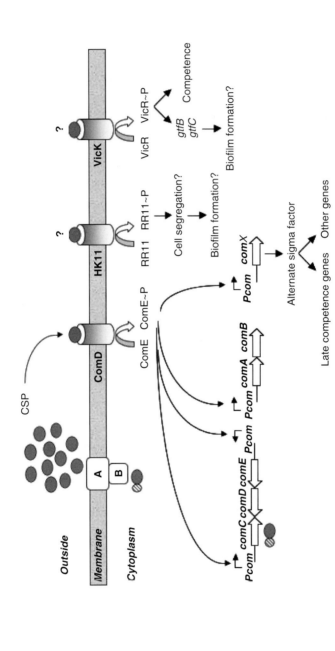

Figure 10.4. Schematic representation of quorum-sensing regulation of genetic competence in *Streptococcus mutans*. See text for details.

free-living counterparts (64). It is likely that the high density of bacterial cells growing in biofilms can facilitate the exchange of intraspecies genetic material by providing optimal conditions for the activation of the CSP-dependent quorum-sensing system. In addition, natural transformation has also been shown to occur in mixed-species biofilms *in vitro*. One of the best examples was the intergeneric transfer of tetracycline resistance in a mixed-species biofilm from *B. subtilis* to a *Streptococcus* sp. grown in a constant-depth biofilm fermentor (99). Moreover, genetic exchange between a Gram-negative (*Treponema denticola*) and a Gram-positive (*S. gordonii*) bacterium has also been demonstrated in an artificial oral biofilm (118). Biofilms may thus function as potential genetic reservoirs, allowing horizontal gene transfer of virulence factors and the dissemination of antibiotic resistance genes.

The first report linking the cell-density-dependent quorum-sensing system to biofilm formation was described in *S. gordonii*. By using transposon mutagenesis, Loo *et al.* (68) found that inactivation of *comD* resulted in a biofilm-defective phenotype. In a recent study, Gilmore *et al.* (34) used real-time PCR to quantify the expression of *S. gordonii* genes known or assumed to be involved in biofilm formation. They demonstrated that the *comD* and *comE* genes encoding the TCSTS required for competence in *S. gordonii* were both upregulated in the biofilm phase. In *S. mutans*, biofilm formation has also been linked to the competence regulon. Li *et al.* (66) demonstrated that inactivation of any of the genes encoding the CSP signaling system resulted in the formation of an abnormal biofilm. In particular, they found that the *comC* null and *comAB* null mutants formed biofilms that lacked the wild-type architecture, whereas the *comD*, *comE* and *comX* null mutants formed biofilms with reduced biomass. The fact that the biofilms formed by the *comC* mutant differed from the *comD*, *comE*, and *comX* mutant biofilms suggested that *S. mutans* may possess at least one alternate CSP and/or receptor (17). Further support for multiple CSP receptors includes the observation that exogenous addition of synthetic CSP into the *comCDE* mutant culture only partly restored the wild-type biofilm architecture (65, 66). Further studies demonstrated a positive correlation between *pcomX* activity in high cell density areas within the biofilm population and competence development (5). Because *S. mutans* relies on a biofilm lifestyle for its survival in the oral cavity, it is not surprising that this bacterium has evolved a quorum-sensing signaling system that helps it to adapt and survive while living in biofilms.

More recently, Petersen *et al.* (92) have demonstrated that CSP promotes the biofilm mode of growth of *Streptococcus intermedius*, thereby enhancing the role of CSP as a "biofilm sensor". *S. intermedius*, a member of the anginosus group, is a bacterium found in the gingival crevice and is associated with abscesses (101). It has been shown that CSP-treated *S. intermedius* cells formed more biofilm than untreated cells, whereas competence development occurred exclusively in the presence of CSP (92).

Other CSP-regulated phenotypes

Within the oral cavity, *S. mutans* is often exposed to transient environmental conditions. For example, during the ingestion of dietary sugars by the host, carbohydrate concentrations can range from 100 mm to as low as 1 μm during sleep. It also encounters pH shifts ranging from above 7.0 to as low as 3.0 (in carious sites). Consequently, the ability of *S. mutans* to tolerate acidic pH is crucial to its survival and pathogenicity. To determine whether cell density can modulate adaptation to acid in *S. mutans*, Li et al. (63) examined the ability of mutants with defects in the ComCDE signaling system to withstand acid challenge. They found that inactivation of the *comCDE* genes resulted in a diminished acid tolerance response, which could be restored to wild-type levels by supplementing with synthetic CSP. These results clearly demonstrated a connection between the CSP-dependent competence signaling pathway and the acid-tolerance response of *S. mutans*.

Another TCSTS, comprising the *hk11* and *rr11* genes, has also been shown to be important for the survival of *S. mutans* at an acidic pH (65, 66). Interestingly, a *com*-box consensus sequence was found in the promoter region of *rr11*, while the deletion of the *hk11* or *rr11* resulted in defects in biofilm formation. A closer examination of the mutant biofilms by scanning electron microscopy revealed that they have a sponge-like architecture composed of cells organized in very long chains, a feature that was previously observed in the biofilm formed by the *comC* null mutant. Hence, it was suggested that *hk11* may act as another CSP receptor, in which interaction could activate another pathway related to cell segregation and ultimately affect the *S. mutans* biofilm architecture (Figure 10.4). However, further studies will be necessary to assign a role to HK11 as a CSP receptor.

In addition to *hk11/rr11*, characterization of the VicK/VicR TCSTS in *S. mutans* revealed that it controls several important physiological properties, including competence development and biofilm formation, in this bacterium (105).The *vicK* and *vicR* (also named covR/S (61)) genes encode a putative HK and putative RR, respectively (9). Based on transformation efficiency assays that were conducted with a *vicR*-overexpressing mutant, it was observed that its natural transformability (without added CSP) was reduced by approximately 100-fold compared with its UA159 wild-type parent. Interestingly, although the addition of synthetic CSP increased the transformation efficiency of the wild-type by approximately 1,000-fold, the transformability of a *vicK*-null mutant and the *vicR*-overexpressing strain were not restored to wild-type levels. Especially, when CSP was added, the *vicK*-deficient mutant showed a 60-fold reduction in transformation efficiency relative to the wild type, thereby suggesting that VicK is responsive to CSP. However, as in the case of HK11, more studies are warranted to assign a definite role for VicK as another receptor for CSP.

Recently, Wen *et al.* (120) examined the *S. mutans* homolog of the bacterial trigger factor, designated RopA, which is a molecular chaperone highly conserved in most bacteria. Interestingly, they found that inactivation of *ropA* affected multiple *S. mutans* phenotypic properties that included decreased stress tolerance, reduced transformation efficiency, and the formation of altered biofilms. Given the established role of the trigger factor in proper folding, processing, and secretion of membrane-associated and extracellular proteins, the authors suggested that RopA may be involved in the processing and/or secretion of proteins, which can affect competence development, biofilm formation, and bacterial stress response in *S. mutans*.

CONCLUSIONS AND FUTURE PERSPECTIVES

Although the process of transformation has been known in streptococci for over 80 years, it is only in the past 10 years that we have understood its molecular underpinnings. In this chapter, we have discussed how quorum sensing in streptococci leads to the development of competence. Interestingly, accumulating evidence suggests that genetic competence is one of many important physiological properties regulated by this elegant mechanism that has been linked to virulence in a plethora of bacterial species. Because many of these quorum-sensing pathways have been associated with critical cell functions, they have provided us with enticing targets for antibacterial therapy.

Synthetic signaling molecules can be used to artificially induce intraspecies communication in streptococci; this can be exploited to design therapeutics to treat or prevent streptococcal infections. Emerging literature discusses the possibility of interfering with quorum sensing as a novel approach to control bacterial pathogenicity (30, 55, 123). The use of quorum-sensing blockers to attenuate virulence, as opposed to conventional therapeutics such as antibiotics, can offer various advantages. Because these blockers can inhibit virulence without necessarily killing bacterial cells, the development of drug-resistant strains can be minimized.

REFERENCES

1 Ajdic, D., W. M. McShan *et al.* 2002. Genome sequence of *Streptococcus mutans* UA159, a cariogenic dental pathogen. *Proc. Natn. Acad. Sci. USA* **99**(22): 14434–9.

2 Alloing, G., C. Granadel *et al.* 1996. Competence pheromone, oligopeptide permease, and induction of competence in *Streptococcus pneumoniae. Molec. Microbiol.* **21**(3): 471–8.

3 Alloing, G., B. Martin *et al.* 1998. Development of competence in *Streptococcus pneumoniae*: pheromone autoinduction and control of quorum-sensing by the oligopeptide permease. *Molec. Microbiol.* **29**(1): 75–83.

4 Alloing, G., M. C. Trombe *et al.* 1990. The ami locus of the gram-positive bacterium *Streptococcus pneumoniae* is similar to binding protein-dependent transport operons of gram-negative bacteria. *Molec. Microbiol.* **4**(4): 633–44.

5 Aspiras, M. B., R. P. Ellen *et al.* 2004. ComX activity of *Streptococcus mutans* growing in biofilms. *FEMS Microbiol. Lett.* **238**(1): 167–74.

6 Avery, O. T., C. M. MacLeod *et al.* 1944. Studies on the chemical nature of the substance inducing transformation of pneumococcal types: induction of trans-formation by a desoxyribonucleic acid fraction isolated from pneumococcus Type III. *J. Exp. Med.* **79**: 137–58.

7 Banas, J. A. 2004. Virulence properties of *Streptococcus mutans. Front. Biosci.* **9**: 1267–77.

8 Bassler, B. L. 2002. Small talk. Cell-to-cell communication in bacteria. *Cell* **109**(4): 421–4.

9 Bhagwat, S. P., J. Nary *et al.* 2001. Effects of mutating putative two-component systems on biofilm formation by *Streptococcus mutans* UA159. *FEMS Microbiol. Lett.* **205**(2): 225–30.

10 Boucher, Y., C. J. Douady *et al.* 2003. Lateral gene transfer and the origins of prokaryotic groups. *A. Rev. Genet.* **37**: 283–328.

11 Campbell, E. A., S. Y. Choi *et al.* 1998. A competence regulon in *Streptococcus pneumoniae* revealed by genomic analysis. *Molec. Microbiol.* **27**(5): 929–39.

12 Cheng, Q., E. A. Campbell *et al.* 1997. The com locus controls genetic transform-ation in *Streptococcus pneumoniae. Molec. Microbiol.* **23**(4): 683–92.

13 Claverys, J. P. and L. S. Havarstein 2002. Extracellular-peptide control of compe-tence for genetic transformation in *Streptococcus pneumoniae. Front. Biosci.* **7**: d1798–814.

14 Claverys, J. P. and B. Martin 1998. Competence regulons, genomics and strepto-cocci. *Molec. Microbiol.* **29**(4): 1126–7.

15 Claverys, J. P., M. Prudhomme *et al.* 2000. Adaptation to the environment: *Streptococcus pneumoniae*, a paradigm for recombination-mediated genetic plasticity? *Molec. Microbiol.* **35**(2): 251–9.

16 Cvitkovitch, D. G. 2001. Genetic competence and transformation in oral streptococci. *Crit. Rev. Oral Biol. Med.* **12**(3): 217–43.

17 Cvitkovitch, D. G., Y. H. Li *et al.* 2003. Quorum-sensing and biofilm formation in Streptococcal infections. *J. Clin. Invest.* **112**(11): 1626–32.

18 Davies, D. G., M. R. Parsek *et al.* 1998. The involvement of cell-to-cell signals in the development of a bacterial biofilm. *Science* **280**(5361): 295–8.

19 de Saizieu, A., C. Gardes *et al.* 2000. Microarray-based identification of a novel *Streptococcus pneumoniae* regulon controlled by an autoinduced peptide. *J. Bacteriol.* **182**(17): 4696–703.

20 Diep, D. B., L. S. Havarstein *et al.* 1994. The gene encoding plantaricin A, a bacteriocin from *Lactobacillus plantarum* C11, is located on the same transcription unit as an agr-like regulatory system. *Appl. Environ. Microbiol.* **60**(1): 160–6.

21 Dowson, C. G., V. Barcus *et al.* 1997. Horizontal gene transfer and the evolution of resistance and virulence determinants in Streptococcus. *Soc. Appl. Bacteriol. Symp. Ser.* **26**: 42S–51S.

22 Dowson, C. G., T. J. Coffey *et al.* 1993. Evolution of penicillin resistance in *Streptococcus pneumoniae*; the role of *Streptococcus mitis* in the formation of a low affinity PBP2B in *S. pneumoniae*. *Molec. Microbiol.* **9**(3): 635–43.

23 Dubnau, D. 1999. DNA uptake in bacteria. *A. Rev. Microbiol.* **53**: 217–44.

24 Echenique, J. R., S. Chapuy-Regaud *et al.* 2000. Competence regulation by oxygen in *Streptococcus pneumoniae*: involvement of *ciaRH* and *comCDE*. *Molec. Microbiol.* **36**(3): 688–96.

25 Echenique, J. R. and M. C. Trombe 2001. Competence repression under oxygen limitation through the two-component MicAB signal-transducing system in *Streptococcus pneumoniae* and involvement of the PAS domain of MicB. *J. Bacteriol.* **183**(15): 4599–608.

26 Erdos, G., S. Sayeed *et al.* 2003. Development and characterization of a pooled *Haemophilus influenzae* genomic library for the evaluation of gene expression changes associated with mucosal biofilm formation in otitis media. *Int. J. Pediatr. Otorhinolaryngol.* **67**(7): 749–55.

27 Fabret, C. and J. A. Hoch 1998. A two-component signal transduction system essential for growth of *Bacillus subtilis*: implications for anti-infective therapy. *J. Bacteriol.* **180**(23): 6375–83.

28 Facklam, R. 2000. What happened to the Streptococci: overview of taxonomic and nomenclature changes. *Clin. Microbiol. Rev.* **15**: 613–30.

29 Ferretti, J. J., D. Ajdic *et al.* 2004. Comparative genomics of streptococcal species. *Ind. J. Med. Res.* **119** (Suppl.): 1–6.

30 Finch, R. G., D. I. Pritchard *et al.* 1998. Quorum-sensing: a novel target for anti-infective therapy. *J. Antimicrob. Chemother.* **42**(5): 569–71.

31 Gasc, A. M., P. Giammarinaro *et al.* 1998. Organization around the *dnaA* gene of *Streptococcus pneumoniae*. *Microbiology* **144** (2): 433–9.

32 Gendron, R., D. Grenier *et al.* 2000. The oral cavity as a reservoir of bacterial pathogens for focal infections. *Microbes Infect.* **2**(8): 897–906.

33 Giammarinaro, P., M. Sicard *et al.* 1999. Genetic and physiological studies of the CiaH-CiaR two-component signal-transducing system involved in cefotaxime resistance and competence of *Streptococcus pneumoniae*. *Microbiology* **145** (8): 1859–69.

34 Gilmore, K. S., P. Srinivas *et al.* (2003). Growth, development, and gene expression in a persistent *Streptococcus gordonii* biofilm. *Infect. Immun.* **71**(8): 4759–66.

35 Gilson, L., H. K. Mahanty *et al.* 1990. Genetic analysis of an MDR-like export system: the secretion of colicin V. *EMBO J.* **9**(12): 3875–94.

36 Grebe, T. W. and J. B. Stock 1999. The histidine protein kinase superfamily. *Adv. Microb. Physiol.* **41**: 139–227.

37 Griffith, F. 1928. The significance of pneumococcal types. *J. Hyg.* **27**: 113–59.

38 Guenzi, E., A. M. Gasc *et al.* 1994. A two-component signal-transducing system is involved in competence and penicillin susceptibility in laboratory mutants of *Streptococcus pneumoniae*. *Molec. Microbiol.* **12**(3): 505–15.

39 Havarstein, L. S. 1998. Bacterial gene transfer by natural genetic transformation. *APMIS* (Suppl.) **84**: 43–6.

40 Havarstein, L. S., G. Coomaraswamy *et al.* 1995. An unmodified heptadecapeptide pheromone induces competence for genetic transformation in *Streptococcus pneumoniae*. *Proc. Natn. Acad. Sci. USA* **92**(24): 11140–4.

41 Havarstein, L. S., D. B. Diep *et al.* 1995. A family of bacteriocin ABC transporters carry out proteolytic processing of their substrates concomitant with export. *Molec. Microbiol.* **16**(2): 229–40.

42 Havarstein, L. S., P. Gaustad *et al.* 1996. Identification of the streptococcal competence-pheromone receptor. *Molec. Microbiol.* **21**(4): 863–9.

43 Havarstein, L. S., R. Hakenbeck *et al.* 1997. Natural competence in the genus Streptococcus: evidence that streptococci can change phenotype by interspecies recombinational exchanges. *J. Bacteriol.* **179**(21): 6589–94.

44 Havarstein, L. S., H. Holo *et al.* 1994. The leader peptide of colicin V shares consensus sequences with leader peptides that are common among peptide bacteriocins produced by gram-positive bacteria. *Microbiology* **140** (9): 2383–9.

45 Havarstein, L. S., and D. A. Morrison 1999. Quorum-sensing and peptide pheromones in streptococcal competence for genetic transformation. In G. M. Dunny & S. C. Winans (eds), *Cell-Cell Signaling in Bacteria*, pp. 9–26. Washington, DC: ASM Press.

46 Hershey, A. D. and M. Chase 1952. Independent functions of viral protein and nucleic acid in growth of bacteriophage. *J. Gen. Physiol.* **36**: 39–56.

47 Hillman, J. D., S. S. Socransky *et al.* 1985. The relationships between streptococcal species and periodontopathic bacteria in human dental plaque. *Arch. Oral. Biol.* **30**(11–12): 791–5.

48 Holo, H., O. Nilssen *et al.* 1991. Lactococcin A, a new bacteriocin from *Lactococcus lactis* subsp. cremoris: isolation and characterization of the protein and its gene. *J. Bacteriol.* **173**(12): 3879–87.

49 Hui, F. M. and D. A. Morrison 1991. Genetic transformation in *Streptococcus pneumoniae*: nucleotide sequence analysis shows comA, a gene required for competence induction, to be a member of the bacterial ATP-dependent transport protein family. *J. Bacteriol.* **173**(1): 372–81.

50 Hui, F. M., L. Zhou *et al.* 1995. Competence for genetic transformation in *Streptococcus pneumoniae*: organization of a regulatory locus with homology to two lactococcin A secretion genes. *Gene* **153**(1): 25–31.

51 Jenkinson, H. F. 2000. Genetics of *Streptococcus sanguis*. In J. I. Rood (ed.), *Gram-Positive Pathogens*, pp. 287–94. Washington, DC: American Society for Microbiology.

52 Jenkinson, H. F., R. A. Baker *et al.* 1996. A binding-lipoprotein-dependent oligopeptide transport system in *Streptococcus gordonii* essential for uptake of hexa- and heptapeptides. *J. Bacteriol.* **178**(1): 68–77.

53 Jenkinson, H. F. and R. J. Lamont 1997. Streptococcal adhesion and colonization. *Crit. Rev. Oral. Biol. Med.* **8**(2): 175–200.

54 Kleerebezem, M. and L. E. Quadri 2001. Peptide pheromone-dependent regulation of antimicrobial peptide production in Gram-positive bacteria: a case of multicellular behavior. *Peptides* **22**(10): 1579–96.

55 Kleerebezem, M., L. E. Quadri *et al.* 1997. Quorum-sensing by peptide pheromones and two-component signal-transduction systems in Gram-positive bacteria. *Molec. Microbiol.* **24**(5): 895–904.

56 Knutsen, E., O. Ween *et al.* 2004. Two separate quorum-sensing systems upregulate transcription of the same ABC transporter in *Streptococcus pneumoniae*. *J. Bacteriol.* **186**(10): 3078–85.

57 Kolenbrander, P. E. 2000. Oral microbial communities: biofilms, interactions, and genetic systems. *A. Rev. Microbiol.* **54**: 413–37.

58 Lange, R., C. Wagner *et al.* 1999. Domain organization and molecular characterization of 13 two-component systems identified by genome sequencing of *Streptococcus pneumoniae*. *Gene* **237**(1): 223–34.

59 Lee, M. S., B. A. Dougherty *et al.* 1999. Construction and analysis of a library for random insertional mutagenesis in *Streptococcus pneumoniae*: use for recovery of mutants defective in genetic transformation and for identification of essential genes. *Appl. Environ. Microbiol.* **65**(5): 1883–90.

60 Lee, M. S. and D. A. Morrison 1999. Identification of a new regulator in *Streptococcus pneumoniae* linking quorum-sensing to competence for genetic transformation. *J. Bacteriol.* **181**(16): 5004–16.

61 Lee, S. F., G. D. Delaney *et al.* 2004. A two-component covRS regulatory system regulates expression of fructosyltransferase and a novel extracellular carbohydrate in *Streptococcus mutans*. *Infect. Immun.* **72**(7): 3968–73.

62 Leonard, C. G. 1973. Early events in development of streptococcal competence. *J. Bacteriol.* **114**(3): 1198–205.

63 Li, Y. H., M. N. Hanna *et al.* 2001. Cell density modulates acid adaptation in *Streptococcus mutans*: implications for survival in biofilms. *J. Bacteriol.* **183**(23): 6875–84.

64 Li, Y. H., P. C. Lau *et al.* 2001. Natural genetic transformation of *Streptococcus mutans* growing in biofilms. *J. Bacteriol.* **183**(3): 897–908.

65 Li, Y. H., P. C. Lau *et al.* 2002. Novel two-component regulatory system involved in biofilm formation and acid resistance in *Streptococcus mutans*. *J. Bacteriol.* **184**(22): 6333–42.

66 Li, Y. H., N. Tang *et al.* 2002. A quorum-sensing signaling system essential for genetic competence in *Streptococcus mutans* is involved in biofilm formation. *J. Bacteriol.* **184**(10): 2699–708.

67 Lina, G., S. Jarraud *et al.* 1998. Transmembrane topology and histidine protein kinase activity of AgrC, the agr signal receptor in *Staphylococcus aureus*. *Molec. Microbiol.* **28**(3): 655–62.

68 Loo, C. Y., D. A. Corliss *et al.* 2000. *Streptococcus gordonii* biofilm formation: identification of genes that code for biofilm phenotypes. *J. Bacteriol.* **182**(5): 1374–82.

69 Lorenz, M. G. and W. Wackernagel 1994. Bacterial gene transfer by natural genetic transformation in the environment. *Microbiol. Rev.* **58**(3): 563–602.

70 Lunsford, R. D. 1998. Streptococcal transformation: essential features and applications of a natural gene exchange system. *Plasmid* **39**(1): 10–20.

71 Lunsford, R. D. and J. London 1996. Natural genetic transformation in *Streptococcus gordonii*: comX imparts spontaneous competence on strain wicky. *J. Bacteriol.* **178**(19): 5831–5.

72 Lunsford, R. D. and A. G. Roble 1997. comYA, a gene similar to comGA of *Bacillus subtilis*, is essential for competence-factor-dependent DNA transformation in *Streptococcus gordonii*. *J. Bacteriol.* **179**(10): 3122–6.

73 Luo, P., H. Li *et al.* 2003. ComX is a unique link between multiple quorum-sensing outputs and competence in *Streptococcus pneumoniae*. *Molec. Microbiol.* **50**(2): 623–33.

74 Luo, P., H. Li *et al.* 2004. Identification of ComW as a new component in the regulation of genetic transformation in *Streptococcus pneumoniae*. *Molec. Microbiol.* **54**(1): 172–83.

75 Luo, P. and D. A. Morrison 2003. Transient association of an alternative sigma factor, ComX, with RNA polymerase during the period of competence for genetic transformation in *Streptococcus pneumoniae*. *J. Bacteriol.* **185**(1): 349–58.

76 Magnuson, R., J. Solomon *et al.* 1994. Biochemical and genetic characterization of a competence pheromone from *B. subtilis*. *Cell* **77**(2): 207–16.

77 Marsh, P. D. 2000. Oral ecology and its impact on oral microbial diversity. In Ellen, R. P. (ed.), *Oral Bacterial Ecology: the Molecular Basis*, pp. 11–65. Wymondham, UK: Horizon Scientific Press.

78 Marsh, P. D. 2004. Dental plaque as a microbial biofilm. *Caries Res.* **38**(3): 204–11.

M. D. SENADHEERA *ET AL.*

79 Martin, B., P. Garcia *et al.* 1995. The *recA* gene of *Streptococcus pneumoniae* is part of a competence-induced operon and controls an SOS regulon. *Dev. Biol. Stand.* **85**: 293–300.

80 Martin, B., M. Prudhomme *et al.* 2000. Cross-regulation of competence pheromone production and export in the early control of transformation in *Streptococcus pneumoniae. Molec. Microbiol.* **38**(4): 867–78.

81 Marugg, J. D., C. F. Gonzalez *et al.* 1992. Cloning, expression, and nucleotide sequence of genes involved in production of pediocin PA-1, and bacteriocin from *Pediococcus acidilactici* PAC1.0. *Appl. Environ. Microbiol.* **58**(8): 2360–7.

82 Mascher, T., D. Zahner *et al.* 2003. The *Streptococcus pneumoniae* cia regulon: CiaR target sites and transcription profile analysis. *J. Bacteriol.* **185**(1): 60–70.

83 McNab, R. and H. F. Jenkinson 1998. Altered adherence properties of a *Streptococcus gordonii* hppA (oligopeptide permease) mutant result from transcriptional effects on cshA adhesin gene expression. *Microbiology* **144** (1): 127–36.

84 Mejean, V. and J. P. Claverys 1993. DNA processing during entry in transformation of *Streptococcus pneumoniae. J. Biol. Chem.* **268**(8): 5594–9.

85 Mitchell, T. J. 2003. The pathogenesis of streptococcal infections: from tooth decay to meningitis. *Nat. Rev. Microbiol.* **1**(3): 219–30.

86 Morrison, D. A., M. C. Trombe *et al.* 1984. Isolation of transformation-deficient *Streptococcus pneumoniae* mutants defective in control of competence, using insertion-duplication mutagenesis with the erythromycin resistance determinant of pAM beta 1. *J. Bacteriol.* **159**(3): 870–6.

87 Mortier-Barriere, I., A. de Saizieu *et al.* 1998. Competence-specific induction of *recA* is required for full recombination proficiency during transformation in *Streptococcus pneumoniae. Molec. Microbiol.* **27**(1): 159–70.

88 Moscoso, M. and J.-P. Claverys 2004. Release of DNA into the medium by competent *Streptococcus pneumoniae*: kinetics, mechanism and stability of the liberated DNA. *Molec. Microbiol.* **54**(3): 783–94.

89 Pakula, R. and W. Walczak 1963. On the nature of competence of transformable streptococci. *J. Gen. Microbiol.* **31**: 125–33.

90 Pestova, E. V., L. S. Havarstein *et al.* 1996. Regulation of competence for genetic transformation in *Streptococcus pneumoniae* by an auto-induced peptide pheromone and a two-component regulatory system. *Molec. Microbiol.* **21**(4): 853–62.

91 Pestova, E. V. and D. A. Morrison 1998. Isolation and characterization of three *Streptococcus pneumoniae* transformation-specific loci by use of a lacZ reporter insertion vector. *J. Bacteriol.* **180**(10): 2701–10.

92 Petersen, F. C., D. Pecharki *et al.* 2004. Biofilm mode of growth of *Streptococcus intermedius* favored by a competence-stimulating signaling peptide. *J. Bacteriol.* **186**(18): 6327–31.

93 Petersen, F. C. and A. A. Scheie 2000. Genetic transformation in *Streptococcus mutans* requires a peptide secretion-like apparatus. *Oral Microbiol. Immunol.* **15**(5): 329–34.

94 Peterson, S., R. T. Cline *et al.* 2000. Gene expression analysis of the *Streptococcus pneumoniae* competence regulons by use of DNA microarrays. *J. Bacteriol.* **182**(21): 6192–202.

95 Peterson, S. N., C. K. Sung *et al.* 2004. Identification of competence pheromone responsive genes in *Streptococcus pneumoniae* by use of DNA microarrays. *Molec. Microbiol.* **51**(4): 1051–70.

96 Pozzi, G., L. Masala *et al.* 1996. Competence for genetic transformation in encapsulated strains of *Streptococcus pneumoniae*: two allelic variants of the peptide pheromone. *J. Bacteriol.* **178**(20): 6087–90.

97 Reichmann, P. and R. Hakenbeck 2000. Allelic variation in a peptide-inducible two-component system of *Streptococcus pneumoniae*. *FEMS Microbiol. Lett.* **190**(2): 231–6.

98 Rimini, R., B. Jansson *et al.* 2000. Global analysis of transcription kinetics during competence development in *Streptococcus pneumoniae* using high density DNA arrays. *Molec. Microbiol.* **36**(6): 1279–92.

99 Roberts, A. P., J. Pratten *et al.* 1999. Transfer of a conjugative transposon, Tn5397 in a model oral biofilm. *FEMS Microbiol. Lett.* **177**(1): 63–6.

100 Robinson, V. L., D. R. Buckler *et al.* 2000. A tale of two components: a novel kinase and a regulatory switch. *Nat. Struct. Biol.* **7**(8): 626–33.

101 Russell, R. R. B. 2000. Pathogenesis of oral streptococci. In J. I. Rood *et al.* (eds), *Gram-Positive Pathogens*, pp. 272–9. Washington, DC: American Society for Microbiology Press.

102 Sablon, E., B. Contreras *et al.* 2000. Antimicrobial peptides of lactic acid bacteria: mode of action, genetics and biosynthesis. *Adv. Biochem. Eng. Biotechnol.* **68**: 21–60.

103 Schroder, H., T. Langer *et al.* 1993. DnaK, DnaJ and GrpE form a cellular chaperone machinery capable of repairing heat-induced protein damage. *EMBO J.* **12**(11): 4137–44.

104 Sebert, M. E., L. M. Palmer *et al.* 2002. Microarray-based identification of *htrA*, a *Streptococcus pneumoniae* gene that is regulated by the CiaRH two-component system and contributes to nasopharyngeal colonization. *Infect. Immun.* **70**(8): 4059–67.

105 Senadheera, D., C. Huang *et al.* 2004. *Streptococcus mutans* covR/S genes control adhesion, biofilm formation and competence development. *J. Dent. Res.* Special Issue (Proceedings from the International Association for Dental Research Annual Meeting), Abstr. 3001.

106 Socransky, S. S. and A. D. Haffajee 2002. Dental biofilms: difficult therapeutic targets. *Periodontol. 2000* **28**: 12–55.

107 Stein, T., S. Borchert *et al.* 2002. Dual control of subtilin biosynthesis and immunity in *Bacillus subtilis*. *Molec. Microbiol.* **44**(2): 403–16.

108 Steinmoen, H., E. Knutsen *et al.* 2002. Induction of natural competence in *Streptococcus pneumoniae* triggers lysis and DNA release from a subfraction of the cell population. *Proc. Natn. Acad. Sci. USA* **99**(11): 7681–6.

109 Steinmoen, H., A. Teigen *et al.* 2003. Competence-induced cells of *Streptococcus pneumoniae* lyse competence-deficient cells of the same strain during cocultivation. *J. Bacteriol.* **185**(24): 7176–83.

110 Stock, A. M., V. L. Robinson *et al.* 2000. Two-component signal transduction. *A. Rev. Biochem.* **69**: 183–215.

111 Strauch, M. A. and J. A. Hoch 1993. Signal transduction in *Bacillus subtilis* sporulation. *Curr. Opin. Genet. Dev.* **3**(2): 203–12.

112 Sturme, M. H., M. Kleerebezem *et al.* 2002. Cell to cell communication by autoinducing peptides in gram-positive bacteria. *Antonie Van Leeuwenhoek* **81**(1–4): 233–43.

113 Throup, J. P., K. K. Koretke *et al.* 2000. A genomic analysis of two-component signal transduction in *Streptococcus pneumoniae*. *Molec. Microbiol.* **35**(3): 566–76.

114 Tomasz, A. 1965. Control of the competent state in Pneumococcus by a hormone-like cell product: an example for a new type of regulatory mechanism in bacteria. *Nature* **208**(6): 155–9.

115 Tomasz, A. and S. M. Beiser 1965. Relationship between the competence antigen and the competence-activator substance in pneumococci. *J. Bacteriol.* **90**(5): 1226–32.

116 Tomasz, A. and R. D. Hotchkiss 1964. Regulation of the transformability of pneumococcal cultures by macromolecular cell products. *Proc. Natn. Acad. Sci. USA* **51**: 480–7.

117 Wagner, C., A. Saizieu Ad *et al.* 2002. Genetic analysis and functional characterization of the *Streptococcus pneumoniae vic* operon. *Infect. Immun.* **70**(11): 6121–8.

118 Wang, B. Y., B. Chi *et al.* 2002. Genetic exchange between *Treponema denticola* and *Streptococcus gordonii* in biofilms. *Oral Microbiol. Immunol.* **17**(2): 108–12.

119 Ween, O., P. Gaustad *et al.* 1999. Identification of DNA binding sites for ComE, a key regulator of natural competence in *Streptococcus pneumoniae*. *Molec. Microbiol.* **33**(4): 817–27.

120 Wen, Z. T., D. G. Suntharaligham *et al.* 2005. Trigger factor in *Streptococcus mutans* is involved in stress tolerance, competence development, and biofilm formation. *Infect. Immunol.* **73**: 219–25.

121 Whittaker, C. J., C. M. Klier *et al.* 1996. Mechanisms of adhesion by oral bacteria. *A. Rev. Microbiol.* **50**: 513–52.

122 Winans, G. M. D. a. S. C. (ed.) 1999. *Cell-Cell Signaling in Bacteria*. Washington, DC: American Society of Microbiology Press.

123 Zhang, L. H. and Y. H. Dong 2004. Quorum-sensing and signal interference: diverse implications. *Molec. Microbiol.* **53**(6): 1563–71.

124 Zhulin, I. B., B. L. Taylor *et al.* 1997. PAS domain S-boxes in Archaea, Bacteria and sensors for oxygen and redox. *Trends Biochem. Sci.* **22**(9): 331–3.

125 Puyet, A., B. Greenberg and S. A. Lacks 1990. Genetic and structural characterization of end A, a membrane-bound nuclease required for transformation of *Streptococcus pneumoniae*. *J. Molec. Biol.* **213**: 727–38.

CHAPTER 11

Signaling by a cell-surface-associated signal during fruiting-body morphogenesis in *Myxococcus xanthus*

Lotte Søgaard-Andersen

Max Planck Institute for Terrestrial Microbiology,
Marburg, Germany

269

INTRODUCTION

Over the past decade the perception of bacterial cells as autonomous individuals, each following their own agenda and not interacting with each other, has been replaced by the view that bacteria interact extensively both within and between species by means of intercellular signal molecules. Each of these signal molecules constitutes part of an information system that is constructed of four parts: the donor cell synthesizing the signal, the signal molecule, the recipient cell, and the output response. As in any other information system, the signal must be tailored to the talents of the recipient. A clear example of a tailor-made signal molecule is the C-signal molecule in *Myxococcus xanthus*. Most intercellular signals identified in bacteria are small (i.e. with a molecular mass of less than 1000 Da), freely diffusible molecules that are part of quorum sensing systems, which help bacterial cells to assess population size (90). However, that is not the case for the C-signal in *Myxococcus xanthus*. The C-signal is a 17 kDa cell-surface-associated protein and is thus non-diffusible, and it helps to guide *M. xanthus* cells into nascent fruiting bodies and to assess their position in a field of cells.

C-signal transmission occurs by a contact-dependent mechanism, i.e. it depends on direct contact between the donor and the receiving cell. The C-signal is used repeatedly during the starvation-induced formation of spore-filled fruiting bodies in *M. xanthus*. Early during fruiting-body formation, the C-signal induces the aggregation of cells into the nascent fruiting bodies. In the late stage of fruiting body formation, the C-signal induces sporulation of cells that have accumulated inside the fruiting bodies. During the same time interval, the C-signal induces the expression of a large number of genes. Why, then, has *M. xanthus* adopted a

Bacterial Cell-to-Cell Communication: Role in Virulence and Pathogenesis, ed. D. R. Demuth and R. J. Lamont. Published by Cambridge University Press. © Cambridge University Press 2005.

cell-surface-associated signal molecule rather than a freely diffusible molecule to induce the two major morphogenetic events in fruiting body formation? *M. xanthus* moves by gliding motility and is faced with the problem that the average gliding speed is only 2–5 µm min^{-1}. It has been argued that this speed, which is only 50–100 times higher than the rate of continental drift, is so low that the directive properties of a diffusible aggregation signal would disperse before *M. xanthus* cells could reorient in a gradient of such a signal. The solution adopted by *M. xanthus* cells to overcome this problem is to use a cell-surface-associated signal. The advantage of this type of signal is that it moves at the same speed as the cells and therefore allows cells to reorient without losing their sense of direction. A second advantage conferred by a signal molecule that hinges on a contact-dependent signal transmission mechanism is that it confers information to the recipient about its position relative to that of other cells in its immediate neighbourhood, i.e. signaling levels are proportional to cell density. Thus, the output response of a cell can be tied in with the position of that cell relative to that of other cells. This property of the C-signaling system is the key to an understanding of one of the hallmarks in fruiting-body formation, the position-specific sporulation of those cells that have accumulated inside the fruiting bodies.

The complete understanding of an intercellular signaling system entails an understanding of how the signal molecule is synthesized, how synthesis is regulated, how the signal is received, and how the reception of the signal is transformed into an output response. In this chapter, these questions are discussed in the context of the C-signaling system in *M. xanthus* and its role in fruiting-body formation.

MULTICELLULARITY AS A SURVIVAL STRATEGY

Myxobacteria are the gold standard for social bacteria, although in strong competition with Actinomycetes. Myxobacteria have adopted a social lifestyle – or multicellularity – as a survival strategy. Myxobacteria are found in the topsoil, where they feed on organic matter and prey on other microorganisms by secreting hydrolytic enzymes and antimicrobials (68). *M. xanthus* forms two morphologically distinct types of biofilm, depending on the nutritional status of the cells (Figure 11.1). In the presence of nutrients, the motile, rod-shaped cells grow and divide. Myxobacteria move by gliding (31, 81); if cells are present on a solid surface they form spreading, cooperatively feeding colonies (13). Cells at the edge of a colony spread coordinately over the surface, forming a thin, film-like structure (Figure 11.1). The developmental program that culminates in

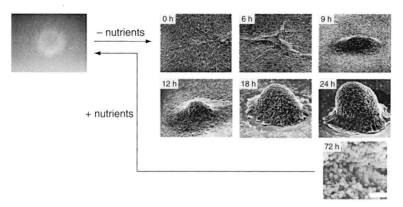

Figure 11.1. The life cycles of *Myxococcus xanthus*. In the panel to the left a vegetative, spreading colony is shown. The diameter of the colony is *c.* 2 cm. In the scanning EM pictures to the right, the different stages of fruiting body formation are shown. In the 72 h panel, a fruiting body has been opened to visualize the spores. Scale bars in 24 and 72 h panels: 5 and 10 μm, respectively. The scanning EM pictures are adapted from Kuner and Kaiser (43).

the formation of the spore-filled fruiting bodies is initiated in response to starvation of cells at a high density on a solid surface (13). Fruiting-body formation proceeds in a relatively stereotyped pattern of morphological events, which are separated in time and space (Figures 11.1, 11.2). The starting point is an initially unstructured lawn of randomly oriented cells on a solid surface. The first explicit signs of fruiting-body formation are evident after 4–6 h with changes in cell behaviour (29) as the cells begin to aggregate to form small aggregation centers (Figure 11.2). A characteristic of the aggregation process is that cells move into the aggregation centers organized as streams of cells, rather than entering the centers as single cells from all directions (61) (Figure 11.2). As more cells enter the aggregation centers, these centers increase in size and eventually become symmetric mounds. By 24 h, the aggregation process is complete and the nascent fruiting bodies each contain *c.* 10^5 densely packed cells (Figures 11.1, 11.2). Inside the nascent fruiting bodies, the rod-shaped, motile cells undergo morphological and physiological differentiation into spherical, non-motile, dormant spores, resulting in mature fruiting bodies (Figure 11.1). Although the aggregation process takes *c.* 24 h to complete, spore maturation is finished *c.* 72 h after the onset of starvation. Although all cells exposed to starvation have the potential to develop into spores (14, 52) only 10–20% of cells differentiate into spores. These cells are specifically those which have accumulated inside the mounds. Up to 30%

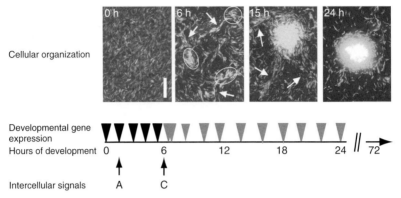

Cellular organization

Developmental gene expression

Hours of development

Intercellular signals

Figure 11.2. Morphogenesis of fruiting bodies in *Myxococcus xanthus*. On the developmental timeline, each triangle represents a gene, which is turned on during fruiting-body formation. Black triangles represent genes that are expressed in all cells; grey triangles represent genes that are expressed only in aggregating and sporulating cells. The times of action of the A and C signals are indicated by the arrows below the timeline. Cellular arrangements during the different stages of fruiting-body formation are shown above the timeline. Cell arrangements are visualized by fluorescence microscopy of green fluorescent protein (GFP)-labeled cells. GFP-labeled wild-type cells were co-developed with non-fluorescent wild-type cells at a ratio of 1 : 40. Images were acquired at the time points indicated. White circles and arrows in the 6 h image indicate aggregation centers and streams, respectively. White arrows in the 15 h image indicate streams of cells entering a nascent fruiting body. Scale bar: 50 μm. The pictures of the cellular arrangements during the different stages of fruiting-body formation are adapted from (28).

remain outside the fruiting bodies. These cells remain rod-shaped and differentiate to a cell type called peripheral rods. Even after extended periods of starvation, peripheral rods do not differentiate into spores (60). Finally, the remaining cells undergo lysis (69).

Fruiting-body formation involves temporally coordinated changes in gene expression in which genes are turned on at specific time points during development (27, 41) (Figure 11.2). Moreover, developmental gene expression is spatially controlled (Figure 11.2). Genes turned on after 6 h are preferentially expressed in aggregating and sporulating cells, whereas genes activated prior to 6 h are expressed in all cells including peripheral rods (30). Finally, genetic and biochemical evidence suggests that fruiting-body formation depends on the exchange of intercellular signals (71) (Figure 11.2).

Myxobacteria are the only bacteria that cope with starvation by forming spore-filled fruiting bodies. This survival strategy is optimally suited to the multicellular life-style of myxobacteria. Each fruiting body consists of *c.* 10^5

spores and is essentially a spore-filled sac. Under the appropriate conditions, the spores in fruiting bodies germinate to give rise to motile, metabolically active vegetative cells. Germination immediately gives rise to a new, spreading, cooperatively feeding colony. Thus, fruiting bodies are optimally designed to ensure that a new vegetative cycle is initiated by a community of cells rather than by a single cell.

GLIDING MOTILITY IN *M. XANTHUS*

The multicellular lifestyle of *M. xanthus* crucially depends on the ability of cells to display active movement. *M. xanthus* cells move by gliding motility, which is the movement of a rod-shaped cell in the direction of the cell's long axis on a surface in the absence of a flagellum (22). Gliding and its regulation constitutes the basis for the formation of spreading colonies and fruiting-body formation. *M. xanthus* has two mechanisms for gliding, adventurous (A)- and social (S)-motility (24). Mutations in both systems (A⁻, S⁻ mutants) result in a non-motile phenotype; mutations that inactivate only one system leave the cells motile by means of the remaining, intact system. The speed of gliding is highly variable for a given cell as it moves across the substrate (Figure 11.3*b*). Periodically, cells stop and then either resume gliding in the same direction or undergo a reversal in which the head becomes the tail and vice versa (5, 28, 29, 82) (Figure 11.3a). S-motility is generally only operational when cells are within contact distance of each other (23), whereas A-motility is operational in single cells (23).

The motility engine that provides the force for S-motility is generated by the retraction of type IV pili (Tfp) (32, 92); for a review on Tfp, see (55). Tfp are polarly localized structures and are normally only present at one pole at a time (32) (Figure 11.3*c*). Extension of Tfp from a cell pole, followed by retraction of Tfp, provides the force for cell movement (58, 76, 85). In addition to Tfp, S-motility depends on LPS O-antigen (6) and on extracellular matrix fibrils (1). Extracellular matrix fibrils are fibers 30 nm thick that coat the cell surface and form an extracellular matrix in which neighboring cells are interconnected (1, 4). The fibrils are made from equal amounts of polysaccharide and protein (3). The contact-dependence of Tfp-dependent motility in *M. xanthus* has been rationalized by the observation that the polysaccharide portion of the extracellular matrix fibrils triggers the retraction of Tfp (51). According to this scheme, S-motility relies on the extension of Tfp from the pole of a cell and attachment to the extracellular matrix fibrils on a nearby cell, followed

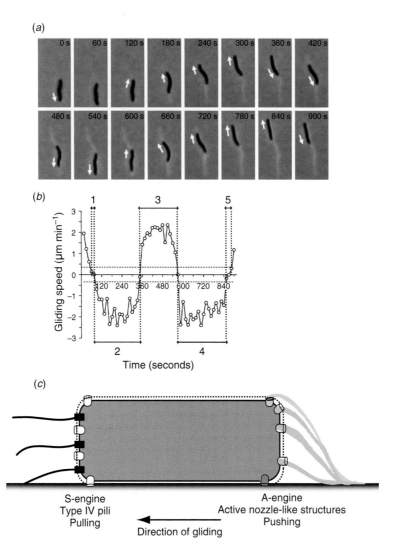

Figure 11.3. Characteristics of *Myxococcus xanthus* gliding motility. (*a*) A sequence of phase-contrast images of a single cell gliding on a solid surface. Images were recorded every 15 s for 900 s. Slime deposited by the cell is evident as a refractile trail; white arrows indicate the direction of gliding. (*b*) The speed profile of the cell in (*a*). The gliding speed was calculated for each 15 s interval between recordings and plotted as a function of recording time. The dashed horizontal lines indicate the detection limit for active movement ($0.35\,\mu m\,min^{-1}$). The dotted vertical lines indicate intervals where the cell showed active movement (intervals 2, 3, and 4) and periods of no active movement (intervals 1 and 5). A change in gliding speed from a negative to a positive value or vice versa indicates a reversal. This cell reversed its direction of gliding at $t = 75$, 345, 570 and 870 s.

by retraction of Tfp and thus the forward movement of the Tfp containing cell. Thus, retraction of Tfp essentially pulls cells forward. Tfp are only observed at one pole at a time (32). Nevertheless, A^-S^+ cells, which move only by means of their S-engine, reverse direction at regular intervals (82), suggesting that the pole at which the Tfp are located changes regularly.

M. *xanthus* cells moving by means of A-motility leave behind a slime trail. Currently, the secreted slime in these slime trails is thought to be the key to an understanding of A-motility. Wolgemuth *et al.* (91) identified (by electron microscopy) nozzle-like structures, which are clustered at both poles of A-motile cells in M. *xanthus* (Figure 11.3c). These structures are similar to the junctional pore complexes used in gliding cyanobacteria to secrete the slime that generates gliding motility in these bacteria (25). Accordingly, it was proposed that the nozzle-like structures in M. *xanthus* are also involved in slime secretion and that slime extrusion generates the force for movement. To account for how slime extrusion could produce sufficient thrust to propel a bacterial cell, Wolgemuth *et al.* (91) proposed that slime is introduced into the nozzle-like structures in a dehydrated form. The subsequent hydration of the slime would cause it to swell and be extruded from the nozzle. Adherence of the hydrated slime to the substrate would provide the slime with a footing that would allow it to push the cell forward. Slime extrusion thus essentially pushes cells forward. Even though the nozzle-like structures are observed at both poles, slime extrusion is only observed at one pole at a time, implying that only one polar cluster of the nozzle-like structures is actively secreting slime at a time (Figure 11.3c). Nevertheless, A^+S^+ cells that move as single cells, and hence only translocate by means of their A-engine, reverse direction regularly (5). This, in turn, implies that the polarity of the active nozzles also switches at regular intervals. The molecular composition of the nozzle-like structures in M. *xanthus* is so far unknown.

Figure 11.3. (cont.)

(c) Longitudinal cross-section of M. *xanthus* cells on a solid surface, showing the arrangement and location of the two engines involved in gliding motility. The dashed and solid lines indicate the outer and inner membrane, respectively. Type IV pili project from the left pole. Black cylinders indicate proteins involved in biogenesis and retraction of type IV pili spanning the cell envelope. White barrels indicate the slime-extruding nozzle-like structures implicated in A-motility. These structures are present at both poles; however, only one cluster at a time is actively secreting slime. In this cell, the cluster at the right pole secretes slime as indicated in grey. Parts (a) and (b) were adapted from (28); (c) was adapted from (78).

Interestingly, homologs of the TolR, TolB, and TolQ proteins have been shown to be important for A-motility (89, 93). In Gram-negative bacteria these proteins, together with the TolA and Pal proteins, form a complex, which is involved in transport processes across the inner and outer membranes (53).

INTERCELLULAR SIGNALING DURING FRUITING-BODY MORPHOGENESIS

Whereas gliding motility and its regulation constitutes the basis for the formation of fruiting bodies, the role of the intercellular signals involved in fruiting-body formation seems to be to coordinate and synchronize the efforts of thousands of cells during the construction process. Genetic evidence suggests that there may be at least five intercellular signals, known as the A- to E-signals, which are involved in fruiting-body formation (71). Initially, mutants deficient in the synthesis of an intercellular signal required for fruiting-body formation were identified in a collection of mutants that were unable to complete fruiting body formation and sporulation (18). A specific characteristic of a group of these mutants was that their development was rescued by co-development with wild-type cells. This rescue is referred to as extracellular complementation (18) to emphasize that it does not involve transfer of genetic material from wild-type cells to mutant cells. In addition it was found that none of the non-autonomous mutants were auxotrophs, suggesting that extracellular complementation did not involve cross-feeding (18). Extracellular complementation experiments involving pairs of non-autonomous mutants lead to the classification of the mutants into four classes referred to as the *asg* (A-signal), *bsg* (B-signal), *csg* (C-signal) and *dsg* (D-signal) mutants, respectively (18). In these co-development experiments, mutants from one class were able to rescue development of a mutant from a different class, i.e. an *asg* mutant could rescue sporulation of a *bsg* mutant and vice versa, whereas mutants belonging to the same class did not result in rescue of development when co-developed (18). Later, a fifth class of non-autonomous mutants referred to as the *esg* (E-signal) mutant was identified (12). Extracellular complementation was explained by suggesting that the inability of a non-autonomous mutant to complete development was caused by the inability to produce an intercellular signal required for development. In the co-development experiments, this missing signal would be provided either by wild-type cells or by a mutant from a different class of non-autonomous mutants (18).

Despite intensive work, only two of the intercellular signals – the A- and C-signals – have been characterized biochemically. Likewise, the A- and C-signals are the only signals whose function has been clarified. The A-signal becomes important for development after 2 h of starvation (44) (Figure 11.2). The A-signal consists of a subset of amino acids, which are produced by the action of extracellular proteases that are activated in response to starvation (45, 65). The A-signal functions as part of a system that measures the density of starving cells: each starving cell produces a constant amount of A-signal and once a threshold concentration of A-signal is reached development proceeds (46). The function of the A-signaling system therefore seems to be to ascertain that fruiting-body morphogenesis is not initiated unless a sufficiently high number of cells are starving. The C-signal comes into action after 6 h of starvation, coinciding with the first signs of morphogenesis (40).

THE C-SIGNAL INDUCES THREE RESPONSES THAT ARE SEPARATED IN TIME AND SPACE

Mutants that cannot synthesize the C-signal are unable to aggregate and sporulate (73) and the expression of the genes that are normally turned on from 6 h is reduced or abolished (40). The C-signal is encoded by the *csgA* gene (73). Consequently, in a *csgA* mutant the developmental program is arrested after 6 h. Development of *csgA* mutant cells can be restored in either of two ways: by co-development with wild-type cells or with cells from either the *asg, bsg, dsg,* or *esg* classes of non-autonomous mutants (18), or by exogenous C-signal (36).

Formally, the developmental defects in a *csgA* mutant and the rescue of these defects by extracellular complementation or by exogenous C-signal only provide evidence that the C-signal is required for fruiting-body formation. These experiments do not provide evidence that the C-signal is a true signal that induces a specific response. To distinguish between these possibilities, Kruse *et al.* (42) overproduced the C-signal early during starvation, the idea being that if the C-signal was a true signal that induced a response(s), then this/these response(s) would be induced earlier in the strain overproducing the signal. In the strain overproducing the C-signal early during development it was observed that aggregation, sporulation, and C-signal-dependent gene expression were indeed induced earlier than in wild-type cells. Thus, this experiment provided the crucial piece of evidence demonstrating that the C-signal induces aggregation, sporulation, and full expression of developmental genes, which are turned on after 6 h.

THE MOLECULAR NATURE OF THE C-SIGNAL

The molecular nature of the C-signal has been controversial. Kim and Kaiser (36, 37) originally purified the C-signal by detergent extraction and biochemical fractionation from starving *M. xanthus* cells on the basis of its ability to rescue development of *csgA* cells. Kim and Kaiser identified the C-signal as a protein with molecular mass *c.* 17 kDa, which is encoded by the *csgA* gene. The understanding that this 17 kDa protein was the C-signal was questioned by several observations that suggested that the CsgA protein might act as an enzyme to produce the C-signal. Firstly, from the sequence of the *csgA* gene, the size of the CsgA protein was predicted to be 25 kDa (47). Secondly, the CsgA protein is homologous to members of the short-chain alcohol dehydrogenase (SCAD) family of intracellular proteins (2, 47). SCAD enzymes contain two conserved sequence motifs, both of which are present in the CsgA protein: an N-terminal motif corresponding to the NAD(P)-coenzyme binding pocket and a more C-terminal motif involved in the catalytic mechanism (63). Consistently, the CsgA protein was observed to bind NAD^+ *in vitro* (47). Thirdly, exogenous CsgA protein purified from *E. coli* restored development of *csgA* cells; however, exogenous CsgA proteins carrying substitutions in either the coenzyme-binding pocket or the catalytic site failed to restore development of *csgA* cells (47). Finally, overproduction of the SocA protein, which is homologous to SCAD enzymes, restored development in a *csgA* mutant *in vivo* (48, 49). However, enzymatic activity of full-length CsgA protein has not been demonstrated.

Recently, these conflicting data were at least partly reconciled. First it was shown that the CsgA protein exists in two forms, one with a molecular mass of *c.* 25 kDa (designated p25), which corresponds to full-length CsgA protein, and one with a molecular mass of *c.* 17 kDa (designated p17) (42), which is similar in size to the C-signal protein purified by Kim and Kaiser (36, 37). The p25 protein is present in vegetative cells and accumulates during fruiting-body formation, whereas p17 is only detected in starving cells (42). By partially purifying the C-signal from starving cells, Lobedanz and Søgaard-Andersen showed that the C-signal co-purifies with p17 (54). In biochemical fractionation experiments, p17 and p25 were both found to be localized to the outer membrane (54). Moreover, by using antibodies directed against the N- and C-terminal of p25, evidence was provided that p17 corresponds to the C-terminal *c.* 17 kDa of p25 (54). It has been found consistently that recombinant p17 proteins, which corresponded to the C-terminal *c.* 17 kDa of p25 and which were purified from *E. coli*, have

C-signal activity (54). Importantly, these recombinant p17 proteins lack the NAD^+ coenzyme-binding pocket and are unable to bind NAD^+ *in vitro* (54). Thus, p17 does not depend on SCAD activity to engage in C-signaling. Rather these data strongly support the idea that p17 is the *bona fide* C-signal.

The precise N-terminus of p17 remains to be determined. Another unsolved question regarding the exact biochemical nature of the C-signal is how p25 and p17 are anchored in the outer membrane. A homology model of p25 revealed that p25 is not an amphiphilic protein and does not have the β-barrel structure typical of outer membrane proteins (54). However, in Triton X-114 phase separation experiments, p17 and p25 both partition to the detergent phase (54). As this partitioning pattern is a characteristic of amphiphilic membrane proteins and lipoproteins, this led to the speculation that p25 and p17 are post-translationally modified with a hydrophobic molecule and that this hydrophobic moiety serves to anchor the proteins in the outer membrane. Neither the chemical nature of this modification nor the position in p17 and p25 that carries the modification have been determined. It should be pointed out that secreted lipoproteins are typically modified on an N-terminal Cys residue. However, p25 does not contain a lipoprotein signal peptide.

SYNTHESIS OF THE C-SIGNAL

To understand a signaling pathway, it is essential to define the mechanisms by which cells produce this signal. Shimkets and co-workers have provided evidence that the *csgA* gene is transcribed from the same promoter during vegetative growth and development (50) and that the translation start codon for p25 synthesis is essential for synthesis of the C-signal (47). These observations argued against the idea that p17 was synthesized from an alternative start codon. A synthesis mechanism for p17 involving translational frameshifts also appeared unlikely, as this mechanism would involve at least two frameshifts. Synthesis mechanism involving transcriptional or translational mechanism having been ruled out by Shimkets and co-workers, Lobedanz and Søgaard-Andersen tested whether p17 synthesis involved proteolytic processing of p25 (54). This hypothesis was tested experimentally by using an *in vitro* protease assay in which total cell extracts were prepared from starving *M. xanthus* cells. When this extract was added to a recombinant p25 protein purified from *E. coli*, a protein with a molecular mass of *c*. 17 kDa was produced. This protein was recognized by the antibodies against the C-terminus of p25, whereas the antibodies against the N-terminus of p25 did not recognize it. Control experiments provided evidence that the recombinant p25 protein does not harbor the protease activity involved in the N-terminal proteolytic processing of p25. By adding protease inhibitors specific for different types of proteases to the cell extract, evidence was obtained that the protease involved in p25 processing is a serine protease. Consistent with the observation that p17 is only detected in developing cells (42), it was observed that the activity of the protease is developmentally regulated. The protease and the corresponding gene remain to be identified.

Several observations – in addition to those already mentioned – still need to be explained before the molecular nature of the C-signal can be declared solved. In particular, the inability of mutant p25 proteins, which carry substitutions in the coenzyme binding pocket or in the catalytic site, to rescue development of csgA cells in the C-signal bioassay (47) is thought-provoking. It has been speculated that the substitutions in the coenzyme binding pocket may interfere with proteolytic processing of p25 and that the substitutions in the catalytic site could interfere with recognition by the C-signal receptor (54).

An unresolved issue is the potential enzymatic activity of p25. Is p25 an enzymatic fossil, which no longer has SCAD activity and which only functions as a precursor for p17? Or does p25 still have enzymatic activity? Clearly, this potential enzymatic activity is not required for fruiting body formation, as development of csgA cells is rescued by exogenous p17. Irrespective of whether p25 has SCAD activity, it is interesting to speculate about why p25 was, at some point in its evolution, adopted to become the precursor for the 17 kDa C-signal protein. An enzyme is optimized to recognize a substrate. This interaction could potentially be exploited to generate a signal–receptor pair in which the signal molecule is the substrate-binding part of the enzyme and the receptor is the original substrate or a protein that carries the original substrate as a secondary modification. It will be interesting to see whether this is the case for the C-signal receptor (cf. below).

C-SIGNAL TRANSMISSION RELIES ON A CONTACT-DEPENDENT MECHANISM

Early on, Kaiser and co-workers observed a conspicuous relationship between cell motility and C-signal transmission. First, it was found that non-motile cells have an altered pattern of developmental gene expression (39). This altered pattern matches exactly the pattern observed in csgA mutants. The similarity in the pattern of developmental expression suggested that motility is required for C-signal transmission. Secondly, it was found that both the donor cell of the C-signal as well as the receiver cell of the C-signal need to be motile in order for C-signal transmission to occur (34). The interpretation of these experiments was that specific cell–cell contacts are needed for C-signal transmission to occur and that motility is required to establish these contacts. This hypothesis was tested in a simple and elegant experimental setup developed by Kim and Kaiser (33). Briefly, an agar plate containing starvation medium was scored with a finely grained sandpaper to create grooves, which were approximately 0.5 to 1 cell length in width. Subsequently, non-motile cells were positioned in the grooves. In the grooves, the rod-shaped cells generally aligned with extensive side-to-side contacts and end-to-end contacts. Amazingly, this simple manipulation of cell position restored C-signal-dependent gene expression and sporulation in the non-motile mutants. Interestingly, cells outside the grooves also aligned in a pattern that established extensive side-to-side contacts. However, these cells neither expressed C-signal dependent genes nor did they sporulate. The defining difference between cells in the grooves and outside the grooves is the frequency of end-to-end contacts (33). Thus, Kim and Kaiser suggested that C-signal transmission depends on direct contacts between adjacent cells and that these contacts were end-to-end contacts (33). Cell motility would be required to establish these contacts (33).

Later, the observation that wild-type cells and *csgA* cells need to be in direct contact in order for extracellular complementation to occur has offered further support to the contact-dependent C-signal transmission mechanism (36). Finally, the findings that CsgA antibodies recognize epitopes, located on the surface of developing wild-type cells (74) and that the 17 kDa C-signal protein is localized to the outer membrane showed that the C-signal is non-diffusible, and thus further supported the idea that C-signal transmission would rely on a contact-dependent mechanism (54). Definitive evidence that C-signal transmission depends on specific end-to-end contacts is still lacking. However, analyses of cell behaviour during rippling, which is a specialized type of organized cell movement that is occasionally observed during the early stages of aggregation, lend support to the notion that C-signal transmission involves end-to-end contacts (70).

THE C-SIGNAL TRANSDUCTION PATHWAY

In order to understand how a single signal can induce three responses, i.e. aggregation, sporulation, and developmental gene expression, which are separated in time and space, it is essential to elucidate the structure of the signal transduction pathway. Several of the components in the C-signal transduction have been identified (Figure 11.4). Most of these components have been identified by using genetic approaches in which developmental mutants were specifically screened for those deficient in C-signal-dependent responses (52, 80).

The receptor of the C-signal has yet to be identified. The first recognized component in the C-signal transduction pathway is the DNA binding response regulator protein FruA, which consists of an N-terminal receiver domain and a C-terminal DNA binding domain (15). *fruA* mutants are unable to aggregate and sporulate and are deficient in the expression of C-signal-dependent genes (15, 26, 62, 78). FruA activity is regulated at the level of transcription as well as at a post-translational level (Figure 11.4). The regulatory mechanism acting at the level of *fruA* transcription ensures the timely induction of *fruA* transcription after 3–6 h of starvation. Several pathways converge to stimulate *fruA* transcription: the early-acting A-signal induces *fruA* transcription by an unknown mechanism (15, 62). The MrpC protein, which is a homolog of the cyclic AMP receptor protein in *E. coli* (84), binds to the *fruA* promoter and induces *fruA* transcription. Finally, the DevT protein directly or indirectly stimulates transcription of *fruA* (7). The regulatory mechanism acting at the post-translational level results in the activation of FruA. Genetic evidence suggests that this activation involves the phosphorylation of a conserved Asp residue in the N-terminal receiver domain (15). In other response regulator proteins, the corresponding Asp residue is phosphorylated by a cognate histidine protein kinase (64). Moreover, genetic evidence suggests that FruA phosphorylation is induced by the C-signal (15, 78). The cognate FruA histidine protein kinase has yet to be identified.

Downstream from phosphorylated FruA the C-signal transduction pathway branches (Figure 11.4). One branch leads to aggregation; the proteins in the Frz chemosensory system act in this branch (78, 80). The Frz proteins constitute a cytoplasmic signal transduction system that controls several gliding-motility parameters including the frequency of gliding reversals (5, 28). The Frz proteins are similar to proteins involved in chemotaxis responses in other bacteria (88). In this branch, the C-signal is an input signal to the Frz system and induces methylation of the FrzCD protein (78), a methyl-accepting chemotaxis protein (57). C-signal-induced methylation of FrzCD occurs via phosphorylation of FruA and depends on the FrzF methyltransferase (15, 78) (Figure 11.5). Presumably C-signaling alters the activity of the Frz system in such a way that aggregation is induced. Genetic evidence suggests that C-signaling inhibits the activity of the Frz system (77). This in turn, alleviates the Frz-dependent inhibition

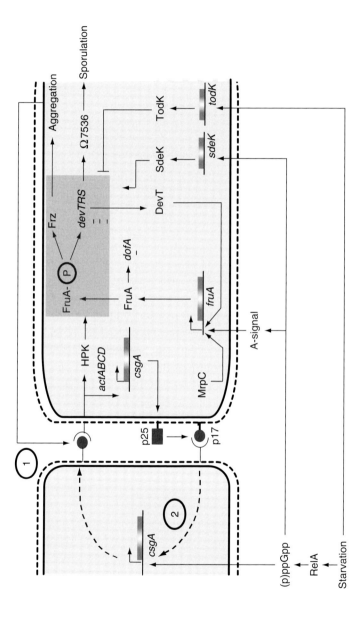

Figure 11.4. Multiple signal transduction pathways converge to control morphogenesis. The schematic illustrates C-signal transmission between two cells. The dashed and continuous lines indicate the outer and inner membrane, respectively. For simplicity, the components in the pathway are only shown in the cell to the right. These components are also present in the cell to the left. In this cell, only *csgA* is shown to illustrate the signal amplification loop labelled "2". The second signal amplification loop is labelled "1". The symbol ⊃– indicates the hypothetical C-signal receptor; HPK indicates the hypothetical FruA histidine protein kinase; and Ω7536 symbolizes the sporulation gene monitored by the Tn5lacΩ7536 insertion. The possible convergence point of the signal transduction pathways defined by the SdeK and TodK kinases with the C-signal transduction pathway is indicated by the shaded box. See the text for a detailed explanation of the pathways.

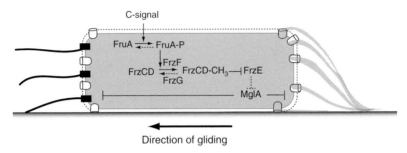

C-signal

FruA ⇌ FruA-P

↓FrzF

FrzCD ⇌ FrzCD-CH₃ ⊣ FrzE

FrzG

MglA ⊣

Direction of gliding

Figure 11.5. The C-signal inhibits polarity switching of the gliding engines in *Myxococcus xanthus*. Longitudinal cross-section of a *M. xanthus* cell on a solid surface, showing the arrangement and location of the two engines involved in gliding motility. The interactions that result in C-signal-dependent inhibition of polarity switching of the two gliding engines during fruiting-body formation are indicated in black. C-signaling results in increased methylation of the FrzCD protein. Genetic evidence suggests that this leads to an inhibition of the FrzE histidine protein kinase. This in turn results in an alleviation of the Frz-dependent inhibition of the MglA protein. Consequently, MglA inhibits polarity switching of the two gliding engines. The figure was adapted from (78).

of the MglA protein, a member of the Ras–Rab–Rho superfamily of small eukaryotic GTPases (21), which controls polarity switching of the A- and S-engines (83). Consequently, MglA inhibits polarity switching of Tfp and the active nozzle-like structures (Figure 11.5). The connection between the C-signal and the Frz system, its interaction with the gliding machinery, and the connection to cell behavior is discussed further below.

The second branch downstream from phosphorylated FruA leads to C-signal-dependent gene expression and sporulation (15, 26) (Figure 11.4). The C-signal and FruA jointly regulate the expression of at least 50 genes (26) including the *devTRS* operon (15). The DevTRS proteins, in turn, are required for the expression of the sporulation gene tagged by the Tn5*lac* Ω7536 insertion (52). Moreover, DevT is required for full expression of the *fruA* gene (7). In addition to regulating the expression of C-signal-dependent genes, FruA regulates the expression of at least eight genes, including the *dofA* gene, in a C-signal independent manner (26) (Figure 11.4). Presumably these genes are induced by unphosphorylated FruA.

A third branch in the C-signal transduction pathway is located upstream from FruA and leads to increased transcription of the *csgA* gene (35) (Figure 11.4). *csgA* is transcribed in vegetative cells and transcription increases approximately four-fold during development (50). This increase in *csgA* transcription is paralleled in the accumulation of p25, which also increases approximately four-fold in response to starvation (42). The increase in *csgA* transcription in response to starvation involves RelA and the stringent response (9) and the four proteins encoded by the *act* operon (17). Specifically, ActA and ActB, both of which are response regulators, are required for full transcription of *csgA* during development whereas ActC and ActD are important for the correct timing of *csgA* transcription. It has been suggested that the Act proteins are specifically involved in the C-signal-dependent increase in *csgA* transcription (17). However, an interaction between C-signal transmission and *act* transcription or the Act protein has yet to be reported. Following *csgA* transcription, p25 accumulates, is exported, and is then processed to produce the 17 kDa C-signal protein. The p17 protein is exposed on the cell surface and may interact with a C-signal receptor on a different cell.

The C-signal transduction pathway is only active during starvation. Starvation sets the pathway in motion by inducing the stringent response (75). The stringent response, in turn, induces *csgA* transcription (9) and A-signal accumulation (20) (Figure 11.4).

The C-signal transduction pathway contains two signal-amplification loops, which ensure that cells engaged in C-signaling are exposed to increased levels of C-signaling during development. In the first loop, C-signaling induces aggregation and thus the accumulation of cells at a higher cell density. As C-signal transmission involves a contact-dependent mechanism, the prediction is that during the aggregation process the level of C-signaling to which an aggregating cell is exposed increases (Figure 11.4, loop 1). In the second amplification loop, C-signaling results in increased *csgA* transcription; this in turn, results in p25 accumulation, which is subsequently processed to p17, which may then engage in C-signaling with a neighboring cell (Figure 11.4, loop 2).

HOW DOES A SINGLE SIGNAL INDUCE THREE RESPONSES SEPARATED IN TIME AND SPACE?

A hallmark in fruiting-body formation is the temporal and spatial coordination of aggregation and sporulation: aggregation precedes sporulation and cells do not initiate sporulation until the aggregation process is complete and cells have accumulated at a high cell density inside the fruiting bodies. With the C-signal inducing aggregation as well as sporulation, the question becomes: how is the C-signal transduction pathway structured to ensure the spatial and temporal coordination of aggregation and sporulation?

Three independent lines of evidence converge to suggest that the three C-signal-dependent responses are induced at different thresholds of C-signaling. Kim and Kaiser added different amounts of exogenous C-signal to starving *csgA* cells and observed that aggregation and early C-signal-dependent gene expression were induced at an intermediate level of C-signaling whereas sporulation and late C-signal-dependent gene expression were induced at a higher level (35). By manipulating expression of the *csgA* gene in vivo, Li *et al.* (50) made similar observations. Finally, by systematically varying the accumulation of the C-signal *in vivo*, Kruse *et al.* (42) observed that increased synthesis of the C-signal early during development resulted in early aggregation, sporulation and C-signal-dependent gene expression. Vice versa, decreased accumulation of the C-signal resulted in delayed aggregation and sporulation. Importantly, it was also observed that overexpression of the C-signal results in uncoupling of aggregation and sporulation, with spores being formed outside fruiting bodies. Thus, the regulated increase in C-signaling levels is a key parameter in the temporal and spatial coordination of aggregation and sporulation.

The current view of how the C-signal ensures the temporal coordination of aggregation and sporulation is as follows (79). The C-signal thresholds in combination with the ordered increase in the level of C-signaling ensure that the C-signal first induces aggregation and early gene expression and then late gene expression and sporulation. According to this model, the C-signal is a timer of developmental events. Moreover, the C-signal is a non-diffusible morphogen as it induces distinct morphogenetic events at distinct thresholds.

The spatial coordination of aggregation, sporulation and gene expression is likely to be a direct consequence of the contact-dependent C-signal transmission mechanism (79). This signal transmission mechanism ensures that the level of C-signaling to which a cell is exposed reflects cell density and thus the position of a cell. The high level of C-signaling that induces late gene expression and sporulation is only obtained in cells that are closely packed inside the nascent fruiting bodies. Consequently, only cells that have accumulated inside the nascent fruiting bodies express late genes and undergo sporulation. The observation that overexpression of *csgA* results in uncoupling of aggregation and sporulation suggests that in wild-type cells the balance between the two C-signal amplification loops is carefully regulated to ensure that the sporulation threshold is explicitly obtained by cells inside the nascent fruiting bodies. Thus, the mechanism of the C-signal essentially allows cells to decode their position with respect to that of other cells and in that way match gene expression, and ultimately sporulation, to their position.

This model also provides an explanation for the spatial control of C-signal-dependent gene expression. C-signal-dependent genes are preferentially expressed in aggregating and sporulating cells whereas they are not expressed in the peripheral rods (30). Peripheral rods are present at a low cell density outside the nascent fruiting bodies. Consequently, they will only infrequently engage in direct contacts with other cells with C-signal transmission. Therefore, they experience a low level of C-signaling that allows neither C-signal-dependent gene expression nor sporulation.

MULTIPLE SIGNAL TRANSDUCTION PATHWAYS CONTROL MORPHOGENESIS

Two additional signal transduction pathways converge with the C-signal transduction pathway to regulate aggregation, sporulation, and gene expression after 6 h (Figure 11.4). Each of the pathways is defined by a specific histidine protein kinase. The SdeK histidine protein kinase is synthesized in a RelA-dependent manner immediately after the initiation of starvation; the SdeK pathway converges with the C-signal transduction pathway to stimulate aggregation, sporulation, and gene expression (16, 66). The pathway defined by the TodK histidine protein kinase inhibits aggregation, sporulation, and gene expression (67). Synthesis of TodK is inhibited by starvation in a RelA-independent manner (67). Genetic evidence suggests that the TodK-dependent inhibition of aggregation, sporulation and gene expression is alleviated after 6–9 h of starvation (67). Interestingly, SdeK and TodK are both predicted to be located in the cytoplasm and they both contain PAS domains in their sensor part. PAS domains have been implicated in sensing changes in redox potential, oxygen, light, small ligands, and overall energy level of a cell and also mediate protein–protein interactions (86). Therefore, it has been proposed that the kinase activities of TodK and SdeK are controlled by intracellular signals, which are informative about the metabolic state of individual cells and which are indicative of continued starvation (67). According to this model, continued starvation of cells results in the accumulation of these signals. This would subsequently trigger an alteration in the activity of the TodK and SdeK kinases, resulting in the alleviation of the inhibitory effect of TodK and stimulating the activating effect of SdeK on the C-signal transduction pathway. In this model, the C-signal transduction pathway is an integration point at which the intercellular signals needed to coordinate the efforts of thousands of cells during fruiting-body formation and intracellular signals reflecting the energy status of individual cells are integrated. The advantage of this kind of integration would be that productive C-signaling, and thus morphogenesis, is strictly coordinated with the energy status of individual cells and only occurs when the energy status of individual cells is appropriate. The SdeK and TodK pathways may also help to make sure that, if nutrients become available during fruiting body formation, then fruiting-body formation is arrested and vegetative growth resumes.

THE C-SIGNAL-DEPENDENT MOTILITY RESPONSE

As fruiting-body formation depends on changes in cell behavior from spreading to aggregation, a complete understanding of the morphogenetic properties of the C-signal entails a description of how this signal molecule alters cell behavior. Using fluorescent time-lapse video microscopy, Jelsbak and Søgaard-Andersen (29) found that, during the aggregation stage of fruiting body formation, the C-signal induces a motility response that includes increases in the gliding speed and in the duration of gliding intervals and decreases in the stop and reversal frequencies. The combined effect of these changes is a switch in the motility behaviour of individual cells from an oscillatory to a unidirectional type of behaviour (Figure 11.6).

In order for *M. xanthus* cells to display directional movement such as that seen during aggregation, they need to be able to regulate their frequency of direction changing. A priori, a stop as well as a reversal could result in a change in the direction of movement. However, as *M. xanthus* cells adhere to each other and to the substratum, they are not buffeted by Brownian motion. Consequently, among the motility parameters that characterize gliding in *M. xanthus*, only a reversal results in a change in the direction of movement. Therefore, among the motility parameters regulated by the C-signal, the key parameter is the reversal frequency (29). The remaining motility parameters regulated by the C-signal may contribute to aggregation by increasing the net distance travelled by a cell per minute (29).

The C-signal dependent decrease in the reversal frequency fits with the genetic and biochemical evidence demonstrating that the C-signal is an input signal to the Frz chemosensory system (78) (Figures 11.5, 11.6). Consistent with the observation that the C-signal induces a decrease in the reversal frequency, methylation of FrzCD correlates with a low reversal frequency (56).

The C-signal-dependent decrease in the reversal frequency suggests that the C-signal inhibits polarity switching of the Tfp and nozzle-like structures involved in S- and A-motility, respectively (77). The net

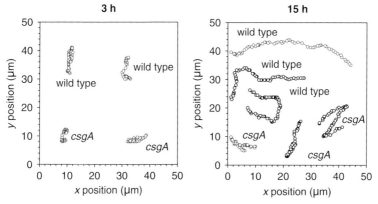

Figure 11.6. Effect of the C-signal on the motility behaviour of *M. xanthus* cells. By means of fluorescent time-lapse video-microscopy the *x*, *y* positions of wild-type cells, which express GFP and which were co-developed with wild-type cells that do not express GFP, and of *csgA* cells, which express GFP and which were co-developed with *csgA* cells that do not express GFP, were recorded every 15 s for a period of 900 s. The trajectories in the left panel show the behaviour of two wild-type cells and of two *csgA* cells starved for 3 h (i.e. 3 h prior to the initiation of C-signaling); the trajectories in the right panel show the behavior of three wild-type cells and of three *csgA* cells starved for 15 h (i.e. 9 h after the initiation of C-signaling). Before the initiation of C-signaling, wild-type and *csgA* cells behave similarly. After the initiation of C-signaling in wild-type cells, these cells display a unidirectional type of behavior. The figure was adapted from (30).

result of this inhibition is that the two motility engines are locked at their respective poles, and consequently C-signaling cells are locked in a unidirectional mode of behavior.

C-SIGNAL-INDUCED AGGREGATION: A MODEL

The ability of C-signaling cells to move a long distance is clearly beneficial during the aggregation stage of fruiting-body formation. However, in order for cells to aggregate they need to move with a sense of direction. The identification of the motility parameters controlled by the C-signal in combination with the contact-dependent C-signal transmission mechanism has allowed the generation of a model for C-signal-induced aggregation (Figure 11.7). The model is composed of three discrete events. The basic event is an end-to-end contact between two cells with C-signal transmission. Cells engaged in end-to-end contact with C-signal transmission gain the ability to travel over longer net distances for as long as they

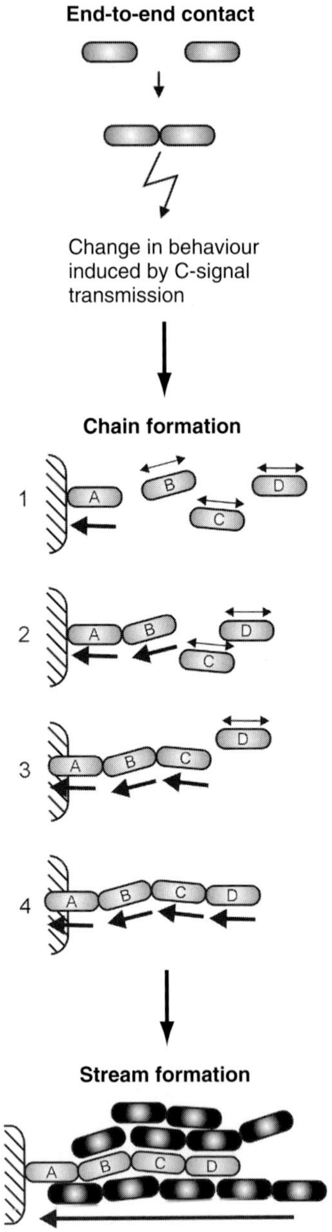

Figure 11.7. A model for C-signal-induced aggregation. The model consists of three elements. The basic element is an end-to-end contact between two cells with C-signal transmission followed by a change in cell behavior. The second element is chain formation, which is predicted to be a consequence of repeated end-to-end contacts with C-signal transmission

maintain the end-to-end contact. In a population of starving cells, this basic event is predicted to result in the second discrete event, chain formation (Figure 11.7). End-to-end contacts will occur frequently at the high cell density required for normal development. This will sequentially result in the recruitment of cells to pairs of cells that are C-signaling and, thus, in the formation of chains of cells. Cells in a chain are moving at the same speed and in the same direction, which is determined by the direction of movement of the cell at the leading end of the chain. The information about which direction to move in is relayed from the leading cell in a chain to the following cell by the direct cell–cell contacts. A reversal of direction of a cell positioned inside a chain would break the chain. However, cell reversals in chains are unlikely based on two observations. First, the cohesiveness of the cells increases during development (72). Second, cells are exposed to increasing levels of C-signaling during aggregation (see above).

The arrangement and movement of cells in chains is predicted to result in the third discrete event, stream formation (Figure 11.7). Movement of cell in a chain may create alignment of neighboring cells. This will result in the formation of secondary chains that are associated with the initiating chain by lateral interactions. This could lead to the formation of the streams of cells that have been observed experimentally (Figure 11.2). Aggregation

---

Figure 11.7. (cont.)

between cells in a field. Consider cell A in scenario 1, which for simplicity is located at the edge of an aggregate and moving towards the aggregate. This cell is moving at a high speed in long gliding intervals and with low stop and reversal frequencies, as it is engaged in C-signaling with cells inside the aggregate. The bold, single-headed arrow indicates this type of behavior. Cells not engaged in C-signaling move at low speeds and with high stop and reversal frequencies, as marked by the double-headed arrows. If cell B establishes an end-to-end contact with cell A (scenario 2), then the two cells will transmit the C-signal to each other. As a consequence, cell B is induced to move with a high speed and low stop and reversal frequencies as long as it maintains the end-to-end contact. As cell A is moving towards the aggregation center, cell B can only move with a high speed and low stop and reversal frequencies if it also moves towards the aggregation center. This sequence of events could be repeated sequentially, adding cell C, etc., thus creating a chain in which cells are moving in the direction determined by the leading cell A (scenarios 3–4). The third element in the model is stream formation. Cell movement in chains is predicted to create alignment of neighboring cells. This will result in the formation of secondary chains of cells (marked by dark gray colour). The cells in these secondary chains are associated with the initiating chain by lateral cell–cell contacts and with other cells in the secondary chain by end-to-end contacts. Together, an initiating chain and its associated secondary chains will make up a stream. The figure was adapted from (28).

centers could form by collisions of streams or by a single stream turning on itself in a spiral movement. Once the leading cell of a stream is trapped inside an aggregation center, the remaining cells in the stream will follow into that center, as the direction of movement of a cell in a stream is determined by the leading cell. In this model C-signal transmission is a local event between two cell ends, which occurs without reference to the global pattern, and the result is a global organization of cells. Therefore, according to this model C-signal-induced aggregation is a self-organizing process.

SIGNAL INTEGRATION DURING FRUITING-BODY MORPHOGENESIS

292

Formation of the spore-filled fruiting bodies is clearly an effective survival strategy in response to starvation. However, it is also costly, as only 10%–20% of cells differentiate into spores. The starvation-induced stringent response and the intercellular A-signal system constitute two checkpoints, which ensure that cells only embark on fruiting-body formation when starvation is severe and gives the impression of being long-lasting. The C-signal comes into action when the cells have passed the tests of the stringent response and the A-signal. The primary functions of the C-signal are to induce and coordinate aggregation, sporulation and full gene expression after 6 h. Thus, the C-signaling system cannot be regarded as a system whose primary function is to monitor starvation. Rather the C-signal is the temporal and spatial coordinator of the different morphogenetic events involved in fruiting-body formation. Regarded in this way, the stringent response and the A- and C-signaling systems essentially act to stimulate fruiting-body formation. Predictably, it is becoming evident that fruiting-body formation is also subject to intensive negative control.

The negative regulators of fruiting-body formation include two cytoplasmic histidine protein kinases (EspA and TodK) (8, 67), a member of the NtrC family of two-component transcriptional regulators (SpdR) (19), the Che3 chemosensory system (38), two serine/threonine protein kinases (Pkn1 and Pkn2) (59, 87), a protein that is homologous to flavin-containing monooxygenases (BcsA) (11), and a protein of unknown biochemical function that modulates the stringent response (SocE) (10). Although the presence of a PAS domain in the sensor part of the TodK kinase hints that it could be involved in monitoring the metabolic state of cells (67), the specific parameters monitored by these systems are still unknown. However, the multitude of negative regulators of fruiting-body formation indicate that cells are continuously monitoring the conditions to evaluate whether or not fruiting-body formation should continue.

From a design point of view, a regulatory circuit that includes the integration of positively as well as negatively acting pathways is highly robust. It endows the system with the capacity to tailor the final decision, i.e. to continue or not to continue fruiting-body formation, to the specific conditions to which the cells are exposed.

CONCLUDING REMARKS

The C-signal is a remarkably versatile signal and we now have a general framework for understanding how a single signal induces three responses that are separated in time and space. However, many questions are still unanswered. So far, the identification of components important for development has to a large extent depended on the often painstaking identification of relevant genes and proteins on a step-by-step basis using genetic approaches. Recently, the *M. xanthus* genome sequence was finished (W. C. Niermann, personal communication). This together with proteome analysis (26) and the construction of an *M. xanthus* DNA microarray (H. Kaplan and D. Kaiser, personal communication) promise to pave the way for the systematic identification of additional important players in the regulatory pathways that drive and coordinate fruiting-body formation.

ACKNOWLEDGEMENTS

I thank Jimmy Jakobsen for carefully reading the manuscript, and Martin Overgaard, Anders Aa. Rasmussen, Sune Lobedanz, Eva Ellehauge and Lars Jelsbak for many helpful discussions.

REFERENCES

1   Arnold, J. W. and L. J. Shimkets 1988. Cell surface properties correlated with cohesion in *Myxococcus xanthus*. *J. Bacteriol*. **170**: 5771–7.
2   Baker, M. E. 1994. *Myxococcus xanthus* C-factor, a morphogenetic paracrine signal, is similar to *Escherichia coli* 3-oxoacyl-[acyl-carrier-protein] reductase and human 17 beta-hydroxysteroid dehydrogenase. *Biochem. J*. **301**: 311–12.
3   Behmlander, R. M. and M. Dworkin 1994. Biochemical and structural analyses of the extracellular matrix fibrils of *Myxococcus xanthus*. *J. Bacteriol*. **176**: 6295–303.
4   Behmlander, R. M. and M. Dworkin 1991. Extracellular fibrils and contact-mediated cell interactions in *Myxococcus xanthus*. *J. Bacteriol*. **173**: 7810–21.
5   Blackhart, B. D. and D. R. Zusman 1985. 'Frizzy' genes of *Myxococcus xanthus* are involved in control of frequency of reversal of gliding motility *Proc. Natn. Acad. Sci. USA* **82**: 8771–4.

6 Bowden, M. G. and H. B. Kaplan 1998. The *Myxococcus xanthus* lipopolysaccharide O-antigen is required for social motility and multicellular development. *Molec. Microbiol.* **30**: 275–84.

7 Boysen, A., E. Ellehauge, B. Julien and L. Søgaard-Andersen 2002. The DevT protein stimulates synthesis of FruA, a signal transduction protein required for fruiting body morphogenesis in *Myxococcus xanthus*. *J. Bacteriol.* **184**: 1540–6.

8 Cho, K. and D. R. Zusman 1999. Sporulation timing in *Myxococcus xanthus* is controlled by the *espAB* locus. *Molec. Microbiol.* **34**: 714–25.

9 Crawford, E. W. and L. J. Shimkets 2000. The *Myxococcus xanthus socE* and *csgA* genes are regulated by the stringent response. *Molec. Microbiol.* **37**: 788–99.

10 Crawford, E. W. and L. J. Shimkets 2000. The stringent response in *Myxococcus xanthus* is regulated by SocE and the CgsA C-signaling protein. *Genes Dev.* **14**: 483–92.

11 Cusick, J. K., E. Hager and R. E. Gill 2002. Characterization of *bcsA* mutations that bypass two distinct signaling requirements for *Myxococcus xanthus* development. *J. Bacteriol.* **184**: 5141–50.

12 Downard, J., S. V. Ramaswamy and K. S. Kil 1993. Identification of *esg*, a genetic locus involved in cell-cell signaling during *Myxococcus xanthus* development. *J. Bacteriol.* **175**: 7762–70.

13 Dworkin, M. 1996. Recent advances in the social and developmental biology of the Myxobacteria. *Microbiol. Rev.* **60**: 70–102.

14 Dworkin, M. and S. M. Gibson 1964. A system for studying microbial morphogenesis: rapid formation of microcysts in *Myxococcus xanthus*. *Science* **146**: 243–4.

15 Ellehauge, E., M. Nørregaard-Madsen and L. Søgaard-Andersen 1998. The FruA signal transduction protein provides a checkpoint for the temporal co-ordination of intercellular signals in *M. xanthus* development. *Molec. Microbiol.* **30**: 807–17.

16 Garza, A. G., J. S. Pollack, B. Z. Harris *et al.* 1998. SdeK is required for early fruiting body development in *Myxococcus xanthus*. *J. Bacteriol.* **180**: 4628–37.

17 Gronewold, T. M. A. and D. Kaiser 2001. The *act* operon controls the level and time of C-signal production for *Myxococcus xanthus* development. *Molec. Microbiol.* **40**: 744–56.

18 Hagen, D. C., A. P. Bretscher and D. Kaiser 1978. Synergism between morphogenetic mutants of *Myxococcus xanthus*. *Dev. Biol.* **64**: 284–96.

19 Hager, E., H. Tse and R. E. Gill 2001. Identification and characterization of *spdR* mutations that bypass the BsgA protease-dependent regulation of developmental gene expression in *Myxococcus xanthus*. *Molec. Microbiol.* **39**: 765–80.

20 Harris, B. Z., D. Kaiser and M. Singer 1998. The guanosine nucleotide (p)ppGpp initiates development and A-factor production in *Myxococcus xanthus*. *Genes Dev.* **12**: 1022–35.

21 Hartzell, P. L. 1997. Complementation of sporulation and motility defects in a prokaryote by a eukaryotic GTPase. *Proc. Natn. Acad. Sci. USA* **94**: 9881–6.

22 Henrichsen, J. 1972. Bacterial surface translocation: a survey and a classification. *Bacteriol. Rev.* **36**: 478–503.

23 Hodgkin, J. and D. Kaiser 1979. Genetics of gliding motility in *Myxococcus xanthus* (Myxobacterales): genes controlling movement of single cells. *Molec. Gen. Genet.* **171**: 167–76.

24 Hodgkin, J. and D. Kaiser 1979. Genetics of gliding motility in *Myxococcus xanthus* (Myxobacteriales): two gene systems control movement. *Molec. Gen. Genet.* **171**: 177–91.

25 Hoiczyk, E. and W. Baumeister 1998. The junctional pore complex, a prokaryotic secretion organelle, is the molecular motor underlying gliding motility in cyanobacteria. *Curr. Biol.* **8**: 1161–8.

26 Horiuchi, T., M. Taoka, T. Isobe, T. Komano and S. Inouye 2002. Role of *fruA* and *csgA* genes in gene expression during development of *Myxococcus xanthus*. Analysis by two-dimensional gel electrophoresis. *J. Biol. Chem.* **277**: 26753–60.

27 Inouye, M., S. Inouye and D. R. Zusman 1979. Gene expression during development of *Myxococcus xanthus*: pattern of protein synthesis. *Dev. Biol.* **68**: 579–91.

28 Jelsbak, L. and L. Søgaard-Andersen 1999. The cell surface-associated intercellular C-signal induces behavioral changes in individual *Myxococcus xanthus* cells during fruiting body morphogenesis. *Proc. Natn. Acad. Sci. USA* **96**: 5031–6.

29 Jelsbak, L. and L. Søgaard-Andersen 2002. Pattern formation by a cell surface-associated morphogen in *Myxococcus xanthus*. *Proc. Natn. Acad. Sci. USA* **99**: 2032–7.

30 Julien, B., A. D. Kaiser and A. Garza 2000. Spatial control of cell differentiation in *Myxococcus xanthus*. *Proc. Natn. Acad. Sci. USA* **97**: 9098–103.

31 Kaiser, D. 2003. Coupling cell movement to multicellular development in myxobacteria. *Nature Rev. Microbiol.* **1**: 45–54.

32 Kaiser, D. 1979. Social gliding is correlated with the presence of pili in *Myxococcus xanthus*. *Proc. Natn. Acad. Sci. USA* **76**: 5952–6.

33 Kim, S. K. and D. Kaiser 1990. Cell alignment required in differentiation of *Myxococcus xanthus*. *Science* **249**: 926–8.

34 Kim, S. K. and D. Kaiser 1990. Cell motility is required for the transmission of C-factor, an intercellular signal that coordinates fruiting body morphogenesis of *Myxococcus xanthus*. *Genes Dev.* **4**: 896–904.

35 Kim, S. K. and D. Kaiser 1991. C-factor has distinct aggregation and sporulation thresholds during *Myxococcus* development. *J. Bacteriol.* **173**: 1722–8.

36 Kim, S. K. and D. Kaiser 1990. C-factor: a cell-cell signaling protein required for fruiting body morphogenesis of *M. xanthus*. *Cell* **61**: 19–26.

37 Kim, S. K. and D. Kaiser 1990. Purification and properties of *Myxococcus xanthus* C-factor, an intercellular signaling protein. *Proc. Natn. Acad. Sci. USA* **87**: 3635–9.

38 Kirby, J. R. and D. R. Zusman 2003. Chemosensory regulation of developmental gene expression in *Myxococcus xanthus*. *Proc. Natn. Acad. Sci. USA* **100**: 2008–13.

39 Kroos, L., P. Hartzell, K. Stephens and D. Kaiser 1988. A link between cell movement and gene expression argues that motility is required for cell-cell signaling during fruiting body development. *Genes Dev.* **2**: 1677–85.

40 Kroos, L. and D. Kaiser 1987. Expression of many developmentally regulated genes in *Myxococcus* depends on a sequence of cell interactions. *Genes Dev.* **1**: 840–54.

41 Kroos, L., A. Kuspa and D. Kaiser 1986. A global analysis of developmentally regulated genes in *Myxococcus xanthus*. *Dev. Biol.* **117**: 252–66.

42 Kruse, T., S. Lobedanz, N. M. S. Berthelsen and L. Søgaard-Andersen 2001. C-signal: A cell surface-associated morphogen that induces and coordinates multicellular fruiting body morphogenesis and sporulation in *M. xanthus*. *Molec. Microbiol.* **40**: 156–68.

43 Kuner, J. M. and D. Kaiser 1982. Fruiting body morphogenesis in submerged cultures of *Myxococcus xanthus*. *J. Bacteriol.* **151**: 458–61.

44 Kuspa, A., L. Kroos and D. Kaiser 1986. Intercellular signaling is required for developmental gene expression in *Myxococcus xanthus*. *Dev. Biol.* **117**: 267–76.

45 Kuspa, A., L. Plamann and D. Kaiser 1992. Identification of heat-stable A-factor from *Myxococcus xanthus*. *J. Bacteriol.* **174**: 3319–26.

46 Kuspa, A., L. Plamann and D. Kaiser 1992. A-signaling and the cell density requirement for *Myxococcus xanthus* development. *J. Bacteriol.* **174**: 7360–9.

47 Lee, B.-U., K. Lee, J. Mendez and L. J. Shimkets 1995. A tactile sensory system of *Myxococcus xanthus* involves an extracellular NAD(P)$^+$-containing protein. *Genes Dev.* **9**: 2964–73.

48 Lee, K. and L. J. Shimkets 1994. Cloning and characterization of the *socA* locus which restores development to *Myxococcus xanthus* C-signaling mutants. *J. Bacteriol.* **176**: 2200–9.

49 Lee, K. and L. J. Shimkets 1996. Suppression of a signaling defect during *Myxococcus xanthus* development. *J. Bacteriol.* **178**: 977–84.

50 Li, S., B.-U. Lee and L. J. Shimkets 1992. *csgA* expression entrains *Myxococcus xanthus* development. *Genes Dev.* **6**: 401–10.

51 Li, Y., H. Sun, X. Ma *et al.* 2003. Extracellular polysaccharides mediate pilus retraction during social motility of *Myxococcus xanthus*. *Proc. Natn. Acad. Sci. USA* **100**: 5443–8.

52 Licking, E., L. Gorski and D. Kaiser 2000. A common step for changing cell shape in fruiting body and starvation-independent sporulation in *Myxococcus xanthus*. *J. Bacteriol.* **182**: 3553–8.

53 Llamas, M. A., J. J. Rodriguez-Herva, R. E. W. Hancock *et al.* 2003. Role of *Pseudomonas putida tol-oprL* gene products in uptake of solutes through the cytoplasmic membrane. *J. Bacteriol.* **185**: 4707–16.

54 Lobedanz, S. and L. Søgaard-Andersen 2003. Identification of the C-signal, a contact-dependent morphogen coordinating multiple developmental responses in *Myxococcus xanthus*. *Genes Dev.* **17**: 2151–61.

55 Mattick, J. S. 2002. Type IV pili and twitching motility. *A. Rev. Microbiol.* **56**: 289–314.

56 McBride, M. J., T. Köhler and D. R. Zusman 1992. Methylation of FrzCD, a methyl-accepting taxis protein of *Myxococcus xanthus*, is correlated with factors affecting cell behaviour. *J. Bacteriol.* **174**: 4246–57.

57 McBride, M. J., R. A. Weinberg and D. R. Zusman 1989. 'Frizzy' aggregation genes of the gliding bacterium *Myxococcus xanthus* show sequence similarities to the chemotaxis genes of enteric bacteria. *Proc. Natn. Acad. Sci. USA* **86**: 424–8.

58 Merz, A. J., M. So and M. P. Sheetz 2000. Pilus retraction powers bacterial twitching motility. *Nature* **407**: 98–102.

59 Munoz, D. J., S. Inouye and M. Inouye 1991. A gene encoding a protein serine/threonine kinase is required for normal development of *M. xanthus*, a gram-negative bacterium. *Cell* **67**: 995–1006.

60 O'Connor, K. A. and D. R. Zusman 1991. Development in *Myxococcus xanthus* involves differentiation into two cell types, peripheral rods and spores. *J. Bacteriol.* **173**: 3318–33.

61 O'Connor, K. A. and D. R. Zusman 1989. Patterns of cellular interactions during fruiting-body formation in *Myxococcus xanthus*. *J. Bacteriol.* **171**: 6013–24.

62 Ogawa, M., S. Fujitani, X. Mao, S. Inouye and T. Komano 1996. FruA, a putative transcription factor essential for the development of *Myxococcus xanthus*. *Molec. Microbiol.* **22**: 757–67.

63 Oppermann, U., C. Filling, M. Hult *et al.* 2003. Short-chain dehydrogenases/reductases (SDR): the 2002 update. *Chem. Biol. Interact.* **143–4**: 247–53.

64 Parkinson, J. S. 1993. Signal transduction schemes of bacteria. *Cell* **73**: 857–71.

65 Plamann, L., A. Kuspa and D. Kaiser 1992. Proteins that rescue A-signal-defective mutants of *Myxococcus xanthus*. *J. Bacteriol.* **174**: 3311–18.

66 Pollack, J. S. and M. Singer 2001. SdeK, a histidine kinase required for *Myxococcus xanthus* development. *J. Bacteriol.* **183**: 3589–96.

67 Rasmussen, A. A. and L. Søgaard-Andersen 2003. TodK, a putative histidine protein kinase, regulates timing of fruiting body morphogenesis in *Myxococcus xanthus*. *J. Bacteriol.* **185**: 5452–64.

68 Reichenbach, H. 1999. The ecology of the myxobacteria. *Environ. Microbiol.* **1**: 15–21.

69 Rosenbluh, A., R. Nir, E. Sahar and E. Rosenberg 1989. Cell-density-dependent lysis and sporulation of *Myxococcus xanthus* in agarose beads. *J. Bacteriol.* **171**: 4923–9.

70 Sager, B. and D. Kaiser 1994. Intercellular C-signaling and the traveling waves of *Myxococcus*. *Genes Dev.* **8**: 2793–804.

71 Shimkets, L. J. 1999. Intercellular signaling during fruiting-body development of *Myxococcus xanthus*. *Au. Rev. Microbiol.* **53**: 525–49.

72 Shimkets, L. J. 1986. Role of cell cohesion in *Myxococcus xanthus* fruiting body formation. *J. Bacteriol.* **166**: 842–88.

73 Shimkets, L. J., R. E. Gill and D. Kaiser 1983. Developmental cell interactions in *Myxococcus xanthus* and the *spoC* locus. *Proc. Natn. Acad. Sci. USA* **80**: 1406–10.

74 Shimkets, L. J. and H. Rafiee 1990. CsgA, an extracellular protein essential for *Myxococcus xanthus* development. *J. Bacteriol.* **172**: 5299–306.

75 Singer, M. and D. Kaiser 1995. Ectopic production of guanosine penta- and tetraphosphate can initiate early developmental gene expression in *Myxococcus xanthus*. *Genes Dev.* **9**: 1633–44.

76 Skerker, J. M. and H. C. Berg 2001. Direct observation of extension and retraction of type IV pili. *Proc. Natn. Acad. Sci. USA* **98**: 6901–4.

77 Søgaard-Andersen, L. 2004. Cell polarity, intercellular signaling and morphogenetic cell movements in *Myxococcus xanthus*. *Curr. Opin. Microbiol.* **7**: 587–93.

78 Søgaard-Andersen, L. and D. Kaiser 1996. C factor, a cell-surface-associated intercellular signaling protein, stimulates the cytoplasmic Frz signal transduction system in *Myxococcus xanthus*. *Proc. Natn. Acad. Sci. USA* **93**: 2675–9.

79 Søgaard-Andersen, L., M. Overgaard, S. Lobedanz *et al.* 2003. Coupling gene expression and multicellular morphogenesis during fruiting body formation in *Myxococcus xanthus*. *Molec. Microbiol.* **48**: 1–8.

80 Søgaard-Andersen, L., F. J. Slack, H. Kimsey and D. Kaiser 1996. Intercellular C-signaling in *Myxococcus xanthus* involves a branched signal transduction pathway. *Genes Dev.* **10**: 740–54.

81 Spormann, A. M. 1999. Gliding motility in bacteria: insights from studies of *Myxococcus xanthus*. *Microbiol. Molec. Biol. Rev.* **63**: 621–41.

82 Spormann, A. M. and A. D. Kaiser 1995. Gliding movements in *Myxococcus xanthus*. *J. Bacteriol.* **177**: 5846–52.

83 Spormann, A. M. and D. Kaiser 1999. Gliding mutants of *Myxococcus xanthus* with high reversal frequencies and small displacements. *J. Bacteriol.* **181**: 2593–601.

84 Sun, H. and W. Shi 2001. Analyses of *mrp* genes during *Myxococcus xanthus* development. *J. Bacteriol.* **183**: 6733–9.

85 Sun, H., D. R. Zusman and W. Shi 2000. Type IV pilus of *Myxococcus xanthus* is a motility apparatus controlled by the *frz* chemosensory system. *Curr. Biol.* **10**: 1143–6.

86 Taylor, B. L. and I. B. Zhulin 1999. PAS domains: internal sensors of oxygen, redox potential, and light. *Microbiol. Molec. Biol. Rev.* **63**: 479–506.

87 Udo, H., M. Inouye and S. Inouye 1996. Effects of overexpression of Pkn2, a transmembrane protein serine/threonine kinase, on development of *Myxococcus xanthus*. *J. Bacteriol.* **178**: 6647–9.

88 Ward, M. J. and D. R. Zusman 1999. Motility in *Myxococcus xanthus* and its role in developmental aggregation. *Curr. Opin. Microbiol.* **2**: 624–9.

89 White, D. J. and P. L. Hartzell 2000. AglU, a protein required for gliding motility and spore maturation of *Myxococcus xanthus*, is related to WD-repeat proteins. *Molec. Microbiol.* **36**: 662–78.

90 Winans, S. C. and B. L. Bassler 2002. Mob psychology. *J. Bacteriol.* **184**: 873–83.

91 Wolgemuth, C., E. Hoiczyk, D. Kaiser and G. Oster 2002. How myxobacteria glide. *Curr. Biol.* **12**: 369–77.

92 Wu, S. S. and D. Kaiser 1995. Genetic and functional evidence that Type IV pili are required for social gliding motility in *Myxococcus xanthus*. *Molec. Microbiol.* **18**: 547–58.

93 Youderian, P., N. Burke, D. White and P. L. Hartzell 2003. Identification of genes required for adventurous gliding motility in *Myxococcus xanthus* with the transposable element *mariner*. *Molec. Microbiol.* **49**: 555–70.

Index

Note: page numbers in *italics* refer to figures and tables.